NASA SP-4009

THE

APOLLO SPACECRAFT

A CHRONOLOGY

VOLUME II
November 8, 1962—September 30, 1964

by

Mary Louise Morse and Jean Kernahan Bays

THE NASA HISTORICAL SERIES

Scientific and Technical Information Office   1973
NATIONAL AERONAUTICS AND SPACE ADMINISTRATION
Washington, D.C.

FOREWORD

This, the second volume of the Apollo Spacecraft Chronology, takes up the story where the first left off, in November 1962. The first volume dealt with the birth of the Apollo Program and traced its early development. The second concerns its teenage period, up to September 30, 1964.

By late 1962 the broad conceptual design of the Apollo spacecraft and the Apollo lunar landing mission was complete. The Administrator formally advised the President of the United States on December 10 that NASA had selected lunar orbit rendezvous over direct ascent and earth orbit rendezvous as the mode for landing on the moon. All major spacecraft contractors had been selected; detailed system design and early developmental testing were under way.

On October 20, 1962, soon after Wally Schirra's six-orbit mission in *Sigma 7*, the first formal overall status review of the Apollo spacecraft and flight mission effort was given to Administrator James E. Webb. The writer of this foreword, who was then the Assistant Director for Apollo Spacecraft Development, recalls George Low, then Director of Manned Spacecraft and Flight Missions under D. Brainerd Holmes, discussing the planning schedule for completion of the Mercury Project in 1963, initiation of Gemini flights in 1964, and the start of Apollo earth orbital flights in 1965. Major design features of the spacecraft and subsystems were discussed and so were facilities, training, flight mission plans, and resources. At the conclusion of the review, Mr. Webb, Dr. Dryden, and Dr. Seamans commented favorably on the overview provided and on the accomplishments and hard planning that had been completed. The chronology of events during the subsequent two years, as summarized herein, provides an interesting comparison with the plans as discussed that day; we came very close to what was planned for 1963 and 1964.

During 1963 formal contract negotiations with the previously selected major spacecraft contractors were completed. In addition most of the contractors for major facilities and support activities on the ground were selected. The latter group included Radio Corporation of America to furnish the spacecraft vacuum test chamber at Houston, Bell Aerosystems for the lunar landing training vehicle, Philco Corporation as prime contractor for the Mission Control Center, Link Division of General Precision, Inc., for the lunar mission simulators, and International Business Machines for the Real Time Computer Complex at Houston's Mission Control Center.

Also in 1963 the Office of Manned Space Flight was reorganized under its new leader, George E. Mueller, to strengthen its systems engineering and integration role in overall management of the Apollo-Saturn Program. In

December Dr. Mueller brought in General Sam Phillips as Deputy Director of the Apollo Program. Soon thereafter Phillips was named Apollo Program Director. A comparable reorganization took place at the Manned Spacecraft Center in Houston as the tempo of spacecraft module design and development increased. At the same time, the prime contractors were selecting and completing negotiations with their subcontractors and suppliers for the thousands of subsystems and components involved. The government-industry team for carrying out the Apollo spacecraft and flight mission related tasks was essentially complete by late 1963. Concurrently, similar activities were proceeding for the Saturn launch vehicles at the Marshall Space Flight Center and for launch site preparation at the Kennedy Space Center, as it was named by President Johnson on November 28, 1963.

Meanwhile, a series of basic program decisions were made; these enabled the spacecraft and lunar landing mission design teams to proceed into detail design. Among these decisions were the following:

• Nominal earth landing would be on the water. This was a change from the original plan which provided for earth landing in either Australia or the southwestern United States. The change was made primarily to take advantage of the softer impact conditions afforded by water landing, although the operational flexibility afforded by ocean landing was an additional favorable factor.

• CSM to LM transposition and docking would be by the free flying mode. This meant that, after injection into translunar trajectory, the crew would detach the CSM from its launch position and would rotate the spacecraft 180° and manually maneuver it into a docked position with the LM.

• The crew would operate the LM from standing position.

• The spacecraft guidance computer would use micrologic design.

• The Lunar Module would have a four-legged, deployable landing gear. This was a change from the original Grumman configuration which had five legs.

• The Lunar Module would be capable of supporting the effective operation of two men on the lunar surface for up to 24 hours, plus 24 hours in flight.

At the same time, rapid progress was made on the development of the spacecraft, on the Saturn launch vehicles, and on the facilities to support them. Typical events in 1963 included:

• The service propulsion prototype engine successfully completed initial firings.

• The first of a number of parachute malfunctions occurred during development drop tests.

• The impact test facility for development and verification of the Command Module landing system at the North American plant in Downey, California, was completed.

• Flight of Saturn SA-4 verified the capability of the Saturn first stage to operate satisfactorily after a simulated in-flight failure of one engine.

FOREWORD

- The Little Joe II launch complex at White Sands was completed and the first Little Joe II test article was launched successfully.
- The LM–1 lunar module mockup was completed.
- Prototype fuel cells were delivered by Pratt & Whitney to North American.
- The first pad abort test was successfully completed at White Sands.
- The J–2 engine successfully completed its initial long duration firing.

The Mercury Program ended with Gordon Cooper's 34-hour earth orbit mission on May 15–16, 1963, the unmanned Lunar Orbiter Project was approved, and scientific guidelines for the Apollo mission were promulgated. A new group of 14 astronauts, including Buzz Aldrin and Mike Collins, who were destined to join Neil Armstrong in the first lunar landing mission, was selected in October 1963.

Dr. Mueller, in the fall of 1963, introduced something that was to have a mighty effect on "landing before this decade is out." It was called "all-up testing." Under the "all-up" concept, launch vehicle and spacecraft development flights were combined, with all elements active and as close to lunar configuration as possible, beginning with the very first flight. This plan replaced the more conventional approach of making initial launch vehicle tests with dummy upper stages and dummy spacecraft.

Because the Saturn I flight program was of an interim non-lunar configuration, it was curtailed and four manned earth orbital flights with the Saturn I launch vehicle were canceled. The Saturn IB development for manned flight was accelerated and all Saturn IB flights, beginning with SA–201, would carry operational spacecraft. Similarly, the Saturn V development flights, beginning with 501, would be in "all-up" configuration and vehicle 501 would be used to obtain reentry data on the Apollo spacecraft. The first manned flight on both the Saturn IB and V would follow two successful unmanned flights, so that the first manned flights could be as early as vehicles 203 and 503 for the IB and V, respectively. This would exploit early successful flight operation of the new launch vehicles by reducing the total number of flights required to qualify the lunar flight configuration for manned operations. The first manned flight on a Saturn V did of course take place on vehicle 503 in December 1968—the successful *Apollo 8* mission.

Another Mueller innovation was the Apollo Executives Group, which first met in the fall of 1963. It brought together senior officials of the major Apollo–Saturn contractors, such as the Presidents of North American, Boeing, and Grumman, with senior NASA Manned Space Flight executives (Mueller, von Braun, Gilruth, and Debus). These periodic meetings proved to be an excellent mechanism for opening lines of communication at the top, for assuring timely top management attention to the most important problems as they arose, and for assuring a coordinated team effort on the many faceted Apollo–Saturn activities. A similar group of Gemini Executives was also set up; there was considerable cross-communication between the two since several of the same organizations were involved in both programs.

During 1964 ground and flight development activity accelerated further and the first of many flight components, the launch escape rockets built by the Lockheed Propulsion Company, successfully completed qualification testing.

In early 1964, the Block II CSM lunar-orbit-rendezvous configuration guidelines were forwarded by NASA to North American, and the Block II mockup was formally reviewed in September of that year. The Block I configuration had been configured before the LOR mode was chosen; as a consequence, it did not have the docking and crew-transfer provisions which, among other changes, were incorporated in the Block II.

The first Gemini mission, a successful unmanned test flight, was launched on April 8. *Ranger VII* provided the first close-up pictures of the moon in July. Project FIRE provided flight data at Apollo reentry speeds, and Saturn I flights SA–5, SA–6, and SA–7 were successfully completed during 1964. SA–7, the seventh straight Saturn I success, provided a functional verification of the Apollo Launch Escape System jettison. The unbroken string of Saturn launch successes presented a far different picture from earlier days when a 50% launch success record was considered exceptional.

In summary, the two years covered by this volume of the Chronology saw the essential completion of the putting together of the Apollo government-industry team, substantial maturing of the design, verification of many essential design features by test, streamlining of the flight program through adoption of the all-up concept, and the acquisition of first data about the lunar surface from the Ranger Program.

As this volume comes to a close, there were still four years to go before the first manned Apollo mission, and nearly five years to the first lunar landing. Many difficulties lay ahead, but the course had been marked and giant strides had been taken along that course.

John H. Disher
Deputy Director, Skylab Program
Former Director, Test Division
Apollo Program Office

CONTENTS

	PAGE
Foreword	iii
List of Illustrations	viii
The Key Events	xi
Preface	xiii
PART I: Defining Contractual Relations; November 8, 1962, Through August 28, 1963	1
PART II: Developing Hardware Distinctions; August 30, 1963, Through April 28–30, 1964	81
PART III: Developing Software Ground Rules; April 29, 1964, Through September 30, 1964	167

Appendixes

1. Glossary of Abbreviations	233
2. Spacecraft Weight by Quarter	235
3. Major Spacecraft Component Manufacturers	239
4. Flight Summary	240
5. Apollo Program Flight Objectives	245
6. Hardware Manufacture and Acceptance	247
7. Funding	249
8. Organizational Charts	250
Index	269

LIST OF ILLUSTRATIONS

All photographs and illustrations are U.S. Government ones except as credited. Persons in the photographs are identified in the captions.

PART I

	PAGE
SA-3 on the pad	4
Little Joe II contract signed	8
Transposition and docking maneuvers	9
Vertical Assembly Building	11
Apollo suit	15
CM boilerplate 3 delivered	16
Simulation of 1/6 g	17
Space Environmental Facility (artist's concept)	20
Cutaway concept of Environmental Chamber	20
Command module with strakes	23
Merritt Island ground-breaking ceremony	26
Crawler-transporter	28
Drop test at NAA	29
Michoud Assembly Facility	31
MSC centrifuge (artist's concept)	33
Little Joe II assembly	34
Hardware recovery possibilities (cartoon)	35
Astronauts in mockup	37
Martin Company "bug"	39
Philco contract signing	40
SA-4 poised for liftoff	43
Command module fabrication	46
Launch escape motor	48
Combustion experiments	49
Hypervelocity Ballistic Range	51
LEM model shown	52
Little Joe II progress checked	54
Lunar personnel propulsion proposal	57

LIST OF ILLUSTRATIONS

	PAGE
Descent engine mockup	60
Skip trajectory	62
Apollo mission simulator	64
Stabilization and control subsystems	66
LEM models	66
Arnold Engineering Development Center facility	67
Frictionless platform	68
BP-6 arrival at WSMR	70
Docking probe and drogue assembly	72
LEM ascent engine	74
Equal-period orbits	75
Little Joe II qualification test	79

PART II

Lunar Orbiter model	83
BP-12 Design Engineering Inspection	87
NASA Administrator examining model	90
Guidance and navigation system mockup	92–93
Lunar Landing Research Facility	97
"Retriever"	99
Optimum LEM descent	100
Prototype space suit	102
Reaction control thrusters	105
BP-6 pad abort test	108
Wind tunnel testing	109
Transonic Wind Tunnel	116
Launch escape motor firing	128
Flight director attitude indicator	134
Launch escape vehicle configuration	135
Apollo spacecraft simulator (Honeywell)	140
Thermal Vacuum Facility	142
Service module mockup	143
Heating experiment	146
Mission Control Center	149
LEM test mockup	150
Evolving spacecraft	152
BP-13 hardware	153
Lunar landing research vehicle	155

THE APOLLO SPACECRAFT: A CHRONOLOGY

	PAGE
Gemini–Titan I launch	156
Space chamber door	157
Project Fire	158

PART III

Communications links	170
BP-12	176
Zero-g mockup	178
Controls and panel displays	179
LEM metal mockup	181
SA-6 space vehicle	182
Apollo flotation collar	184
Docking concept	185
Simplified mockup	187
Gas storage system	191
J-2 engine assembly line	193
BP-15 stack	197
Lunar TV camera	198
Crawler track	199
Ranger VII photos	203
Water-cooled garment	205
Electroluminescent lighting	206
Thermal overgarments	207
BP-15 transported to pad	208
Honeycomb installation	210
Suit test	211
CM uprighting system	212
Weightless familiarization	214
Astronaut mobility test	216
Flight directors	219
Flight kit assembly	220
LEM foot pad	223
BP-15	226
Service module propulsion test	228
Block II mockup	229

THE KEY EVENTS

1962

During December: Manned Spacecraft Center (MSC) prepared the preliminary lunar landing mission design.

1963

January 2: Radio Corporation of America awarded contract for two large vacuum chambers at MSC for space environmental testing.
January 28: Philco Corporation selected as prime contractor for the Mission Control Center (MCC) at MSC.
February 18: Signed definitive contract with General Dynamics/Convair for Little Joe II test vehicles.
March 11: NASA and Grumman Aircraft Engineering Corporation formalized a definitive contract for development of the lunar excursion module (LEM).
July 12: International Business Machines awarded definitive contract for the realtime computer complex at MSC's MCC.
August 14: Signed definitive contract worth $938.4 million with North American Aviation, Inc. (NAA), for command and service modules (CSM) on a cost-plus-fixed-fee basis.
August 30: NASA approved the Lunar Orbiter program.
November 7: Apollo Pad Abort Mission 1, using command module (CM) boilerplate (BP) 6, conducted at White Sands Missile Range (WSMR).
November 22: Preliminary ground rules selected for the Spacecraft Development Test Program and lunar landing sites.

1964

January 21: NAA presented a block change concept for the Block II CSM design for lunar missions.
March 9: MSC assigned funds and responsibility for developing scientific instruments for lunar exploration.
April 28–30: NAA held basic mockup inspection and review for the Block II CSM.
May 4: Program mission objectives and ground rules specified by the Apollo Mission Planning Task Force.
May 13: First flight test of Little Joe II using CM BP-12 at WSMR.
July 28: Ranger VII televised pictures of lunar surface up to impact.
September 14: Ground rules firmly defined for LEM guidance and control system.
September 18: Apollo Mission A-102, using BP-15 for the CSM and SA-7 for the launch vehicle, confirmed compatibility of the Saturn Block II and CSM as well as the launch escape vehicle system.
September 30: NAA conducted formal inspection and review of the Block II CSM mockup.

PREFACE

Project Apollo, America's program to land men on the moon, aimed at what surely will be recorded as one of the epochal achievements of mankind. For any insight into the significance of this "giant leap," it is essential to reckon with the technology and to appreciate the hard work—and the sacrifices—that made *Apollo 11* possible.

This, the second volume of *The Apollo Spacecraft: A Chronology*, tells a part of this story. It follows the precedents and format of the first in the series (NASA SP-4009). The third volume, now nearly completed, will chronicle developments within Project Apollo through detailed hardware design and early ground and flight testing. A fourth will cover the development phase, recovery from the Apollo 204 fire, the first lunar landing flight, and the lunar exploration phase of the program.

By this series of documented resource books, the authors have hoped to provide a tool for further historical studies of Project Apollo and for attempts to understand scientific and technological change during the decade of the Sixties. Our immediate aim has been to serve not only the needs of scholarship and management, but also the "average American" who might wish to probe behind the headlines of space news.

Largely because our research has relied most heavily on records held at NASA's Manned Spacecraft Center (MSC) in Houston, Texas, the title of this series indicates its bias toward *spacecraft* development. Many NASA chronologies and historical monographs—some completed and some in progress—analyze, describe, and interpret various other aspects of American aeronautical and astronautical progress. In manned space flight, for example, *This New Ocean: A History of Project Mercury* (NASA SP-4101) has been written and a history of Project Gemini is nearing completion. Perhaps someday the full complexity of the interrelated technological and scientific activities of Project Apollo may be synthesized more meaningfully. But for now we have presented a skeletal outline of events that affected conceptual design and early engineering work on both hardware and software for the Apollo spacecraft.

Part I, "Defining Contractual Relations," deals mostly with establishing government-industry working arrangements and preliminary hardware design once the prime contractors were selected. Part II, "Developing Hardware Distinctions," characterizes the period of late 1963 and early 1964 as a time of technological transition. And Part III, "Developing Software Ground Rules," tells schematically how, during much of 1964, preliminary mission planning and ground tests led toward the freezing of hardware design and the movement toward flight testing.

As in previous chronologies for Mercury, Gemini, and Apollo's beginnings, the primary sources used here are NASA and industry correspondence and reports. The materials should serve as a foundation for many analytical monographs and, eventually, a narrative history of the whole Apollo program. The available documents are so plentiful and comprehensive that the primary historical problem has been one of selection. The text that follows has been edited downward in size several times, but we trust our critical readers to point out its worst sins of omission and commission. Measurements for the most part were originally in the English system, then converted to metric.

The authors of this volume worked with MSC historians James M. Grimwood and Ivan D. Ertel by virtue of a NASA contract (NAS 9-6331) with the University of Houston's Department of History. Professors James A. Tinsley and Loyd S. Swenson, Jr., provided academic support, while NASA historians Eugene M. Emme, Frank W. Anderson, Jr., and William D. Putnam encouraged the processes of research and revision toward publication. Archivists in Washington, notably Lee D. Saegesser, at other NASA Centers, and in Houston, particularly Billie D. Rowell, have been immensely helpful. Courtney G. Brooks and Sally D. Gates edited the final comment edition, and Corinne L. Morris prepared the manuscript copy. Ertel illustrated the text, while Anderson and Carrie E. Karegeannes shepherded the work through the publication process. To these and many other informants, readers, librarians, and historians the authors and editors of this series are indebted.

M. L. M.
J. K. B.

December 1, 1971

PART I

Defining Contractual Relations

November 8, 1962, through August 28, 1963

PART I

The Key Events

1962

November 16: Saturn–Apollo 3 (SA–3) launch marks first full-weight liftoff of Saturn C–1 rocket.

December 4: Contract for Vertical Assembly Building at Cape Canaveral let to a consortium of four New York architectural engineering firms.

During December: Manned Spacecraft Center (MSC) prepared the preliminary lunar landing mission design.

1963

January 2: Contract let to Radio Corporation of America for two large vacuum chambers at MSC for space environmental testing.

January 18: Contract let to Bell Aerosystems Company for two lunar landing research vehicles by Flight Research Center.

January 28: Philco Corporation selected as prime contractor for the Mission Control Center (MCC) at MSC.

February 8: Definitive contract let to Raytheon Company for command module (CM) onboard digital computer.

February 13: MSC reorganized Apollo Spacecraft Project Office.

February 18: Definitive contract let to General Dynamics/Convair for the Little Joe II test vehicle.

February 20: NASA reorganized the Office of Manned Space Flight.

March 11: Definitive contract formalized between NASA and Grumman Aircraft Engineering Corporation for the Lunar Excursion Module.

March 13: First long-duration static test of Saturn SA–5 first stage.

March 28: Saturn SA–4 launched in successful test of engine-out capability.

April 10: Contract let to Link Division, General Precision, Inc., for lunar mission simulators.

May 3: First of series of qualification drop tests for the earth landing system conducted at El Centro, Calif.

May 15–16: Last flight of Mercury: Cooper in *Faith 7*.

June 14–19: *Vostok V* and *VI* tandem flights.

During June: Most CM subsystem designs frozen.

July 12: Definitive contract let to International Business Machines for the realtime computer complex at MSC's MCC.

August 5: First static firing test of Saturn S–IV stage for SA–5 conducted by Douglas Aircraft Company in Sacramento, Calif.

August 14: Definitive contract with North American Aviation, Inc., for command and service modules signed on a cost-plus-fixed-fee basis for $938.4 million.

August 28: First Little Joe II launched at White Sands Missile Range, N. Mex.

PART I

Defining Contractual Relations

November 8, 1962, through August 28, 1963

1962
November
8

"Not one or two men will make the landing on the moon, but, figuratively, the entire Nation." That is how NASA's Deputy Administrator, Hugh L. Dryden, described America's commitment to Apollo during a speech in Washington, D.C. "What we are buying in our national space program," Dryden said, "is the knowledge, the experience, the skills, the industrial facilities, and the experimental hardware that will make the United States first in every field of space exploration. . . . The investment in space progress is big and will grow, but the potential returns on the investment are even larger. And because it concerns us all, scientific progress is everyone's responsibility. Every citizen should understand what the space program really is about and what it can do."

U.S. Congress, House, Committee on Science and Astronautics, *Astronautical and Aeronautical Events of 1962*, 88th Cong., 1st Sess. (June 12, 1963), pp. 235–36.

9

The Manned Spacecraft Center (MSC) and the Raytheon Company came to terms on the definitive contract for the Apollo spacecraft guidance computer. (See February 8, 1963.)

Manned Space Flight [MSF] Management Council Meeting, November 27, 1962, Agenda Item 2, p. 3.

13

North American Aviation, Inc., selected the Aerospace Electrical Division of Westinghouse Electric Corporation to build the power conversion units for the command module (CM) electrical system. The units would convert direct current from the fuel cells to alternating current.

Aviation Daily, November 13, 1962, p. 71.

15

The Aerojet-General Corporation reported completion of successful firings of the prototype service propulsion engine. The restartable engine, with an ablative thrust chamber, reached thrusts up to 21 500 pounds. [Normal thrust rating for the service propulsion engine is 20 500.]

Aviation Daily, November 15, 1962, p. 89; *Aviation Week and Space Technology*, 77 (November 19, 1962), p. 40.

THE APOLLO SPACECRAFT: A CHRONOLOGY

The Saturn–Apollo 3 vehicle on its launch pedestal ready for the countdown. This photo was taken three days before liftoff and presented an unusual view of the Pad 34 area and the proximity of the Atlantic Ocean.

1962

November

16

Saturn-Apollo 3 (Saturn C–1, later called Saturn I) was launched from the Atlantic Missile Range. Upper stages of the launch vehicle were filled with 23 000 gallons of water to simulate the weight of live stages. At its peak altitude of 167 kilometers (104 miles), four minutes 53 seconds after launch, the rocket was detonated by explosives upon command from earth. The water was released into the ionosphere, forming a massive cloud of ice particles several miles in diameter. By this experiment, known as "Project Highwater," scientists had hoped to obtain data on atmospheric physics, but poor telemetry made the results questionable. The flight was the third straight success for the Saturn C–1 and the first with maximum fuel on board.

> MSFC Historical Office, *History of the George C. Marshall Space Flight Center From July 1 Through December 31, 1962* (MHM–6), Vol. I, p. 193; MSFC, "Saturn SA-3 Flight Evaluation," MPR-SAT-63-1, January 8, 1963, Vol. I, pp. 8, 151; *The Washington Post*, November 17, 1962; *The New York Times*, November 17, 1962.

PART I: DEFINING CONTRACTUAL RELATIONS

1962

November

17

Four Navy officers were injured when an electrical spark ignited a fire in an altitude chamber, near the end of a 14-day experiment at the U.S. Navy Air Crew Equipment Laboratory, Philadelphia, Pa. The men were participating in a NASA experiment to determine the effect on humans of breathing pure oxygen for 14 days at simulated altitudes.

> Edward L. Michel, George B. Smith, Jr., Richard S. Johnston, *Gaseous Environment Considerations and Evaluation Programs Leading to Spacecraft Atmosphere Selection*, NASA Technical Note, TN D-2506 (1965), p. 5.

19

About 100 Grumman Aircraft Engineering Corporation and MSC representatives began seven weeks of negotiations on the lunar excursion module (LEM) contract. After agreeing on the scope of work and on operating and coordination procedures, the two sides reached fiscal accord. Negotiations were completed on January 3, 1963. Eleven days later, NASA authorized Grumman to proceed with LEM development. (See March 11, 1963.)

> MSC, "Project Apollo Quarterly Status Report No. 2 for Period Ending December 31, 1962," p. 21; "Project Apollo Quarterly Status Report No. 3 for Period Ending March 31, 1963," p. 1; NASA Contract No. NAS 9-1100, "Project Apollo Lunar Excursion Module Development Program," January 14, 1963; Clyde B. Bothmer, memorandum for distribution, "Minutes of the Fourteenth Meeting of the Management Council held on Tuesday, January 29, 1963, at the Launch Operations Center, Cocoa Beach, Florida," with enclosure: subject as above, p. 3.

19

North American defined requirements for the command and service modules (CSM) stabilization and control system.

> North American Aviation, Inc. [hereafter cited as NAA], "Apollo Monthly Progress Report," SID 62-300-8, November 30, 1962, p. 52.

20

NASA invited ten industrial firms to submit bids by December 7 for a contract to build a control center at MSC and to integrate ground operational support systems for Apollo and the rendezvous phases of Gemini. On January 28, 1963, NASA announced that the contract had been awarded to the Philco Corporation, a subsidiary of the Ford Motor Company.

> NASA News Release 63-14, "Philco to Develop Manned Flight Control Center at Houston," January 28, 1963; *Aviation Daily*, November 20, 1962, p. 111.

23

A Goddard Space Flight Center report summarizing recommendations for ground instrumentation support for the near-earth phases of the Apollo missions was forwarded to the Apollo Task Group of the NASA Headquarters Office of Tracking and Data Acquisition (OTDA). This report presented a preliminary conception of the Apollo network.

The tracking network would consist of stations equipped with 9-meter (30-foot) antennas for near-earth tracking and communications and of stations having 26-meter (85-foot) antennas for use at lunar distances. A unified S-band system, capable of receiving and transmitting voice, telemetry, and television on a single radio-frequency band, was the basis of the network operation.

1962

November

On March 12, 1963, during testimony before a subcommittee of the House Committee on Science and Astronautics, Edmond C. Buckley, Director of OTDA, described additional network facilities that would be required as the Apollo program progressed. Three Deep Space Instrumentation Facilities with 26-meter (85-foot) antennas were planned: Goldstone, Calif. (completed); Canberra, Australia (to be built); and a site in southern Europe (to be selected). Three new tracking ships and special equipment at several existing network stations for earth-orbit checkout of the spacecraft would also be needed.

> Goddard Space Flight Center, Tracking and Data Systems Directorate, "A Ground Instrumentation Support Plan for the Near-Earth Phases of Apollo Missions," November 23, 1962; U.S. Congress, House, Subcommittee on Applications and Tracking of the Committee on Science and Astronautics, *1964 NASA Authorization*, Hearings, 88th Cong., 1st Sess. (1963), pp. 2795-2801.

26

At a news conference in Cleveland, Ohio, during the 10-day Space Science Fair there, NASA Deputy Administrator Hugh L. Dryden stated that inflight practice at orbital maneuvering was essential for lunar missions. He believed that landings would follow reconnaissance of the moon by circumlunar and near-lunar-surface flights.

> *The Plain Dealer*, Cleveland, November 27, 1962.

27

NASA awarded a $2.56 million contract to Ling-Temco-Vought, Inc. (LTV), to develop the velocity package for Project Fire, to simulate reentry from a lunar mission. An Atlas D booster would lift an instrumented payload (looking like a miniature Apollo CM) to an altitude of 122 000 meters (400 000 feet). The velocity package would then fire the reentry vehicle into a minus 15 degree trajectory at a velocity of 11 300 meters (37 000 feet) per second. On December 17, Republic Aviation Corporation, developer of the reentry vehicle, reported that design was 95 percent complete and that fabrication had already begun.

> *Wall Street Journal*, November 27, 1962; LTV, Chance Vought Corporation, Astronautics Div., "Fire Velocity Package," (undated), pp. 1-1, 11-4; *Aviation Week and Space Technology*, 77 (December 17, 1962), pp. 53, 55, 57.

27

MSC officials met with representatives of Jet Propulsion Laboratory (JPL) and the NASA Office of Tracking and Data Acquisition (OTDA). They discussed locating the third Deep Space Instrumentation Facility (DSIF) in Europe instead of at a previously selected South African site. (See Volume I of this chronology [NASA SP-4009], September 13, 1960.) JPL had investigated several European sites and noted the communications gap for each. MSC stated that a coverage gap of up to two hours was undesirable but not prohibitive. JPL and OTDA agreed to place the European station where the coverage gap would be minimal or nonexistent. However, the existence of a communications loss at a particular location would not be an overriding factor against a site which promised effective technical and logistic support and political stability. MSC agreed that this was a reasonable approach.

<small>Memorandum, Gerald M. Truszynski, NASA, for file, "Meeting at MSC on Location of DSIF Station," December 3, 1964.</small>

MSC released a sketch of the space suit assembly to be worn on the lunar surface. It included a portable life support system which would supply oxygen and pressurization and would control temperature, humidity, and air contaminants. The suit would protect the astronaut against solar radiation and extreme temperatures. The helmet faceplate would shield him against solar glare and would be defrosted for good visibility at very low temperatures. An emergency oxygen supply was also part of the assembly.

Four days earlier, MSC had added specifications for an extravehicular suit communications and telemetry (EVSCT) system to the space suit contract with Hamilton Standard Division of United Aircraft Corporation. The EVSCT system included equipment for three major operations:

(1) Full two-way voice communication between two astronauts on the lunar surface, using the transceivers in the LEM and CM as relay stations

(2) Redundant one-way voice communication capability between any number of suited astronauts

(3) Telemetry of physiological and suit environmental data to the LEM or CM for relay to earth via the S-band link.

[The EVSCT contract was awarded to International Telephone and Telegraph (ITT) Corporation's Kellogg Division. (See March 26, 1963.)]

<small>Memorandum, Ralph S. Sawyer, MSC, to Crew Systems Div., Attn: James V. Correale, "Extravehicular Suit Communications and Telemetry System Specifications," November 23, 1962; MSC News Release, "Project Apollo Space Suits," November 26, 1962; *The Evening Star*, Washington, November 28, 1962; *The Houston Post*, November 27, 1962.</small>

Representatives of Hamilton Standard and International Latex Corporation (ILC) met to discuss mating the portable life support system to the ILC space suit configuration. As a result of mockup demonstrations and other studies, over-the-shoulder straps similar to those in the mockup were substituted for the rigid "horns."

<small>Hamilton Standard, "Monthly Progress Report through November 30, 1962, for Apollo Space Suit Assembly," PR-2-11-62, Item 7.2.</small>

MSC Director Robert R. Gilruth reported to the Manned Space Flight (MSF) Management Council that formal negotiations between NASA and North American on the Apollo spacecraft development contract would begin in January 1963. He further informed the council that the design release for all Apollo systems, with the exception of the space suit, was scheduled for mid-1963; the suit was scheduled for January 1964.

<small>MSF Management Council Meeting, November 27, 1962, Agenda Item 2, pp. 2-3 [and supplemental page].</small>

A $6 million contract for the Little Joe II launch vehicle went to General Dynamics/Convair of San Diego, Calif., November 28, 1962. J. H. Famme, president of General Dynamics/Convair, signed the contract. Observing were, left to right, J. Harris, contracts manager for Little Joe II at Convair; C. D. Sword, Apollo procurement chief; and J. B. Hurt, Convair's Little Joe project manager. Convair had previously been awarded a letter contract to design and manufacture the Little Joe II test vehicle. (See May 11, 1962, entry, *The Apollo Spacecraft: A Chronology*, Volume I).

1962 **November**	AC Spark Plug Division of General Motors Corporation assembled the first CM inertial reference integrating gyro (IRIG) for final tests and calibration. Three IRIGs in the CM navigation and guidance system provided a reference from which velocity and attitude changes could be sensed. Delivery of the unit was scheduled for February 1963. (See February 11, 1963.)

"Apollo Quarterly Status Report No. 2," p. 13.

During the Month	North American completed a study of CSM–LEM transposition and docking. During a lunar mission, after the spacecraft was fired into a trajectory toward the moon, the CSM would separate from the adapter section containing the LEM. It would then turn around, dock with the LEM, and pull the second vehicle free from the adapter. The contractor studied three methods of completing this maneuver: free fly-around, tethered fly-around, and mechanical repositioning. Of the three, the company recommended the free fly-around, based on NASA's criteria of minimum weight, simplicity of

PART I: DEFINING CONTRACTUAL RELATIONS

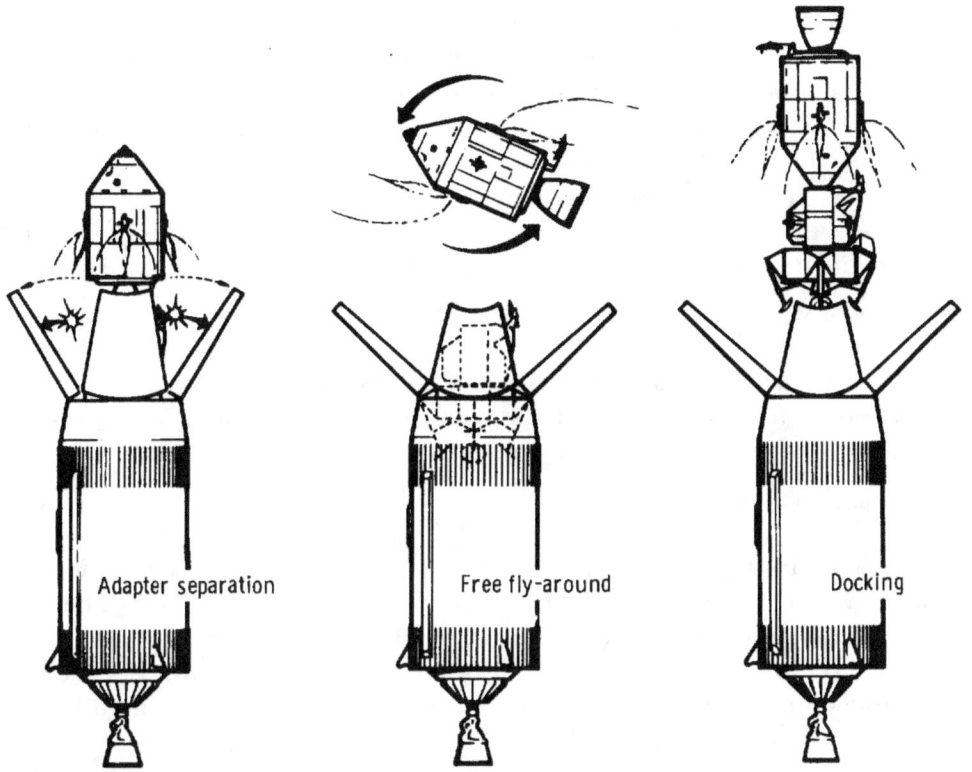

Three phases of activity in the line drawing indicate the techniques of the free fly-around method of the docking exercise between the CSM and the LEM.
—Line drawing by NAA.

design, maximum docking reliability, minimum time of operation, and maximum visibility.

1962
November

Also investigated was crew transfer from the CM to the LEM, to determine the requirements for crew performance and, from this, to define human engineering needs. North American concluded that a separate LEM airlock was not needed but that the CSM oxygen supply system's capacity should be increased to effect LEM pressurization.

On November 29, North American presented the results of docking simulations, which showed that the free flight docking mode was feasible and that the 45-kilogram (100-pound) service module (SM) reaction control system engines were adequate for the terminal phase of docking. The simulations also showed that overall performance of the maneuver was improved by providing the astronaut with an attitude display and some form of alignment aid, such as probe.

MSC, "Abstract of Proceedings, Flight Technology Systems Meeting No. 12, November 27, 1962," November 30, 1962; "Apollo Monthly Progress Report," SID 62-300-8, pp. 11-14.

1962

November

During the Month

North American reported several problems involving the CM's aerodynamic characteristics; their analysis of CM dynamics verified that the spacecraft could—and on one occasion did—descend in an apex-forward attitude. The CM's landing speed then exceeded the capacity of the drogue parachutes to reorient the vehicle; also, in this attitude, the apex cover could not be jettisoned under all conditions. During low-altitude aborts, North American went on, the drogue parachutes produced unfavorable conditions for main parachute deployment. (See January 18, 1963.)

"Apollo Monthly Progress Report," SID 62-300-8, p. 77.

During the Month

Extensive material and thermal property tests indicated that a Fiberglas honeycomb matrix bonded to the steel substructure was a promising approach for a new heatshield design for the CM. (See February 1, 1963.)

Ibid., pp. 143–144.

During the Month

Collins Radio Company selected Motorola, Inc., Military Electronics Division, to develop and produce the spacecraft S-band transponder. The transponder would aid in tracking the spacecraft in deep space; also, it would be used to transmit and receive telemetry signals and to communicate between ground stations and the spacecraft by FM voice and television links. The formal contract with Motorola was awarded in mid-February 1963.

Also, Collins awarded a contract to the Leach Corporation for the development of command and service module (CSM) data storage equipment. The tape recorders must have a five-hour capacity for collection and storage of data, draw less than 20 watts of power, and be designed for in-flight reel changes.

Ibid., p. 89; NAA, "Apollo Facts," RBO070163, (undated), pp. 43–44.

During the Month

MSC awarded a $222 000 contract to the Air Force Systems Command for wind tunnel tests of the Apollo spacecraft at its Arnold Engineering Development Center, Tullahoma, Tenn.

Aviation Week and Space Technology, 77 (November 12, 1962), p. 81.

During the Month

North American made a number of changes in the layout of the CM:

• Putting the lithium hydroxide canisters in the lower equipment bay and food stowage compartments in the aft equipment bay
• Regrouping equipment in the left-hand forward equipment bay to make pressure suit disconnects easier to reach and to permit a more advanced packaging concept for the cabin heat exchanger
• Moving the waste management control panel and urine and chemical tanks to the right-hand equipment bay
• Revising the aft compartment control layout to eliminate the landing impact attenuation system and to add tie rods for retaining the heatshield

PART I: DEFINING CONTRACTUAL RELATIONS

1962
November

• Preparing a design which would incorporate the quick release of the crew hatch with operation of the center window (drawings were released, and target weights and criteria were established)

• Redesigning the crew couch positioning mechanism and folding capabilities

• Modifying the footrests to prevent the crew's damaging the sextant.

"Apollo Monthly Progress Report," SID 62-300-8, pp. 36, 71-72, 102, 104, 195.

December 3

The MSC Apollo Spacecraft Project Office (ASPO) outlined the photographic equipment needed for Apollo missions. This included two motion picture cameras (16- and 70-mm) and a 35-mm still camera. It was essential that the camera, including film loading, be operable by an astronaut wearing pressurized gloves. On February 25, 1963, NASA informed North American that the cameras would be government furnished equipment.

Memorandum, Charles W. Frick, MSC, to Office of Asst. Dir. for Information and Control Systems, Attn: Instrumentation and Electronic Systems Div., "Cameras for Apollo Spacecraft," December 3, 1962; letter, H. P. Yschek, MSC, to NAA, Space and Information Systems Div., "Contract Change Authorization No. Twenty-Six," February 25, 1963.

The U.S. Army Corps of Engineers, acting for NASA, awarded a $3.332 million contract to four New York architectural engineering firms to design

At left is an artist's concept of the Vertical Assembly Building at Merritt Island; below, the construction in progress as of July 31, 1964.

1962

December

the Vertical Assembly Building (VAB) at Cape Canaveral. The massive VAB became a space-age hangar, capable of housing four complete Saturn V launch vehicles and Apollo spacecraft where they could be assembled and checked out. The facility would be 158.5 meters (520 feet) high and would cost about $100 million to build. Subsequently, the Corps of Engineers selected Morrison-Knudson Company, Perini Corp., and Paul Hardeman, Inc., to construct the VAB.

<small>Orlando Sentinel, December 5, 1962; MSC, Space News Roundup, January 9, 1963, p. 6; The Kennedy Space Center Story (KSC, 1969), pp. 19–20.</small>

4

The first test of the Apollo main parachute system, conducted at the Naval Air Facility, El Centro, Calif., foreshadowed lengthy troubles with the landing apparatus for the spacecraft. One parachute failed to inflate fully, another disreefed prematurely, and the third disreefed and inflated only after some delay. No data reduction was possible because of poor telemetry. North American was investigating.

<small>MSF Management Council Minutes, December 18, 1962, p. 2; NAA, "Apollo Monthly Progress Report," SID 62-300-9, January 15, 1963, p. 20.</small>

5

At a meeting held at Massachusetts Institute of Technology (MIT) Instrumentation Laboratory, representatives of MIT, MSC, Hamilton Standard Division, and International Latex Corporation examined the problem of an astronaut's use of optical navigation equipment while in a pressurized suit with helmet visor down. MSC was studying helmet designs that would allow the astronaut to place his face directly against the helmet visor; this might avoid an increase in the weight of the eyepiece. In February 1963, Hamilton Standard recommended adding corrective devices to the optical system rather than adding corrective devices to the helmet or redesigning the helmet. In the same month, ASPO set 52.32 millimeters (2.06 inches) as the distance of the astronaut's eye away from the helmet. MIT began designing a lightweight adapter for the navigation instruments to provide for distances of up to 76.2 millimeters (3 inches).

<small>"Apollo Quarterly Status Report No. 2," p. 9; Hamilton Standard Div., "Minutes of Space Suit Navigation System Optical Interface Meeting," HSER 2582-2, December 5, 1962, pp. 1–2.</small>

5

The General Electric Policy Review Board, established by the MSF Management Council, held its first meeting. On February 9, the General Electric Company (GE) had been selected by NASA to provide integration analysis (including booster-spacecraft interface), ensure reliability of the entire space vehicle, and develop and operate a checkout system. The Policy Review Board was organized to oversee the entire GE Apollo effort.

<small>Memorandum, James E. Sloan, NASA, to Wernher von Braun, Kurt H. Debus, and Robert R. Gilruth, "General Electric Policy Review Board," December 6, 1962; draft, "General Electric Policy Review Board Charter," December 4, 1962; memorandum, Sloan to Gilruth and Walter C. Williams, "Charter of Policy Review Board for General Electric Manned Lunar Landing Program Effort," January 8, 1963 (charter enclosed).</small>

PART I: DEFINING CONTRACTUAL RELATIONS

With NASA's concurrence, North American released the Request For Proposals on the Apollo mission simulator. A simulated CM, an instructor's console, and a computer complex now supplanted the three part-task trainers originally planned. An additional part-task trainer was also approved. A preliminary report describing the device had been submitted to NASA by North American. The trainer was scheduled to be completed by March 1964.

1962
December
8

"Apollo Quarterly Status Report No. 2," p. 34; NAA, "Apollo Monthly Progress Report," SID 62-300-12, May 1, 1963, p. 2.

NASA Administrator James E. Webb, in a letter to the President, explained the rationale behind the Agency's selection of lunar orbit rendezvous (rather than either direct ascent or earth orbit rendezvous) as the mode for landing Apollo astronauts on the moon. (See Volume I, July 11, 1962.) Arguments for and against any of the three modes could have been interminable: "We are dealing with a matter that cannot be conclusively proved before the fact," Webb said. "The decision on the mode . . . had to be made at this time in order to maintain our schedules, which aim at a landing attempt in late 1967."

10

John M. Logsdon, "NASA's Implementation of the Lunar Landing Decision," (HHN-81), August 1969, pp. 85, 87.

NASA authorized North American's Columbus, Ohio, Division to proceed with a LEM docking study.

11

TWX, J. F. Leonard, NAA, to NASA, [Attn:] D. B. Cherry, December 14, 1962.

The first static firing of the Apollo tower jettison motor, under development by Thiokol Chemical Corporation, was successfully performed.

11

"Apollo Monthly Progress Report," SID 62-300-9, p. 14; "Apollo Quarterly Status Report No. 2," p. 6.

Northrop Corporation's Ventura Division, prime contractor for the development of sea-markers to indicate the location of the spacecraft after a water landing, suggested three possible approaches:

12

(1) A shotgun shell type that would dispense colored smoke
(2) A floating, controlled-rate dispenser (described as an improvement on the current water-soluble binder method)
(3) A floating panel with relatively permanent fluorescent qualities.

Northrop Ventura recommended the first method, because it would produce the strongest color and size contrast and would have the longest life for its weight.

Memorandum, W. E. Oller, Northrop Ventura, to MSC, Attn: P. Armitage, "NAS 9-482, Status of Remainder of Program," December 12, 1962.

MSC officials, both in Houston and at the Preflight Operations Division at Cape Canaveral, agreed on a vacuum chamber at the Florida location to test

13

1962
December

spacecraft systems in a simulated space environment during prelaunch checkout.

> Memorandum, A. D. Mardel, MSC, to Distribution, "Minutes of meeting on NASA AMR Vacuum Chamber requirements," December 14, 1962.

15 The first working model of the crew couch was demonstrated during an inspection of CM mockups at North American. As a result, the contractor began redesigning the couch to make it lighter and simpler to adjust. Design investigation was continuing on crew restraint systems in light of the couch changes. An analysis of acceleration forces imposed on crew members during reentry at various couch back and CM angles of attack was nearing completion.

> "Apollo Quarterly Status Report No. 2," pp. 9, 10; NASA-Resident Apollo Spacecraft Project Office (RASPO/NAA), "Consolidated Activity Report . . . , December 1, 1962–January 5, 1963," p. 3.

18 MSC Director Robert R. Gilruth reported to the MSF Management Council that tests by Republic Aviation Corporation, the U.S. Air Force School of Aerospace Medicine (SAM) at Brooks Air Force Base, Tex., and the U.S. Navy Air Crew Equipment Laboratory (ACEL) at Philadelphia, Pa., had established that, physiologically, a spacecraft atmosphere of pure oxygen at 3.5 newtons per square centimeter (five pounds per square inch absolute [psia]) was acceptable. During the separate experiments, about 20 people had been exposed to pure oxygen environments for periods of up to two weeks without showing adverse effects. Two fires had occurred, one on September 10 at SAM and the other on November 17 at ACEL. The cause in both cases was faulty test equipment. On July 11, NASA had ordered North American to design the CM for 3.5 newtons per square centimeter (5-psia), pure-oxygen atmosphere.

> MSF Management Council Minutes, December 18, 1962, p. 3; "Apollo Quarterly Status Report No. 2," p. 11; "Abstract of Proceedings, Crew Systems Meeting No. 13, December 18, 1962," December 20, 1962.

19 NASA announced that *Ranger VI* (see Volume I, August 29, 1961) would be used for intensive reliability tests. Resultant improvements would be incorporated into subsequent spacecraft (numbers VII–IX), delaying the launchings of those vehicles by "several months." The revised schedule was based on recommendations by a Board of Inquiry headed by Cdr. Albert J. Kelley (USN), Director of Electronics and Control in the NASA Office of Advanced Research and Technology. (See Volume I, October 18, 1962.) The Kelley board, appointed by NASA Space Sciences Director Homer E. Newell after the *Ranger V* flight, consisted of officials from NASA Headquarters, five NASA Centers, and Bellcomm, Inc. The board concluded that increased reliability could be achieved through spacecraft design and construction modifications and by more rigorous testing and checkout. (See January 30, 1964.)

PART I: DEFINING CONTRACTUAL RELATIONS

The Washington Post, December 20, 1962; *The Evening Star*, Washington, December 20, 1962; U.S. Congress, House, Subcommittee on Space Sciences and Advanced Research and Technology of the Committee on Science and Astronautics, *1964 NASA Authorization*, Hearings on H. R. 5466, 88th Cong., 1st Sess. (1963), pp. 1597–1598.

1962
December

MSC prognosticated that, during landing, exhaust from the LEM's descent engine would kick up dust on the moon's surface, creating a dust storm. Landings should be made where surface dust would be thinnest.

20

NASA Project Apollo Working Paper No. 1052, "A Preliminary Analysis of the Effects of Exhaust Impingement on the Lunar Surface During the Terminal Phases of Lunar Landing," December 20, 1962.

North American delivered CM boilerplate (BP) 3, to Northrop Ventura, for installation of an earth-landing system. BP–3 was scheduled to undergo parachute tests at El Centro, Calif., during early 1963.

21

RASPO/NAA, "Consolidated Activity Report . . . , December 1, 1962–January 5, 1963," p. 7.

The Minneapolis-Honeywell Regulator Company submitted to North American cost proposal and design specifications on the Apollo stabilization and

26

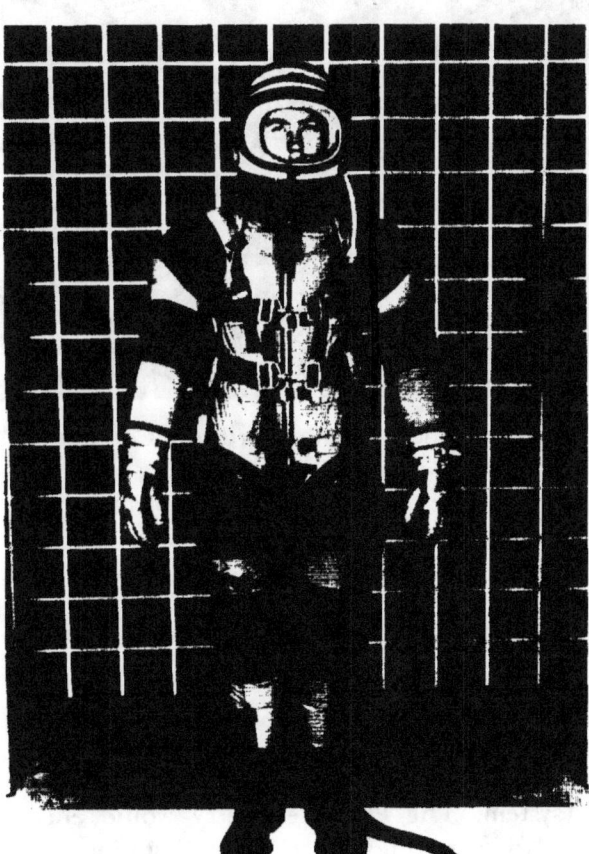

Many changes were made in space suits before the Apollo suit was finally developed in flight configuration. Joe Kosmo, Crew Systems Division, is shown on December 20, 1962, modeling a suit with constant volume joints constructed of restrained bellows.

CM boilerplate (BP) 3 was being off-loaded at the Northrop Corporation's Ventura Division plant at Newbury Park, Calif.

—Northrop photo.

1962
December

control system, based upon the new Statement of Work drawn up on December 17.

"Apollo Quarterly Status Report No. 2," p. 16.

28

North American selected Radiation, Inc., to develop the CM pulse code modulation (PCM) telemetry system. The PCM telemetry would encode spacecraft data into digital signals for transmission to ground stations. The $4.3 million contract was officially announced on February 15, 1963.

PART I: DEFINING CONTRACTUAL RELATIONS

"Apollo Monthly Progress Report," SID 62-300-9, p. 20; NAA, "Apollo Facts," RBC/070163, (undated), pp. 44-45; *Space Business Daily*, February 26, 1963, p. 243.

1962
December
28

Lockheed Propulsion Company successfully static fired four launch escape system pitch-control motors. In an off-the-pad or low-altitude abort, the pitch-control motor would fix the trajectory of the CM after its separation from the launch vehicle.

"Apollo Monthly Progress Report," SID 62-300-9, p. 14; NAA, "Quarterly Reliability Status Report," SID 62-557-4, January 31, 1964, pp. 242, 246.

North American's Rocketdyne Division completed the first test firings of the CM reaction control engines.

28

Ralph B. Oakley, *Historical Summary, S&ID Apollo Program* (NAA, Space and Information Systems Div., January 20, 1966), p. 8; "Apollo Monthly Progress Report," SID 62-300-9, p. 13.

Langley Research Center conducted studies to determine what problems might be encountered by an astronaut wearing a space suit and walking on the lunar surface where the lunar force of gravity is only one sixth of that on earth. In this laboratory device, a system of slings supported most of the weight of the man and allowed him to walk and jump under conditions simulating lunar gravity. Here a scientist was being prepared for tests in the facility.

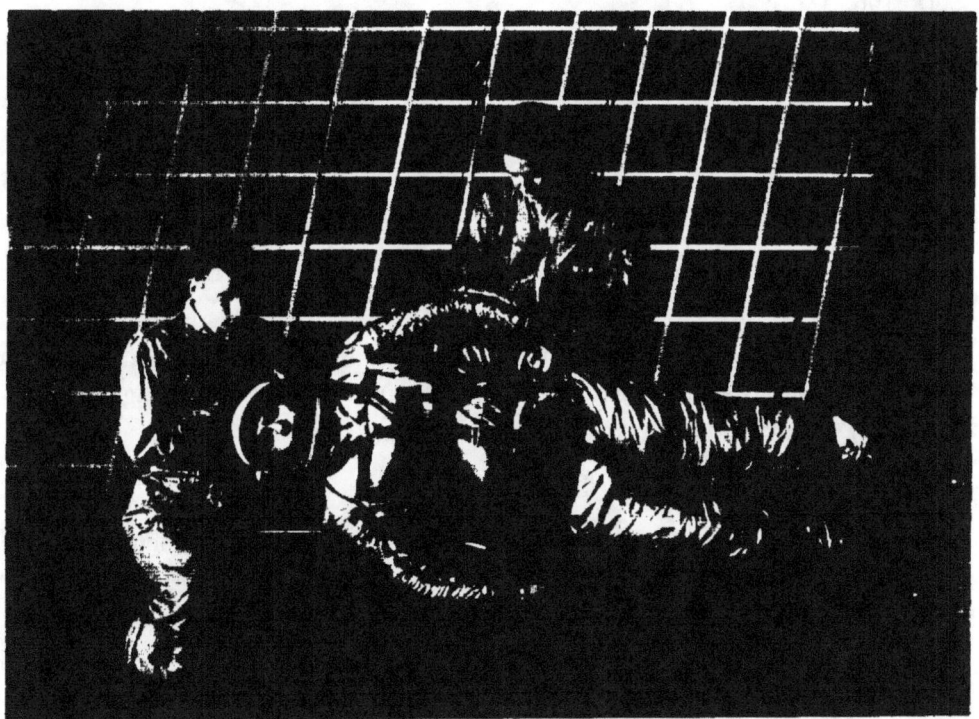

1962

December

During the Month

MSC prepared the Project Apollo lunar landing mission design. This plan outlined ground rules, trajectory analyses, sequences of events, crew activities, and contingency operations. It also predicted possible planning changes in later Apollo flights.

"Apollo Quarterly Status Report No. 2," p. 4.

During the Month

In the first of a series of reliability-crew safety design reviews on all systems for the CM, North American examined the spacecraft's environmental control system (ECS). The Design Review Board approved the overall ECS concept, but made several recommendations for further refinement. Among these were:

• The ECS should be made simpler and the system's controls should be better marked and located.
• Because of the pure oxygen environment, all flammable materials inside the cabin should be eliminated.
• Sources of possible atmospheric contamination should be further reviewed, with emphasis upon detecting and controlling such toxic gases inside the spacecraft.

"Quarterly Reliability Status Report," SID 62-557-4.

During the Month

NASA and General Dynamics/Convair (GD/C) began contract negotiations on the Little Joe II launch vehicle, which was used to flight-test the Apollo launch escape system. The negotiated cost was nearly $6 million. GD/C had already completed the basic structural design of the vehicle. (See February 18, 1963.)

General Dynamics, Convair Div., *Little Joe II Test Launch Vehicle, NASA Project Apollo: Final Report*, GDC-66-042 (May 1966), Vol. I, pp. 1-2, 1-4, 4-2, 4-3.

During the Month

North American reported three successful static firings of the launch escape motor. The motor would pull the CM away from the launch vehicle if there were an abort early in a mission.

"Apollo Quarterly Status Report No. 2," p. 6; "Quarterly Reliability Status Report," SID 62-557-4, p. 242.

During the Month

MSC reported that the general arrangement of the CM instrument panel had been designed to permit maximum manual control and flight observation by the astronauts.

"Apollo Quarterly Status Report No. 2," pp. 8, 9.

During the Month

MSC Flight Operations Division examined the operational factors involved in Apollo water and land landings. Analysis of some of the problems leading to a preference for water landing disclosed that:

• Should certain systems on board the CM fail, the spacecraft could land as far as 805 kilometers (500 miles) from the prime recovery area. This contingency could be provided for at sea, but serious difficulties might be encountered on land.

- Because Apollo missions might last as long as two weeks, weather forecasting for the landing zone probably would be unreliable.
- Hypergolic fuels were to remain on board the spacecraft through landing. During a landing at sea, the bay containing the tanks would flood and seawater would neutralize the liquid fuel or fumes from damaged tanks. On land, the possibility of rupturing the tanks was greater and the danger of toxic fumes and fire much more serious.
- Should the CM tumble during descent, the likelihood of serious damage to the spacecraft was less for landings on water.
- On land, obstacles such as rocks and trees might cause serious damage to the spacecraft.
- The spacecraft would be hot after reentry. Landing on water would cool the spacecraft quickly and minimize ventilation problems.
- The requirements for control during reentry were less stringent in a sea landing, because greater touchdown dispersions could be allowed.
- Since the CM must necessarily be designed for adequate performance in a water landing (all aborts during launch and most contingencies required a landing at sea), the choice of water as the primary landing surface could relieve some constraints in spacecraft design. (See February 1 and March 5, 1963; February 25, 1964.)

Memorandum, Christopher C. Kraft, Jr., MSC, to Mgr., ASPO, "Review of Operational Factors Involved in Water and Land Landings," undated (ca. December 1962).

1962

December

The contract for the development and production of the CSM C-band transponder was awarded to American Car and Foundry Industries, Inc., by Collins Radio Company. The C-band transponder was used for tracking the spacecraft. Operating in conjunction with conventional, earth-based, radar equipment, it transmitted response pulses to the Manned Space Flight Network.

"Apollo Quarterly Status Report No. 2," p. 18; "Apollo Monthly Progress Report," SID 62-300-9, p. 10.

During the Month

Grumman agreed to use existing Apollo components and subsystems, where practicable, in the LEM. This promised to simplify checkout and maintenance of spacecraft systems.

MSC, "Contract Implementation Plan, Lunar Excursion Module, Project Apollo," November 11, 1962, p. 5; *Aviation Week and Space Technology*, 78 (January 14, 1963), p. 39.

During the Quarter

MSC awarded a $3.69 million contract to the Radio Corporation of America (RCA) Service Company to design and build two vacuum chambers at MSC. The facility was used in astronaut training and spacecraft environmental testing. Using carbon arc lamps, the chambers simulated the sun's intensity,

1963

January

2

At the left is an artist's concept of the MSC space environmental simulation laboratory, showing the complete facility. Below is a drawing of the larger chamber, including the position of simulated solar sources.

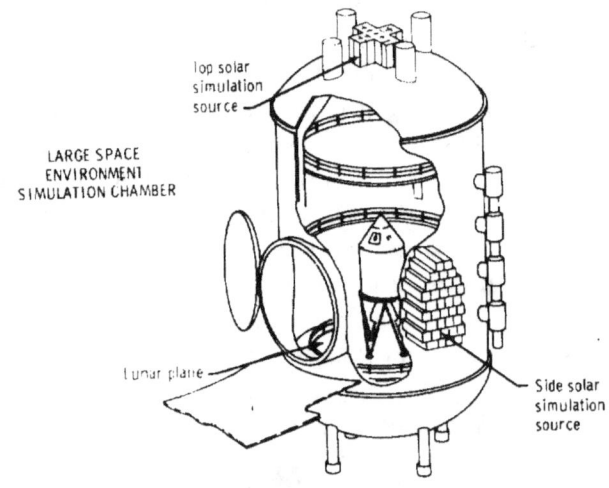

1963

January

permitting observation of the effects of solar heating encountered on a lunar mission. At the end of July, MSC awarded RCA another contract (worth $3 341 750) for these solar simulators.

> MSC Release 63-1, "Contract Awarded to RCA Services Company" [January 2, 1963]; MSC, "Consolidated Activity Report for the Office of the Director, Manned Space Flight, July 21–August 17, 1963," p. 3.

8

After studying the present radar coverage provided by ground stations for representative Apollo trajectories, North American recommended that existing C-band radars be modified to increase ranging limits. The current capability for tracking to 920 kilometers (500 nautical miles), while satisfactory for near-earth trajectories, was wholly inadequate for later Apollo missions. Tracking capability should be extended to 59 000 kilometers (32 000 nautical miles), North American said; and to improve tracking accuracy, transmitter power and receiver sensitivity should be increased.

> Memorandum, C. H. Feltz, NAA, to MSC, Attn: J. T. Markley, "Contract No. NAS 9–150, Research and Development for Project Apollo Spacecraft, C-Band Coverage Preliminary Report," January 8, 1963.

PART I: DEFINING CONTRACTUAL RELATIONS

Joseph F. Shea, Director of the Office of Systems in NASA's Office of Manned Space Flight (OMSF), briefed MSC officials on the nature and scope of NASA's contract with Bellcomm for systems engineering support. Also, Shea familiarized them with the organization and operation of the Office of Systems *vis-a-vis* Bellcomm. [Bellcomm, a separate corporation formed by American Telephone and Telegraph and Western Electric early in 1962, specifically at NASA's request, furnished engineering support to the overall Apollo program.] Bellcomm's studies, either in progress or planned, included computer support, environmental hazards, mission safety and reliability, communications and tracking, trajectory analyses, and lunar surface vehicles.

1963

January

8

Memorandum, Paul E. Purser, MSC, to Distribution, "Operations of OMSF Office of Systems and Bellcomm," January 14, 1963.

MSC and OMSF agreed that an unmanned Apollo spacecraft must be flown on the Saturn C-1 before a manned flight. SA-10 was scheduled to be the unmanned flight and SA-111, the first manned mission.

10

Memorandum, John H. Disher, NASA, to MSC, Attn: Paul E. Purser, "Review of Apollo Quarterly Status Report No. 2," January 23, 1963.

The MSC Flight Operations Division's Mission Analysis Branch analyzed three operational procedures for the first phase of descent from lunar orbit:

16

(1) The first was a LEM-only maneuver. The LEM would transfer to an orbit different from that of the CSM but with the same period and having a pericynthion of 15 240 meters (50 000 feet). After one orbit and reconnaissance of the landing site, the LEM would begin descent maneuvers.

(2) The second method required the entire spacecraft (CSM/LEM) to transfer from the initial circular orbit to an elliptical orbit with a pericynthion of 15 240 meters (50 000 feet).

(3) The third technique involved the LEM's changing from the original 147-kilometer (80-nautical-mile) circular orbit to an elliptic orbit having a pericynthion of 15 240 meters (50 000 feet). The CSM, in turn, would transfer to an elliptic orbit with a pericynthion of 65 kilometers (30 nautical miles). This would enable the CSM to keep the LEM under observation until the LEM began its descent to the lunar surface.

Comparisons of velocity changes and fuel requirements for the three methods showed that the second technique would use much more fuel than the others and, therefore, was not recommended for further consideration.

[Apocynthion and pericynthion are the high and low points, respectively, of an object in orbit around the moon (as, for example, a spacecraft sent from earth). Apolune and perilune also refer to these orbital parameters, but these latter two words apply specifically to an object launched from the moon itself.]

Memorandum, Stephen Huzar, MSC, to Chief, FOD, "Comparison of Fuel Requirements for Three Near-Moon Orbital Techniques Associated With the Planning of the Lunar Landing Mission," January 16, 1963.

1963

16–February 15 North American awarded Airborne Instruments Laboratory, a division of Cutler-Hammer, Inc., a contract for the CM recovery antenna system.

> NAA, "Apollo Monthly Progress Report," SID 62-300-10, March 1, 1963, p. 3.

16–February 15 Representatives of North American, Langley Research Center, Ames Research Center, and MSC discussed CM reentry heating rates. They agreed on estimates of heating on the CM blunt face, which absorbed the brunt of reentry, but afterbody heating rates were not as clearly defined. North American was studying Project Mercury flight data and recent Apollo wind tunnel tests to arrive at revised estimates.

> "Apollo Quarterly Status Report No. 3," p. 33; "Apollo Monthly Progress Report," SID 62-300-10, p. 7.

17 Christopher C. Kraft, Jr., of MSC's Flight Operations Division (FOD), advised ASPO that the digital up-data link being developed for the Gemini program appeared acceptable for Apollo as well. In late October 1962, representatives of FOD and ASPO had agreed that an independent up-data link (a means by which the ground could feed current information to the spacecraft's computer during a mission) was essential for manned Apollo flights. Kraft proposed that the Gemini-type link be used for Apollo as well, and on June 13 MSC ordered North American to include the device in the CM.

> Memorandum, Christopher C. Kraft, Jr., MSC, to Mgr., ASPO, "Apollo Up-Data Link," January 17, 1963; letter, H. P. Yschek, MSC, to NAA, Space and Information Systems Div., "Contract Change Authorization No. Fifty-Four," June 13, 1963.

17 President John F. Kennedy sent his budget request for Fiscal Year 1964 to Congress. The President recommended a NASA appropriation of $5.712 billion, $3.193 billion of which was for manned space flight. Apollo received a dramatic increase—$1.207 billion compared with $435 million the previous year. NASA Administrator James E. Webb nonetheless characterized the budget, about half a billion dollars less than earlier considered, as one of "austerity." While it would not appreciably speed up the lunar landing timetable, he said, NASA could achieve the goal of placing a man on the moon within the decade.

> *The Houston Post*, January 18, 1963.

18 Two aerodynamic strakes were added to the CM to eliminate the danger of a hypersonic apex-forward trim point on reentry. [During a high-altitude launch escape system (LES) abort, the crew would undergo excessive g forces if the CM were to trim apex forward. During a low-altitude abort, there was the potential problem of the apex cover not clearing the CM. (See November 1962.)] The strakes, located in the yaw plane, had a maximum span of one foot and resulted in significant weight penalties. The size of the strakes had to be increased later because of changes in the CM which moved the center of gravity forward and because of the additional ablative

PART I: DEFINING CONTRACTUAL RELATIONS

A command module with strakes is shown inside the clean room at NAA just after de-mating from the service module following combined systems checkout. The strakes may be observed at either side of the command module, just above the aft heatshield line.　　　　　　　　　　—NAA photo.

material needed to combat the increased heating of the strakes during reentry. Removal of the strakes would cause a major redesign to permit the apex cover to be jettisoned in the low angle-of-attack (apex forward) region. In the summer of 1963, however, MSC and North American representatives agreed that the strakes should be removed and an apex-mounted flap be added. The flap could be jettisoned with the LES tower during normal missions and retained with the CM during a LES abort.

North American then suggested a "tower flap dual mode" approach. This concept incorporated fixed surfaces at the upper end of the LES tower which would be exposed to the air stream after jettison of the expended rocket casing. For aborts below 9140 meters (30 000 feet), the jettison motor would pull away the expended motor casing, the LES tower, and apex cover. The contractor carried out extensive wind tunnel tests of this configuration and reported to MSC during October that a 0.5941-square-meter (920-square-inch) planer flap located in the upper bay of the LES, coupled with a more favorable CM center of gravity, would be required to solve the reentry problem.

1963

January

1963

January

An independent investigation of deployable aerodynamic surfaces, or canards, at the forward end of the LES rocket motor was also being conducted. These canards would act as lifting surfaces to destabilize the LES and cause it to reorient the spacecraft to a heatshield-forward position. (See November 12, 1963, February 7 and 25, 1964.)

> "Apollo Monthly Progress Report," SID 62-300-9, p. 6; *ibid.*, SID 62-300-10, p. 5; *ibid.*, SID 62-300-11, April 1, 1963, p. 7; *ibid.*, SID 62-300-12, p. 8; *ibid.*, SID 62-300-15, August 1, 1963, p. 5; *ibid.*, SID 62-300-16, September 1, 1963, p. 8; *ibid.*, SID 62-300-17, October 1, 1963, p. 5; *ibid.*, SID 62-300-18, November 1, 1963, p. 3; *ibid.*, SID 62-300-19, December 1, 1963, p. 5; *ibid.*, SID 62-300-20, January 1, 1964, p. 5; *ibid.*, SID 62-300-21, February 1, 1964, p. 3; *ibid.*, SID 62-300-23, April 1, 1964, p. 3; "ASPO Weekly Activity Report, September 19-25, 1963," p. 3; "ASPO Weekly Activity Report, September 26-October 2, 1963," p. 2; "ASPO Status Report For Period Ending October 16, 1963"; "ASPO Status Report For Period October 16-November 12, 1963"; "ASPO Status Report For Period December 18-January 14, 1964"; "ASPO Status Report For Week Ending December 4, 1963"; "ASPO Status Report For Week Ending December 17, 1963"; "ASPO Status Report For Week Ending January 7, 1964"; "Monthly ASPO Status Report For Period January 16-February 12, 1964"; "Apollo Quarterly Status Report No. 3," p. 32; "Apollo Quarterly Status Report No. 4 for Period Ending June 30, 1963," p. 28; "Apollo Quarterly Status Report No. 5 for Period Ending September 30, 1963," p. 40; "Apollo Quarterly Status Report No. 6 for Period Ending December 31, 1963," p. 37; MSC, "Weekly Activity Report for the Office of the Director, Manned Space Flight, June 30-July 6, 1963," p. 4; "Minutes of NASA-NAA Technical Management Meeting, February 25, 1964"; Oakley, *Historical Summary, S&ID Apollo Program*, p. 12.

18

NASA's Flight Research Center (FRC) announced the award of a $3.61 million contract to Bell Aerosystems Company of Bell Aerospace Corporation for the design and construction of two manned lunar landing research vehicles. The vehicles would be able to take off and land under their own power, reach an altitude of about 1220 meters (4000 feet), hover, and fly horizontally. A fan turbojet engine would supply a constant upward push of five-sixths the weight of the vehicle to simulate the one-sixth gravity of the lunar surface. Tests would be conducted at FRC.

> *Astronautics and Aeronautics, 1963* (NASA SP-4004), p. 17; *Daily Press*, Newport News, Va., January 13, 1963; *Wall Street Journal*, January 22, 1963; *Aviation Daily*, January 24, 1963, p. 161.

23

The Hamilton Standard space suit contract was amended to include supplying space suit communications and telemetry equipment. (See November 27, 1962.)

> Hamilton Standard, "Monthly Progress Report for the Period of January 1 through 31, 1963, for Apollo Space Suit Assembly," PR-4-1-63, p. 1.

24

The first evaluation of crew mobility in the International Latex Corporation (ILC) pressure suit was conducted at North American to identify interface problems. Three test subjects performed simulated flight tasks inside a CM mockup. CM spatial restrictions on mobility were shown. Problems involving suit sizes, crew couch dimensions, and restraint harness attachment, adjustment, and release were appraised. Numerous items that conflicted

PART I: DEFINING CONTRACTUAL RELATIONS

with Apollo systems were noted and passed along to ILC for correction in the continuing suit development program. (See March 26–28.)

1963
January

"Project Apollo Spacecraft, Test Program Weekly Activities Report (Period, 21 January 1963 through 27 January 1963)," p. 6.

MSC announced new assignments for the seven original astronauts: L. Gordon Cooper, Jr., and Alan B. Shepard, Jr., would be responsible for the remaining pilot phases of Project Mercury; Virgil I. Grissom would specialize in Project Gemini; John H. Glenn, Jr., would concentrate on Project Apollo; M. Scott Carpenter would cover lunar excursion training; and Walter M. Schirra, Jr., would be responsible for Gemini and Apollo operations and training. As Coordinator for Astronaut Activities, Donald K. Slayton would maintain overall supervision of astronaut duties.

26

Specialty areas for the second generation were: trainers and simulators, Neil A. Armstrong; boosters, Frank Borman; cockpit layout and systems integration, Charles Conrad, Jr.; recovery system, James A. Lovell, Jr.; guidance and navigation, James A. McDivitt; electrical, sequential, and mission planning, Elliot M. See, Jr.; communications, instrumentation, and range integration, Thomas P. Stafford; flight control systems, Edward H. White II; and environmental control systems, personal equipment, and survival equipment, John W. Young.

MSC Fact Sheet No. 113, "Specialized Assignments for MSC Astronauts and Flight Crew Personnel," January 26, 1963; *The Washington Post*, January 27, 1963.

NASA announced the selection of the Philco Corporation as prime contractor for the Mission Control Center (MCC) at MSC. To be operational in mid-1964, MCC would link the spacecraft with ground controllers at MSC through the worldwide tracking network.

28

NASA News Release 63-14, "Philco to Develop Manned Flight Mission Control Center at Houston," January 28, 1963; *Wall Street Journal*, January 29, 1963.

Following a technical conference on the LEM electrical power system (EPS), Grumman began a study to define the EPS configuration. Included was an analysis of EPS requirements and of weight and reliability for fuel cells and batteries. Total energy required for the LEM mission, including the translunar phase, was estimated at 61.3 kilowatt-hours. Upon completion of this and a similar study by MSC, Grumman decided upon a three-cell arrangement with an auxiliary battery. Capacity would be determined when the EPS load analysis was completed. (See March 7.)

28

"Apollo Quarterly Status Report No. 3," pp. 27–28.

Grumman and NASA announced the selection of four companies as major LEM subcontractors:

30

(1) Rocketdyne for the descent engine (see February 13)
(2) Bell Aerosystems Company for the ascent engine (see February 25)

25

Ground was broken for the MSC Operations and Checkout Building at Merritt Island January 28, 1963. Participants were, left to right, Walter C. Williams, Director of Flight Operations, MSC; G. Merritt Preston, Director of Pre-Flight Operations Division, MSC; Kurt H. Debus, Director, Launch Operations Center; D. Brainerd Holmes, Director, NASA Office of Manned Space Flight; Wernher von Braun, Director, Marshall Space Flight Center; Col. H. R. Parfitt, District Engineer, U.S. Army; and Col. E. Richardson, U.S. Air Force.

1963

January

(3) The Marquardt Corporation for the reaction control system (see March 11)

(4) Hamilton Standard for the environmental control system (see March 4).

MSC News Release 63-14, January 30, 1963; *Aviation Daily*, January 30, 1963, p. 210; *Wall Street Journal*, January 31, 1963.

During the Month

MSC awarded a contract to Chance Vought Corporation for a study of guidance system techniques for the LEM in an abort during lunar landing.

NASA News Release 63-41, "January Contracts," March 4, 1963.

February 1

NASA authorized North American to extend until June 10 the CM heatshield development program. This gave the company time to evaluate and recommend one of the three ablative materials still under consideration. The materials were subjected to tests of thermal performance, physical and mechanical properties, and structural compatibility with the existing heatshield substructure. North American sought also to determine the manufacturing feasibility of placing the materials in a Fiberglas honeycomb matrix bonded to a steel substructure. (See November 1962.)

Letter, H. P. Yschek, MSC, to NAA, Space and Information Systems Div., "Contract Change Authorization No. Thirteen, Revision 2," March 11, 1963.

PART I: DEFINING CONTRACTUAL RELATIONS

1963
February
1

Walter C. Williams, MSC's Associate Director, defined the Center's criteria on the location of earth landing sites for Gemini and Apollo spacecraft: site selection as well as mode of landing (i.e., land versus water) for each mission should be considered separately. Constraints on trajectory, landing accuracy, and landing systems must be considered, as well as lead time needed to construct landing area facilities. Both Gemini and Apollo flight planning had to include water as well as land landing modes. (See December 1962.) Although the Apollo earth landing system was designed to withstand the shock of coming down on varying terrains, some experience was necessary to verify this capability. Because of the complexity of the Apollo mission and because the earth landing system did not provide a means of avoiding obstacles, landing accuracy was even more significant for Apollo than for Gemini. With so many variables involved, Williams recommended that specific landing locations for future missions not be immediately designated. (See March 5 and February 25, 1964.)

> Memorandum, Walter C. Williams, MSC, to NASA Headquarters, Attn: OMSF, "Designation of Landing Sites for Projects Gemini and Apollo," February 1, 1963.

6

Aerojet-General Corporation, Sacramento, Calif., began full-scale firings of a service propulsion engine with a redesigned injector baffle.

> MSC, "Consolidated Activity Report for the Office of the Director, Manned Space Flight, January 27–February 23, 1963," p. 56.

7

NASA announced a simplified terminology for the Saturn booster series: Saturn C–1 became "Saturn I," Saturn C–1B became "Saturn IB," and Saturn C–5 became "Saturn V."

> MSC Fact Sheet No. 136, "NASA Simplifies Names of Saturn Launch Vehicles," February 7, 1963.

8

MSC issued a definitive contract for $15 029 420 to the Raytheon Company, Space and Information Systems Division, to design and develop the CM onboard digital computer. The contract was in support of the MIT Instrumentation Laboratory, which was developing the Apollo guidance and navigation systems. Announcement of the contract was made on February 11.

> MSC, "Consolidated Activity Report for the Office of the Director, Manned Space Flight, January 27–February 23, 1963," p. 29; MSC News Release 63-18, February 11, 1963; *Missiles and Rockets*, 12 (February 18, 1963), p. 42.

11

The first inertial reference integrating gyro produced by AC Spark Plug was accepted by NASA and delivered to the MIT Instrumentation Laboratory. (See November 1962.)

> MSC, "Consolidated Activity Report for the Office of the Director, Manned Space Flight, January 27–February 23, 1963," p. 57.

A completed crawler-transporter at Merritt Island.

1963

February

12

NASA selected the Marion Power Shovel Company to design and build the crawler-transport, a device to haul the Apollo space vehicle (Saturn V, complete with spacecraft and associated launch equipment) from the Vertical Assembly Building to the Merritt Island, Fla., launch pad, a distance of about 5.6 kilometers (3.5 miles). The crawler would be 39.6 meters (130 feet) long, 35 meters (115 feet) wide, and 6 meters (20 feet) high, and would weight 2.5 million kilograms (5.5 million pounds). NASA planned to buy two crawlers at a cost of $4 to $5 million each. Formal negotiations began on February 20 and the contract was signed on March 29.

Saturn Illustrated Chronology (MHR-3, August 10, 1964), p. 73; NASA News Release 63-27, "Marion to Build NASA Crawler," February 12, 1963.

13

In a reorganization of ASPO, MSC announced the appointment of two deputy managers. Robert O. Piland, deputy for the LEM, and James L. Decker, deputy for the CSM, would supervise cost, schedule, technical design, and production. J. Thomas Markley was named Special Assistant to the Apollo Manager, Charles W. Frick. Also appointed to newly created positions were Caldwell C. Johnson, Manager, Spacecraft Systems Office, CSM; Owen E. Maynard, Acting Manager, Spacecraft Systems Office, LEM; and David W. Gilbert, Manager, Spacecraft Systems Office, Guidance and Navigation.

MSC News Release 63-27, February 13, 1963.

13

Grumman began discussions with Rocketdyne on the development of a throttleable LEM descent engine. Engine specifications (helium injected, 10:1 thrust variation) had been laid down by MSC. (See May 1.)

MSC, "Consolidated Activity Report for the Office of the Director, Manned Space Flight, January 27–February 23, 1963," p. 57; "Apollo Quarterly Status Report No. 3," p. 25.

PART I: DEFINING CONTRACTUAL RELATIONS

The North American Apollo impact test facility at Downey, Calif., was completed. This facility consisted mainly of a large pool with overhead framework and mechanisms for hydrodynamic drop tests of the CM. Testing at the facility began with the drop of boilerplate 3 on March 11.

1963

February

15

<small>Oakley, *Historical Summary, S&ID Apollo Program*, p. 8; "Apollo Monthly Progress Report," SID 62-300-11, pp. 10, 21.</small>

NASA issued a definitive contract for $6 322 643 to General Dynamics/Convair for the Little Joe II test vehicle. (See May 11, 1962, Vol. I.) A number of changes defined by contract change proposals were incorporated into the final document:

- Four instead of five vehicles to be manufactured and delivered
- Launching from White Sands Missile Range (WSMR), N.M., instead of Cape Canaveral
- Additional support equipment, better definition of vehicle design, and responsibility for launch support.

18

<small>*Little Joe II Test Launch Vehicle, NASA Project Apollo: Final Report*, Vol. I, pp. 1-2, 1-4; MSC, "Consolidated Activity Report for the Office of the Director, Manned Space Flight, January 27-February 23, 1963," p. 28.</small>

A boilerplate spacecraft is dropped in the impact test facility at NAA's Downey, Calif., plant. The tower was 43.6 meters (143 feet) high, the pendulum pivot was 38.1 meters (125 feet), and maximum impact velocity was 12.2 meters (40 feet) per second vertical and 15.2 meters (50 feet) per second horizontal. —NAA photo.

1963

February 18

North American selected Bell Aerosystems Company to provide propellant tanks for the CSM reaction control system. These tanks were to be the "positive expulsion" type (i.e., fuel and oxidizer would be contained inside flexible bladder; pressure against one side of the device would force the propellant through the RCS lines).

"Apollo Monthly Progress Report," SID 62-300-10, p. 3; *Aviation Daily*, February 18, 1963, p. 312.

19

North American shipped CM boilerplate 19 to Northrop Ventura for use as a parachute test vehicle.

MSC, "Consolidated Activity Report for the Office of the Director, Manned Space Flight, January 27–February 23, 1963," p. 55.

20

At a meeting of the MSC–MSFC Flight Mechanics Panel, it was agreed that Marshall would investigate "engine-out" capability (i.e., the vehicle's performance should one of its engines fail) for use in abort studies or alternative missions. Not all Saturn I, IB, and V missions included this engine-out capability. Also, the panel decided that the launch escape system would be jettisoned ten seconds after S–IV ignition on Saturn I launch vehicles. (See March 28.)

MSC, "Consolidated Activity Report for the Office of the Director, Manned Space Flight, January 27–February 23, 1963," p. 58.

20

In a reorganization of OMSF, Director D. Brainerd Holmes appointed Joseph F. Shea as Deputy Director for Systems and George M. Low as Deputy Director for Programs. All major OMSF directorates had previously reported directly to Holmes. In the new organizational structure, Director of Systems Studies William A. Lee, Director of Systems Engineering John A. Gautraud, and Director of Integration and Checkout James E. Sloan would report to Shea. Director of Launch Vehicles Milton W. Rosen, Director of Space Medicine Charles H. Roadman, and the Director of Spacecraft and Flight Missions (then vacant) would report to Low. William E. Lilly, Director of Administration, would provide administrative support in both major areas.

NASA News Release 63-32, "Holmes Names Two Deputies," February 20, 1963; *The Washington Post*, February 21, 1963.

21

MSC issued a Request for Proposals (due by March 13) for a radiation altimeter system. Greater accuracy than that provided by available radar would be needed during the descent to the lunar surface, especially in the last moments before touchdown. Preliminary MSC studies had indicated the general feasibility of an altimeter system using a source-detector-electronics package. After final selection and visual observation of the landing site, radioactive material would be released at an altitude of about 30 meters (100 feet) and allowed to fall to the surface. The detector would operate in conjunction with electronic circuitry to compute the spacecraft's altitude.

PART I: DEFINING CONTRACTUAL RELATIONS

Studies were also under way at MSC on the possibility of using laser beams for range determination.

1963
February

> Memorandum, George W. Brandon, MSC, to Asst. Dir. for Information and Control Systems, "Request for Proposal, Low Level Radiation Altimeter System," November 13, 1962; *Aviation Daily*, February 21, 1963, p. 335.

The MSC Lunar Surface Experiments Panel held its first meeting. This group was formed to study and evaluate lunar surface experiments and the adaptability of Surveyor and other unmanned probes for use with manned missions.

24–March 23

> MSC, "Consolidated Monthly Activity Report for the Office of the Director, Manned Space Flight, February 24–March 23, 1963," p. 44.

Grumman began initial talks with the Bell Aerosystems Company on development of the LEM ascent engine. Complete specifications were expected by March 2.

25

> MSC, "Consolidated Activity Report for the Office of the Director, Manned Space Flight, January 27–February 23, 1963," p. 28.

MSC ordered North American to provide batteries, wholly independent of the main electrical system in the CM, to fire all pyrotechnics aboard the spacecraft.

25

> Letter, H. P. Yschek, MSC, to NAA, Space and Information Systems Div., "Contract Change Authorization No. Twenty-Eight," February 25, 1963.

NASA announced the signing of a formal contract with The Boeing Company for the S-IC (first stage) of the Saturn V launch vehicle, the largest

25

Aerial view of the Michoud Operations Plant, New Orleans, La.

1963

February

rocket unit under development in the United States. The $418 820 967 agreement called for the development and manufacture of one ground test and ten flight articles. Preliminary development of the S-IC, which was powered by five F-1 engines, had been in progress since December 1961 under a $50 million interim contract. Booster fabrication would take place primarily at the Michoud Operations Plant, New Orleans, La., but some advance testing would be done at MSFC and the Mississippi Test Operations facility.

<small>NASA News Release 63-37, "NASA Contracts with Boeing for Saturn V Booster," February 25, 1963; *Aviation Daily*, February 27, 1963, p. 361.</small>

26

Two aerospace technologists at MSC, James A. Ferrando and Edgar C. Lineberry, Jr., analyzed orbital constraints on the CSM imposed by the abort capability of the LEM during the descent and hover phases of a lunar mission. Their study concerned the feasibility of rendezvous should an emergency demand an immediate return to the CSM.

Ferrando and Lineberry found that, once abort factors are considered, there exist "very few" orbits that are acceptable from which to begin the descent. They reported that the most advantageous orbit for the CSM would be a 147-kilometer (80-nautical-mile) circular one.

<small>Memorandum, James A. Ferrando and Edgar C. Lineberry, Jr., to Chief, Flight Operations Div., "The Influence of LEM Abort Capability Upon the Selection of the Command Module Lunar Orbit," February 26, 1963.</small>

26

NASA selected Ford, Bacon, and Davis, Inc., to design MSC's flight acceleration facility, including a centrifuge capable of spinning a simulated CM and its crew at gravity forces equal to those experienced in space flight.

<small>*Space Business Daily*, February 26, 1963, p. 243; *Aviation Daily*, February 26, 1963, p. 358.</small>

27

Aviation Daily reported an announcement by Frank Canning, Assistant LEM Project Manager at Grumman, that a Request for Proposals would be issued in about two weeks for the development of an alternate descent propulsion system. Because the descent stage presented what he called the LEM's "biggest development problem," Canning said that the parallel program was essential.

<small>*Aviation Daily*, February 27, 1963, p. 362.</small>

27

The Apollo Mission Planning Panel held its organizational meeting at MSC. The panel's function was to develop the lunar landing mission design, coordinate trajectory analyses for all Saturn missions, and develop contingency plans for all manned Apollo missions.

Membership on the panel included representatives from MSC, MSFC, NASA Headquarters, North American, Grumman, and MIT, with other NASA Centers being called on when necessary. By outlining the most

PART I: DEFINING CONTRACTUAL RELATIONS

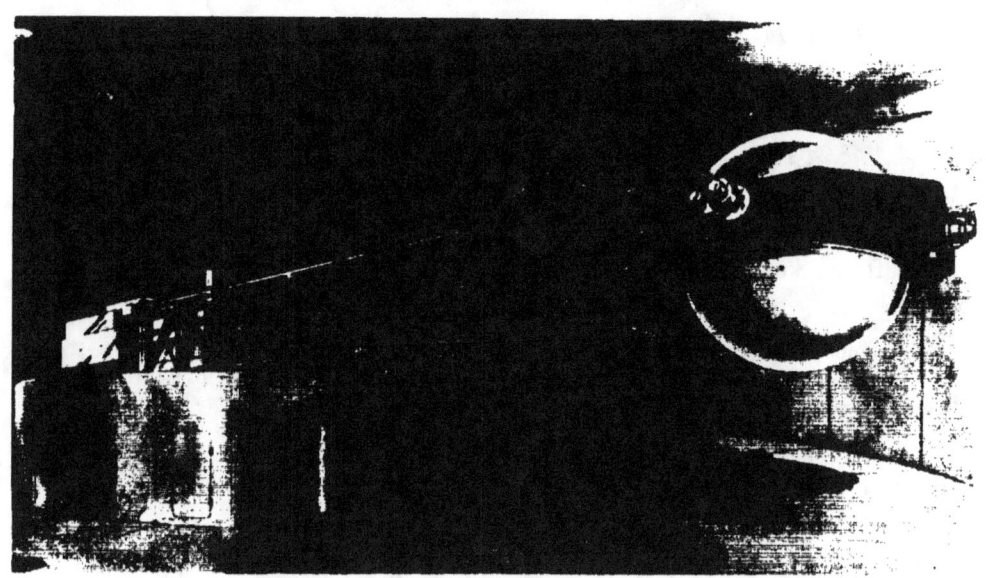

An artist's concept of the MSC flight acceleration facility.

accurate mission plan possible, the panel would ensure that the spacecraft could satisfy Apollo's anticipated mission objectives. Most of the panel's influence on spacecraft design would relate to the LEM, which was at an earlier stage of development than the CSM. The panel was not given responsibility for preparing operational plans to be used on actual Apollo missions, however.

MSC, "Minutes of Meeting on Apollo Mission Planning Panel Organization Meeting, February 27, 1963," March 7, 1963.

Elgin National Watch Company received a subcontract from North American for the design and development of central timing equipment for the Apollo spacecraft. [This equipment provided time-correlation of all spacecraft time-sensitive events. Originally, Greenwich Mean Time was to be used to record all events, but this was later changed. (See August 30–September 5, 1963.)]

Chicago Tribune, February 27, 1963; Wall Street Journal, February 28, 1963.

Grumman began fabrication of a one-tenth scale model of the LEM for stage separation tests. In launching from the lunar surface, the LEM's ascent engine fires just after pyrotechnic severance of all connections between the two stages, a maneuver aptly called "fire in the hole."

Also, Grumman advised that, from the standpoint of landing stability, a five-legged LEM was unsatisfactory. Under investigation were a number

1963
February

27

During
the
Month

1963

February

of landing gear configurations, including retractable legs. (See April 17 and May 20–22.)

> Grumman Aircraft Engineering Corporation [hereafter cited as GAEC], "Monthly Progress Report No. 1, LPR-10-1, March 10, 1963," pp. 5, 6, 8.

During the Month

NASA amended the GE contract, authorizing the company's Apollo Support Department to proceed with the PACE program. (See March 25, 1964.) [PACE (prelaunch automatic checkout equipment) would be used for spacecraft checkout. It would be computer-directed and operated by remote control.]

> GE, "Support Program Monthly Progress Report, February 1963," NASw-410-MR-2. [NOTE: Use of the acronym "PACE" was subsequently dropped at the insistence of a company claiming prior rights to the name.]

March 4

Grumman began initial discussions with Hamilton Standard on the development of the LEM environmental control system.

> MSC, "Consolidated Activity Report for the Office of the Director, Manned Space Flight, January 27–February 23, 1963," p. 57; "Consolidated Monthly Activity Report for the Office of the Director, Manned Space Flight, February 24–March 23, 1963," p. 8.

The first photos released by General Dynamics/Convair in San Diego, Calif., of the assembly of the Little Joe II launch vehicle included this one showing the thrust bulkhead. This was the lowest section of the vehicle, designed to secure the seven solid-fuel rockets in the inner chambers. This launch vehicle was used in the qualification test flight in August 1963.
—General Dynamics photo.

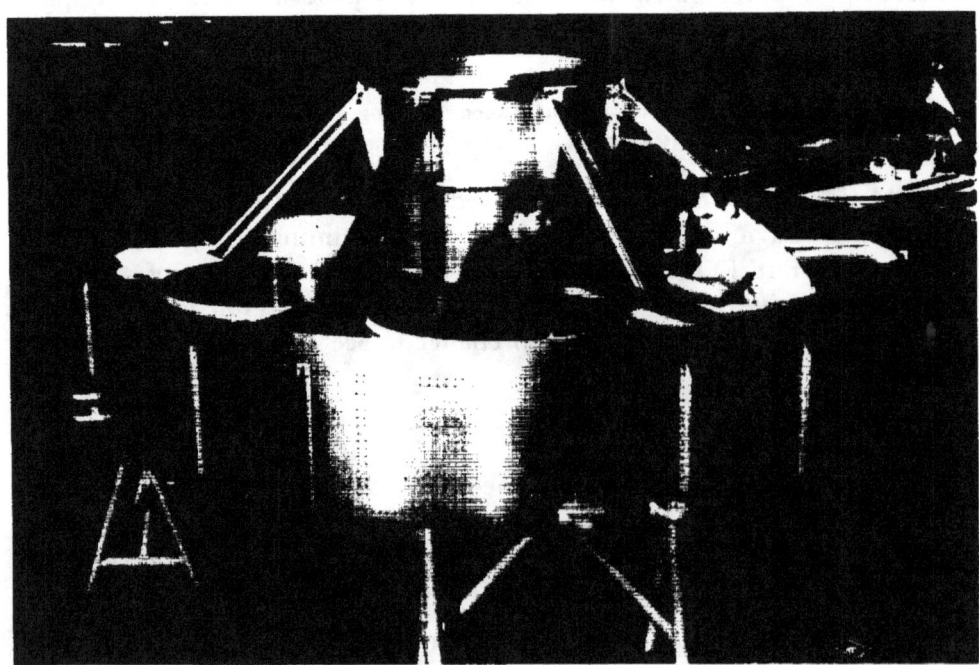

PART I: DEFINING CONTRACTUAL RELATIONS

Many modes of recovering the spacecraft or launch vehicle stages were considered at various times in the early years of the Apollo program. All these considerations prompted the accompanying cartoon, produced by an unidentified artist, incorporating the use of rotors, parasail, glidesail, and solid parachutes to return a happy first stage back to earth while two feathered friends hitched a ride and wondered how it worked.

As a parallel to the existing Northrop Ventura contract, and upon authorization by NASA, North American awarded a contract for a solid parachute program to the Pioneer Parachute Company. [A solid parachute is one with solid (unbroken) gores; the sole opening in the canopy is a vent at the top. Ringsail parachutes (used on the Northrop Ventura recovery system) have slotted gores. In effect, each panel formed on the gores becomes a "sail."] (See June 28.)	**1963** March 4

"Apollo Quarterly Status Report No. 3," p. 18; letter, H. P. Yschek, MSC, to NAA, Space and Information Systems Div., "Contract Change Authorization No. Twenty-Seven," February 25, 1963.

MSC "acquired" under a loan agreement an amphibious landing craft from the Army. Equipment to retrieve Apollo boilerplate spacecraft and other objects used in air drops and flotation tests was installed. The vessel, later named the *Retriever*, arrived at its Seabrook, Tex., docking facility late in June.	4

MSC News Release 63–38, "MSC Acquires Test Vehicle," March 4, 1963; MSC, *Space News Roundup*, June 26, 1963, p. 1.

1963

March 5

MSC awarded a $67 000 contract to The Perkin-Elmer Corporation to develop a carbon dioxide measurement system, a device to measure the partial carbon dioxide pressure within the spacecraft's cabin. Two prototype units were to be delivered to MSC for evaluation. About seven months later, a $249 000 definitive contract for fabrication and testing of the sensor was signed. (See May 6.)

> MSC, "Consolidated Monthly Activity Report for the Office of the Director, Manned Space Flight, February 24–March 23, 1963," p. 30; "Consolidated Activity Report for the Office of the Director, Manned Space Flight, September 22–October 19, 1963," p. 47.

5

NASA announced an American agreement with Australia, signed on February 26, that permitted the space agency to build and operate several new tracking stations "down under." A key link in the Jet Propulsion Laboratory's network of Deep Space Instrumentation Facilities would be constructed in Tidbinbilla Valley, 18 kilometers (11 miles) southwest of Canberra. Equipment at this site included a 26-meter (85-foot) parabolic dish antenna and electronic equipment for transmitting, receiving, and processing radio signals from spacecraft. Tracking stations would be built also at Carnarvon and Darwin.

> NASA News Release 63–47, "NASA to Establish Deep Space Tracking Facility in Australia," March 5, 1963; *Aviation Daily*, March 8, 1963, p. 52.

5

The Mission Analysis Branch (MAB) of MSC's Flight Operations Division cited the principal disadvantages of the land recovery mode for Apollo missions. (See February 1.) Of primary concern was the possibility of landing in an unplanned area and the concomitant dangers involved. For water recovery, the main disadvantages were the establishment of suitable landing areas in the southern hemisphere and the apex-down flotation problem. MAB believed no insurmountable obstacles existed for either approach. (See February 25, 1964.)

> Memorandum, John Bryant, MSC, to Chief, FOD, "Operational Considerations in the Selection of Primary Land or Sea Return Areas for Apollo," March 5, 1963.

6

North American completed construction of Apollo boilerplate (BP) 9, consisting of launch escape tower and CSM. It was delivered to MSC on March 18, where dynamic testing on the vehicle began two days later. On April 8, BP-9 was sent to MSFC for compatibility tests with the Saturn I launch vehicle.

> MSC, "Consolidated Monthly Activity Report for the Office of the Director, Manned Space Flight, February 24–March 23, 1963," p. 50; Oakley, *Historical Summary, S&ID Apollo Program*, p. 8; *Birmingham Post-Herald*, April 5, 1963; *The Huntsville Times*, April 9, 1963; *The Birmingham News*, April 9, 1963.

6

The first Block I Apollo pulsed integrating pendulum accelerometer, produced by the Sperry Gyroscope Company, was delivered to the MIT Instrumentation Laboratory. [Three accelerometers were part of the guid-

PART I: DEFINING CONTRACTUAL RELATIONS

ance and navigation system. Their function was to sense changes in spacecraft velocity.]

<small>MSC, "Consolidated Monthly Activity Report for the Office of the Director, Manned Space Flight, February 24–March 23, 1963," p. 53.</small>

Grumman representatives presented their technical study report on power sources for the LEM. (See January 28.) They recommended three fuel cells in the descent stage (one cell to meet emergency requirements), two sets of fluid tanks, and two batteries for peak power loads. For industrial competition to develop the power sources, Grumman suggested Pratt and Whitney Aircraft and GE for the fuel cells, and Eagle-Picher, Electrical Storage Battery, Yardney, Gulton, and Delco-Remy for the batteries.

<small>"Activity Report, RASPO/GAEC, 3/3/63–3/9/63" (undated), pp. 1–2.</small>

North American moved CM boilerplate (BP) 6 from the manufacturing facilities to the Apollo Test Preparation Interim Area at Downey, Calif.

During a visit to NAA during March 1963, Astronauts M. Scott Carpenter, John H. Glenn, Jr., and Walter M. Schirra, Jr., took time out to "try the spacecraft on for size." The spacecraft mockup was one of the items inspected as they toured the NAA spacecraft facilities at Downey, Calif.

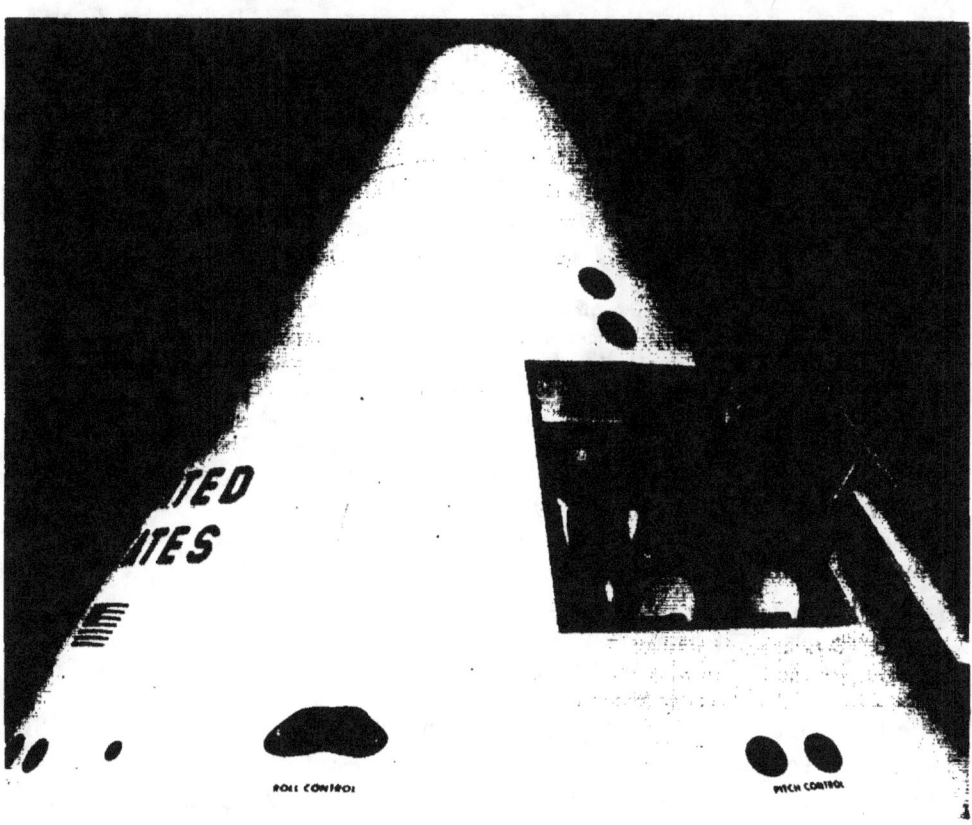

1963
March

7

8

1963
March

During the next several weeks, BP-6 was fitted with a pad adapter, an inert launch escape system, and a nose cone, interstage structure, and motor skirt. (See July 1-2 and November 7.)

> MSC, "Postlaunch Memorandum Report for Apollo Pad Abort I," November 13, 1963, pp. A1-1 through A1-5.

10 Grumman presented its first monthly progress report on the LEM. In accordance with NASA's list of high-priority items, principal engineering work was concentrated on spacecraft and subsystem configuration studies, mission plans and test program investigations, common usage equipment surveys, and preparation for implementing subcontractor efforts.

> "Monthly Progress Report No. 1," LPR-10-1, p. 4.

11 Grumman completed its first "fire-in-the-hole" model test. (See February 1963.) Even though preliminary data agreed with predicted values, they nonetheless planned to have a support contractor, the Martin Company, verify the findings.

> "Activity Report, RASPO/GAEC, 3/10/63-3/16/63" (undated), p. 2.

11 NASA announced signing of the contract with Grumman for development of the LEM. (See November 19, 1962.) Company officials had signed the document on January 21 and, following legal reviews, NASA Headquarters had formally approved the agreement on March 7. Under the fixed-fee contract (NAS 9-1100) ($362.5 million for costs and $25.4 million in fees) Grumman was authorized to design, fabricate, and deliver nine ground test and 11 flight vehicles. The contractor would also provide mission support for Apollo flights. MSC outlined a developmental approach, incorporated into the contract as "Exhibit B, Technical Approach," that became the "framework within which the initial design and operational modes" of the LEM were developed.

> NASA-MSC, "Lunar Excursion Module, Project Apollo, Exhibit B, Technical Approach, Contract NAS 9-1100," December 20, 1962, p. 1; MSF Management Council Meeting, January 29, 1963, Agenda Item 3, "MSC Status Report," pp. 23, 26; MSF Management Council Minutes, January 29, 1963, p. 3; MSC, "Consolidated Monthly Activity Report for the Office of the Director, Manned Space Flight, February 24-March 23, 1963," p. 29; "Apollo Quarterly Status Report No. 3," p. 1; NASA News Release 63-51, "Contract Signed to Develop Lunar Excursion Module," March 11, 1963.

11 Grumman began early contract talks with the Marquardt Corporation for development of the LEM reaction control system.

> MSC, "Consolidated Activity Report for the Office of the Director, Manned Space Flight, January 27-February 23, 1963," p. 57; "Consolidated Monthly Activity Report for the Office of the Director, Manned Space Flight, February 24-March 23, 1963," p. 7.

13 The first stage of the Saturn SA-5 launch vehicle was static fired at MSFC for 144.44 seconds in the first long-duration test for a Block II S-I. The cluster of eight H-1 engines produced 680 thousand kilograms (1.5 million

PART I: DEFINING CONTRACTUAL RELATIONS

pounds) of thrust. An analysis disclosed anomalies in the propulsion system. In a final qualification test two weeks later, when the engines were fired for 143.47 seconds, the propulsion problems had been corrected.

1963

March

MSFC Historical Office, *History of the George C. Marshall Space Flight Center from January 1 through June 30, 1963* (MHM–7), Vol. I, pp. 21–22; *The Huntsville Times*, March 14, 1963.

A bidders' conference was held at Grumman for a LEM mechanically throttled descent engine to be developed concurrently with Rocketdyne's helium injection descent engine. (See February 27.) Corporations represented were Space Technology Laboratories; United Technology Center, a division of United Aircraft Corporation; Reaction Motors Division, Thiokol Chemical Corporation; and Aerojet-General Corporation. Technical and cost proposals were due at Grumman on April 8.

14

"Activity Report, RASPO/GAEC, 3/10/63–3/16/63" (undated), p. 1.

Homer E. Newell, Director of NASA's Office of Space Sciences, summarized results of studies by Langley Research Center and Space Technology Laboratories on an unmanned lunar orbiter spacecraft. These studies had been prompted by questions of the reliability and photographic capabilities of such spacecraft. Both studies indicated that, on a five-shot program, the

14

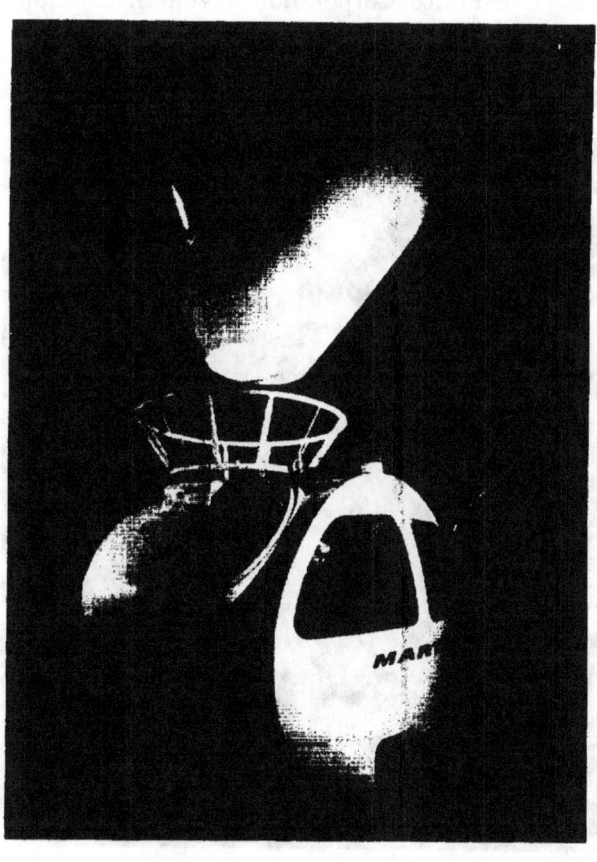

A model of a two-man lunar landing "bug" developed by Martin Company of Baltimore simulated the critical rendezvous and docking portions of a manned mission to the moon. The two-man "bug" featured a round docking ring into which a simulated nose cone of an Apollo spacecraft could be inserted as part of a rendezvous and docking maneuver. The "bug" could move up to about 25 degrees about three axes of motion—roll, pitch, and yaw. The nose cone of the simulated Apollo spacecraft approaching a docking maneuver.
—Martin Company photo.

1963

March

probability was .93 for one and .81 for two successful missions; they also confirmed that the spacecraft would be capable of photographing a landed Surveyor to assist in Apollo site verification.

> Memorandum, Newell, NASA, to Dir., OMSF, "Questions on the unmanned lunar orbiter," March 14, 1963, with four enclosures; Bruce K. Byers, "Lunar Orbiter: a Preliminary History" (HHN-71), August 1969, pp. 21-22.

20

John A. Hornbeck, president of Bellcomm, testified before the House Committee on Science and Astronautics' Subcommittee on Manned Space Flight concerning the nature and scope of Bellcomm's support for NASA's Apollo program. In answer to the question as to how Bellcomm would decide "which area would be the most feasible" for a lunar landing, Hornbeck replied, ". . . the safety of the landing—that will be the paramount thing." He said that his company was studying a number of likely areas, but would "not recommend a specific site at the moment." Further, "Preliminary studies . . . suggest that the characteristics of a 'good' site for early exploration might be (1) on a lunar sea, (2) 10 miles [16 kilometers] from a continent, and (3) 10 miles [16 kilometers] from a postmarial crater." This type of site, Hornbeck said, would permit the most scientific activity

Oscar T. Simpson, General Manager of Philco Corporation's Western Development Laboratories, signed the definitive contract for Philco's activities in equipping the Mission Control Center. Looking on was James Stroup, a contracting officer for MSC.

PART I: DEFINING CONTRACTUAL RELATIONS

practicable, and would enable NASA's planners to design future missions for even greater scientific returns.

1963
March

U.S. Congress, House, Subcommittee on Manned Space Flight of the Committee on Science and Astronautics, *1964 NASA Authorization*, Hearings on H.R. 5466 (Superseded by H.R. 7500), [No. 3] Part 2(a), 88th Cong., 1st Sess. (1963), p. 378.

MSC awarded the Philco Corporation a definitive contract (worth almost $33.8 million) to provide flight information and flight control display equipment (with the exception of the realtime computer complex) for the Mission Control Center at MSC. NASA Headquarters approved the contract at the end of the month.

21

MSC, "Consolidated Monthly Activity Report for the Office of the Director, Manned Space Flight, February 24–March 23, 1963," p. 29; "Apollo Quarterly Status Report No. 3," p. 49; *Space Business Daily*, April 4, 1963, p. 432.

General Dynamics/Convair completed structural assembly of the first launcher for the Little Joe II test program. During the next few weeks, electrical equipment installation, vehicle mating, and checkout were completed. The launcher was then disassembled and delivered to WSMR on April 25, 1963.

25

Little Joe II Test Launch Vehicle, NASA Project Apollo: Final Report, Vol. I, pp. 1–4 and 1–6.

North American analyzed lighting conditions in the CM and found that glossy or light-colored garments and pressure suits produced unsatisfactory reflections on glass surfaces. A series of tests were planned to define the allowable limits of reflection on windows and display panel faces to preclude interference with crew performance.

25–31

"Project Apollo Spacecraft Test Program, Weekly Activity Report (Period 25 March 1963 through 31 March 1963)," p. 5.

Hamilton Standard Division awarded a contract to ITT/Kellogg for the design and manufacture of a prototype extravehicular suit telemetry and communications system to be used with the portable life support system. (See November 27, 1962.)

26

Memorandum, Michael B. Luse, MSC, to Crew Systems Division, Attn: M. I. Radnofsky, "Extra-Vehicular Suit Telemetry and Communication System," March 11, 1964.

MSC announced the beginning of CM environmental control system tests at the AiResearch Manufacturing Company simulating prelaunch, ascent, orbital, and reentry pressure effects. Earlier in the month, analysis had indicated that the CM interior temperature could be maintained between 294 K (70° F) and 300 K (80° F) during all flight operations, although prelaunch temperatures might rise to a maximum of 302 K (84° F).

26

"Apollo Monthly Progress Report," SID 62-300-11, p. 12; MSC News Release 63-61, March 26, 1963.

41

1963
March
26-28

A meeting was held at North American to define CM-space suit interface problem areas. (See January 24.) Demonstrations of pressurized International Latex suits revealed poor crew mobility and task performance inside the CM, caused in part by the crew's unavoidably interfering with one another.

Other items received considerable attention: A six-foot umbilical hose would be adequate for the astronaut in the CM. The location of spacecraft water, oxygen, and electrical fittings was judged satisfactory, as were the new couch assist handholds. The astronaut's ability to operate the environmental control system (ECS) oxygen flow control valve while couched and pressurized was questionable. Therefore, it was decided that the ECS valve would remain open and that the astronaut would use the suit control valve to regulate the flow. It was also found that the hand controller must be moved about nine inches forward.

> Memorandum, J. F. Saunders, Jr., RASPO/NAA, to L. McMillion, MSC, "Data Transmittal," April 5, 1963, with enclosures: Agenda and Minutes of Meeting, "Command Module—Space Suit Interface Meeting No. 4, NAA, Downey—26, 27, 28 March 1963."

27

The Apollo Mission Planning Panel (see February 27) set forth two firm requirements for the lunar landing mission. First, both LEM crewmen must be able to function on the lunar surface simultaneously. MSC contractors were directed to embody this requirement in the design and development of the Apollo spacecraft systems. Second, the panel established duration limits for lunar operations. These limits, based upon the 48-hour LEM operation requirement, were 24 hours on the lunar surface and 24 hours in flight on one extreme, and 45 surface hours and 3 flight hours on the other. Grumman was directed to design the LEM to perform throughout this range of mission profiles.

> MSC, "Abstract of Meeting on Apollo Mission Planning Meeting No. 1, March 27, 1963," March 29, 1963; memorandum, Robert V. Battey, MSC, to Action Committee, "Errata to Abstract of Mission Planning Panel Meeting No. 1," April 1, 1963.

28

NASA launched Saturn SA-4 from Cape Canaveral. The S-I Saturn stage reached an altitude of 129 kilometers (80 statute miles) and a peak velocity of 5906 kilometers (3660 miles) per hour. This was the last of four successful tests for the first stage of the Saturn I vehicle. After 100 seconds of flight, No. 5 of the booster's eight engines was cut off by a preset timer. That engine's propellants were rerouted to the remaining seven, which continued to burn. This experiment confirmed the "engine-out" capability that MSFC engineers had designed into the Saturn I. (See February 20.)

> *Saturn Illustrated Chronology*, pp. 76-77; *History of Marshall . . . January 1-June 30, 1963*, Vol. I, pp. 16-18.

PART I: DEFINING CONTRACTUAL RELATIONS

The Saturn SA-4 stood on Pad 34, Cape Canaveral, shortly before liftoff.

North American selected two subcontractors to build tankage for the SM: Allison Division of General Motors Corporation to fabricate the fuel and oxidizer tanks; and Airite Products, Inc., those for helium storage.

"Apollo Monthly Progress Report," SID 62-300-11, p. 3.

RCA completed a study on ablative versus regenerative cooling for the thrust chamber of the LEM ascent engine. Because of low cooling margins available with regenerative cooling, Grumman selected the ablative

1963
March
During the Month

During the Month

1963

March

method, which permitted the use of either ablation or radiation cooling for the nozzle extension. (See September 19–October 16.)

"Apollo Quarterly Status Report No. 3," p. 26; GAEC, "Monthly Progress Report No. 2," LPR–10–2, April 10, 1963, p. 12.

During the Month

Grumman met with representatives of North American, Collins Radio Company, and Motorola, Inc., to discuss common usage and preliminary design specifications for the LEM communications system. These discussions led to a simpler design for the S-band receiver and to modifications to the S-band transmitter (required because of North American's design approach).

"Monthly Progress Report No. 2," LPR–10–2, p. 15.

During the Month

MSC sent MIT and Grumman radar configuration requirements for the LEM. The descent equipment would be a three-beam doppler radar with a two-position antenna. Operating independently of the primary guidance and navigation system, it would determine altitude, rate of descent, and horizontal velocity from 7000 meters (20 000 feet) above the lunar surface. The LEM rendezvous radar, a gimbaled antenna with a two-axis freedom of movement, and the rendezvous transponder mounted on the antenna would provide tracking data, thus aiding the LEM to intercept the orbiting CM. The SM would be equipped with an identical rendezvous radar and transponder.

"Apollo Quarterly Status Report No. 3," p. 23.

During the Quarter

MSC reported that preliminary plans for Apollo scientific instrumentation had been prepared with the cooperation of NASA Headquarters, Jet Propulsion Laboratory, and the Goddard Space Flight Center. The first experiments would not be selected until about December 1963, allowing scientists time to prepare proposals. Prime consideration would be given to experiments that promised the maximum return for the least weight and complexity, and to those that were man-oriented and compatible with spacecraft restraints. Among those already suggested were seismic devices (active and passive), and instruments to measure the surface bearing strength, magnetic field, radiation spectrum, soil density, and gravitational field. MSC planned to procure most of this equipment through the scientific community and through other NASA and government organizations.

Ibid., p. 30.

During the Quarter

To provide a more physiologically acceptable load factor orientation during reentry and abort, MSC was considering revised angles for the crew couch in the CM. To reduce the couch's complexity, North American had pro-

PART I: DEFINING CONTRACTUAL RELATIONS

posed adjustments which included removable calf pads and a movable head pad. (See April 3.)

Ibid., p. 6.

MSC reported that stowage of crew equipment, some of which would be used in both the CM and the LEM, had been worked out. Two portable life support systems and three pressure suits and thermal garments were to be stowed in the CM. Smaller equipment and consumables would be distributed between modules according to mission phase requirements.

Ibid., p. 22.

Grumman began "Lunar Hover and Landing Simulation IIIA," a series of tests simulating a LEM landing. Crew station configuration and instrument panel layout were representative of the actual vehicle.

Through this simulation, Grumman sought primarily to evaluate the astronauts' ability to perform the landing maneuver manually, using semiautomatic as well as degraded attitude control modes. Other items evaluated included the flight control system parameters, the attitude and thrust controller configurations, the pressure suit's constraint during landing maneuvers, the handling qualities and operation of LEM test article 9 as a free-flight vehicle, and manual abort initiation during the terminal landing maneuver.

GAEC, "Final Report: Lunar Landing Simulation IIIA," LED-770-4, April 1, 1964, p. 1.

The Soviet Union announced the successful launch of the *Lunik IV* probe toward the moon. The 1412-kilogram (3135-pound) spacecraft's mission was not immediately disclosed, but Western observers speculated that an instrumented soft landing was planned. On April 6, at 4:26 a.m. Moscow time, *Lunik IV* passed within 8499 kilometers (5281 miles) of the moon. The Soviet news agency, Tass, reported that data had been received from the spacecraft throughout its flight and that radio communication would continue for a few more days.

The Washington Post, April 3 and 5, 1963; *The New York Times*, April 3, 1963; *The Sunday Star*, Washington, April 7, 1963.

Charles W. Frick resigned as ASPO Manager and Robert O. Piland was named Acting ASPO Manager.

MSC Announcement 178, "New Assignment of Personnel," April 3, 1963.

At a North American design review, NASA representatives expressed a preference for a fixed CM crew couch. This would have the advantages of

1963
March

During the Quarter

April
1

2

3

3

A number of "boilerplate" command modules were fabricated as test vehicles early in the Apollo program. Above, inspecting the assembly of one on April 1, 1963, were George Lemke, left, NASA resident Apollo project manager, and John Paup, vice president of NAA's Space and Information Systems Division and Apollo project manager. At right, an Apollo command module heatshield skin is pre-fitted with metal honeycomb at NAA's Downey plant, as J. W. Fleetwood, right, manager of Apollo manufacturing, and general foreman B. E. Dean discuss the project. —NAA photos.

simplified design, elimination of couch adjustments by the crew, and better placement of the astronauts to withstand reentry loads. NASA authorized North American to adopt the concept following a three-week study by the company to determine whether a favorable center of gravity could be achieved without a movable couch.

1963
April

Use of the fixed couch required relocation of the main and side display panels and repositioning of the translational and rotational hand controllers. During rendezvous and docking operations, the crew would still have to adjust their normal body position for proper viewing.

"Apollo Monthly Progress Report," SID 62-300-12, p. 11; *ibid.*, SID 62-300-13, June 1, 1963, pp. 1, 7-8.

North American awarded a $9.5 million letter contract to the Link Division of General Precision, Inc., for the development and installation of two spacecraft simulators, one at MSC and the other at the Launch Operations Center. Except for weightlessness, the trainers would simulate the entire lunar mission, including sound and lighting effects. (See December 8, 1962.)

10

"Apollo Quarterly Status Report No. 4," p. 40; "Apollo Monthly Progress Report," SID 62-300-12, p. 2; *Aviation Daily*, May 1, 1963, p. 1.

Wesley E. Messing, MSC WSMR Operations Manager, notified NASA, North American, and General Dynamics/Convair (GD/C) that Phase I of the range's launch complex was completed. GD/C and North American could now install equipment for the launch of boilerplate 6 and the Little Joe II vehicle.

10

TWX, Messing to MSC (Attn: W. C. Williams and R. O. Piland), NASA Hqs (Attn: G. M. Low), GD/C (Attn: J. B. Hurt), and NAA, S&ID (Attn: J. L. Pearce), April 10, 1963.

North American chose Simmonds Precision Products, Inc., to design and build an electronic measurement and display system to gauge the service propulsion system propellants. Both a primary and a backup system were required by the contract, which was expected to cost about $2 million.

16–May 15

"Apollo Monthly Progress Report," SID 62-300-13, p. 2; *Space Business Daily*, June 26, 1963, p. 824.

On the basis of wind tunnel tests and analytical studies, North American recommended a change in the planned test of the launch escape system (LES) using boilerplate 22. In an LES abort, the contractor reported, 18 300 meters (60 000 feet) was the maximum altitude at which high dynamic pressure had to be considered. Therefore North American proposed an abort simulation at that altitude, where maximum dynamic pressures were reached, at a speed of Mach 2.5.

16–May 15

The abort test would demonstrate two possibly critical areas:

(1) Any destablizing effect of large LES motor plumes on the CM

1963

April

(2) The ability of the CM's reaction control system to arrest CM rotation following tower jettison.

"Apollo Quarterly Status Report No. 4," pp. 28, 29; "Apollo Monthly Progress Report," SID 62-300-13, p. 5; MSC, "Postlaunch Report for Apollo Mission A-003" (BP-22) (June 28, 1965), p. 2-1; memorandum, J. D. Reed, MSC, to Distr., "Meeting on BP-22 Test Objectives and Trajectories, June 30, 1964," July 2, 1964.

16–May 15

North American simplified the CM water management system by separating it from the freon system. A 4.5-kilogram (10-pound) freon tank was installed in the left-hand equipment bay. Waste water formed during prelaunch and boost, previously ejected overboard, could now be used as an emergency coolant. The storage capacity of the potable water tank was reduced from 29 to 16 kilograms (64 to 36 pounds) and the tank was moved to the lower equipment bay to protect it from potential damage during landing. These and other minor changes caused a reduction in CM weight and an increase in the reliability of the CM's water management system.

"Apollo Quarterly Status Report No. 4," p. 7; "Apollo Monthly Progress Report," SID 62-300-13, p. 13.

17

At a mechanical systems meeting at MSC, customer and contractor achieved a preliminary configuration freeze for the LEM. After "considerable discussion," Grumman agreed to begin designing systems and subsystems based on this configuration, bearing in mind that certain unresolved areas (the docking system scanning telescope location and function, and the outcome

A Lockheed Propulsion Company technician tightened bolts on an inert Apollo escape motor in preparation for its delivery to NAA. —Lockheed photo.

of visibility studies) would have a substantial effect on the final configuration. Several features of the design of the two stages were agreed upon:

• Descent—four cylindrical propellant tanks (two oxidizer and two fuel); four-legged deployable landing gear (see February)

• Ascent—a cylindrical crew cabin (about 234 centimeters [92 inches] in diameter) and a cylindrical tunnel (pressurized) for equipment stowage; an external equipment bay.

GAEC, "Monthly Progress Report No. 3," LPR-10-6, May 10, 1963, pp. 3, 4, 7-8.

1963
April

Combustion experiments for Project Apollo produced striking results at Honeywell's Aeronautical Division in Minneapolis. Over 115 different materials which might be used in or on the Honeywell-developed Apollo stabilization and control system were tested to determine ease of ignition, duration and effect of burning, and toxicity of burn-products in a typical space capsule atmosphere. Test specimens of materials ranging from wire insulation, tapes and tying cords to foams, casting resins, and plastics were suspended in a bell jar containing a 100-percent oxygen atmosphere. An electrically heated coil generated temperatures from 1073 K to 1173 K (800 to 900 degrees C). Here, technician Bill Williams closely observed the explosive effect of overheating an epoxy adhesive sample. —Minneapolis-Honeywell photo.

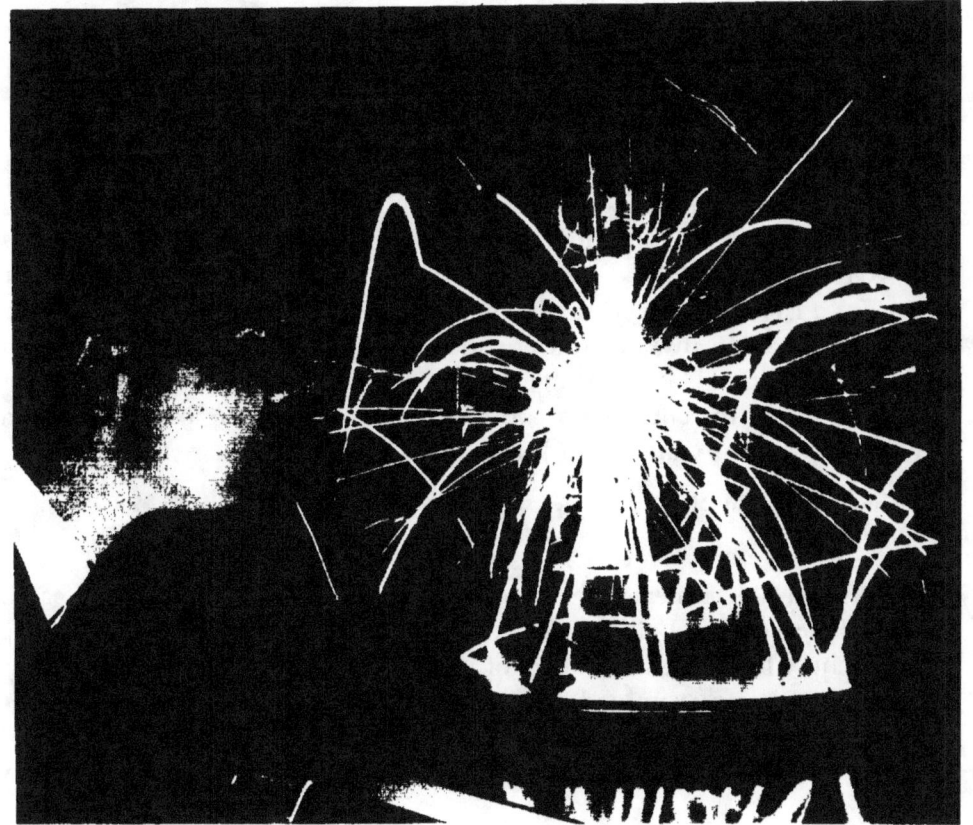

THE APOLLO SPACECRAFT: A CHRONOLOGY

1963

April 18

North American signed a $6 million definitive contract with Lockheed Propulsion Company for the development of solid propellant motors for the launch escape system. Work on the motors had begun on February 13, 1962, when Lockheed was selected.

<small>"Apollo Facts," p. 38; *Space Business Daily*, June 27, 1963, p. 834.</small>

25–26

At ASPO's request, Wayne E. Koons of the Flight Operations Division visited North American to discuss several features of spacecraft landing and recovery procedures. Koon's objective, in short, was to recommend a solution when ASPO and the contractor disagreed on specific points, and to suggest alternate courses when the two organizations agreed. A question had arisen about a recovery hoisting loop. Neither group wanted one, as its installation added weight and caused design changes. In another area, North American wanted to do an elaborate study of the flotation characteristics of the CM. Koons recommended to ASPO that a full-scale model of the CM be tested in an open-sea environment.

There were a number of other cases wherein North American and ASPO agreed on procedures which simply required formal statements of what would be done. Examples of these were:
- Spacecraft reaction control fuel would be dumped before landing (in both normal and abort operations)
- The "peripheral equipment bay" would be flooded within 10 minutes after landing
- Location aids would be dye markers and recovery antennas.

<small>Memorandum, W. E. Koons, MSC, to Chief, Flight Operations Div., "Report of visit to NAA, S&ID, Downey, Calif., 25–26 April 1963," May 7, 1963.</small>

30

The Apollo Spacecraft Mission Trajectory Sub-Panel discussed earth parking orbit requirements for the lunar mission. The maximum number of orbits was fixed by the S–IVB's 4.5-hour duration limit. Normally, translunar injection (TLI) would be made during the second orbit. The panel directed North American to investigate the trajectory that would result from injection from the third, or contingency, orbit. The contractor's study must reckon also with the effects of a contingency TLI upon the constraints of a free return trajectory and fixed lunar landing sites.

<small>Minutes of Second Meeting of the Apollo Spacecraft Mission Trajectory Sub-Panel, April 30, 1963.</small>

During the Month

NASA issued a technical note reporting that scientists at Ames Research Center Hypervelocity Ballistic Range, Moffett Field, Calif., were conducting experiments simulating the impact of micrometeoroids on the lunar surface. The experimenters examined the threat of surface debris, called secondary ejecta, that would be thrown from resultant craters. Data indicated that secondary particles capable of penetrating an astronaut's space suit nearly

PART I: DEFINING CONTRACTUAL RELATIONS

The Hypervelocity Ballistic Missile Range at Ames Research Center launched projectiles at speeds up to 27 360 kilometers (17 000 miles) per hour to simulate micrometeoroids in outer space.

equaled the number of primary micrometeoroids. Thus the danger of micrometeoroid impact to astronauts on the moon may be almost double what was previously thought.

    Donald E. Gault, Eugene M. Shoemaker, and Henry J. Moore, *Spray Ejected From the Lunar Surface by Meteoroid Impact*, NASA TN D-1767, April 1963, p. 1; *Aviation Week and Space Technology*, 78 (January 14, 1964), pp. 54–55, 57, 59.

NASA and General Dynamics/Convair (GD/C) negotiated a second Little Joe II launch vehicle contract. (See February 18.) For an additional $337 456, GD/C expanded its program to include the launch of a qualification test vehicle before the scheduled Apollo tests. This called for an accelerated production schedule for the four launch vehicles and their pair of launchers. An additional telemetry system and an instrumentation transmitter system were incorporated in the qualification test vehicle, which was equipped with a simulated payload. At the same time, NASA established earlier launch dates for the first two Apollo Little Joe II missions.

    Little Joe II Test Launch Vehicle, NASA Project Apollo: Final Report, Vol. I, p. 4–3.

1963
April

During
the
Month

THE APOLLO SPACECRAFT: A CHRONOLOGY

1963

April

During the Month

Grumman reported to MSC the results of studies on common usage of communications. Television cameras for the two spacecraft would be identical (see May 2); the LEM transponder would be as similar as possible to that in the CSM.

"Monthly Progress Report No. 3," LPR-10-6, p. 21.

During the Month

Grumman recommended that the LEM reaction control system (RCS) be equipped with dual interconnected tanks, separately pressurized and employing positive expulsion bladders. The design would provide for an emergency supply of propellants from the main ascent propulsion tanks. The RCS oxidizer to fuel ratio would be changed from 2.0:1 to 1.6:1. MSC approved both of these changes.

Ibid., p. 20; "Apollo Quarterly Status Report No. 3," p. 20.

Examining a one-eighth scale model of the LEM are, left to right, Congressman George P. Miller, Chairman of the House Committee on Science and Astronautics; Joseph M. Gavin, Grumman vice president; and Robert S. Mullaney, Grumman Apollo Program Manager. —Grumman photo.

PART I: DEFINING CONTRACTUAL RELATIONS

1963
May

1

Grumman reported that it had advised North American's Rocketdyne Division to go ahead with the lunar excursion module descent engine development program. Negotiations were complete and the contract was being prepared for MSC's review and approval. The go-ahead was formally issued on May 2. (See January 30, February 13, and November 21.)

> MSC, "Consolidated Activity Report for the Office of the Director, Manned Space Flight, April 28–May 18, 1963," p. 32; "Apollo Quarterly Status Report No. 4," p. 21; GAEC, "Monthly Progress Report No. 4," LPR-10-7, June 10, 1963, p. 2.

2

NASA, North American, Grumman, and RCA representatives determined the alterations needed to make the CM television camera compatible with that in the LEM: an additional oscillator to provide synchronization, conversion of operating voltage from 115 AC to 28 DC, and reduction of the lines per frame from 400 to 320.

> NAA, "Apollo Monthly Progress Report," SID 62-300-14, July 1, 1963, p. 9.

3

At El Centro, Calif., Northrop Ventura conducted the first of a series of qualification tests for the Apollo earth landing system (ELS). The test article, CM boilerplate 3, was dropped from a specially modified Air Force C-133. The test was entirely successful. The ELS's three main parachutes reduced the spacecraft's rate of descent to about 9.1 meters (30 feet) per second at impact, within acceptable limits.

> MSC News Release 63-85, May 3, 1963; "Apollo Monthly Progress Report," SID 62-300-13, p. 10.

6

NASA authorized North American to procure carbon dioxide sensors as part of the environmental control system instrumentation on early spacecraft flights. (See March 5.)

> Letter, H. P. Yschek, MSC, to NAA, Space and Information Systems Div., "Contract Change Authorization No. Forty-Three," May 6, 1963.

6

Astronauts M. Scott Carpenter, Walter M. Schirra, Jr., Neil A. Armstrong, James A. McDivitt, Elliot M. See, Jr., Edward H. White II, Charles Conrad, Jr., and John W. Young participated in a study in LTV's Manned Space Flight Simulator at Dallas, Tex. Under an MSC contract, LTV was studying the astronauts' ability to control the LEM manually and to rendezvous with the CM if the primary guidance system failed during descent. (See September and October 10, 1963, and April 24, 1964.)

> MSC News Release 63-81, May 6, 1963.

7

MSC announced a reorganization of ASPO:
    Acting Manager: Robert O. Piland
    Deputy Manager, Spacecraft: Robert O. Piland
    Assistant Deputy Manager for CSM: Caldwell C. Johnson
    Deputy Manager for System Integration: Alfred D. Mardel
    Deputy Manager, LEM: James L. Decker

A NASA team inspected progress on Little Joe II in San Diego, Calif., May 6, 1963. Left to right, Walter C. Williams, MSC Deputy Director; Acting Apollo Project Manager Robert O. Piland; Convair Little Joe II Program Manager J. B. Hurt; and James C. Elms, MSC Deputy Director.

1963

May

Manager, Spacecraft Systems Office: David W. Gilbert
Manager, Project Integration Office: J. Thomas Markley

MSC Announcement No. 193, "Reorganization of the Apollo Spacecraft Project Office," May 7, 1963.

PART I: DEFINING CONTRACTUAL RELATIONS

The first meeting of the LEM Flight Technology Systems Panel was held at MSC. The panel was formed to coordinate discussions on all problems involving weight control, engineering simulation, and environment. The meeting was devoted to a review of the status of LEM engineering programs.

    Memorandum, Gerald L. Hunt, MSC, to Chief, Flight Operations Div., "LEM Flight Technology System Meeting No. 1," May 20, 1963, with enclosures.

May 1963

10

MSC Director Robert R. Gilruth announced a division of management responsibilities between operations and development within MSC. Walter C. Williams, Deputy Director for Mission Requirements and Flight Operations, would develop mission plans and rules, crew training, ground support and mission control complexes, and would manage all MSC flight operations. At the same time, he would serve as Director of Flight Operations in the NASA Headquarters OMSF with complete mission authority during flight tests of Mercury, Gemini, and Apollo. James C. Elms, Deputy Director for Development and Programs, would manage all MSC manned space flight projects and would plan, organize, and direct MSC administrative and technical support.

    MSC News Release 63-88, May 10, 1963.

10

NASA Associate Administrator Robert C. Seamans, Jr., directed that a Communications and Tracking Steering Panel and a Working Group be organized. They would develop specifications, performance requirements, and implementation plans for the Manned Space Flight Network in support of the Apollo flight missions.

    Memorandum, Robert C. Seamans, Jr., NASA, to Director, Office of Manned Space Flight, et al., "Functional organization to develop specifications, performance requirements and implementation plans for the Manned Space Flight Network," May 10, 1963.

10

Grumman selected Space Technology Laboratories (STL) to develop and fabricate a mechanically throttled descent engine for the LEM, paralleling Rocketdyne's effort. (See February 27 and March 14.) Following NASA and MSC concurrence, Grumman began negotiations with STL on June 1.

    MSC, "Consolidated Activity Report for the Office of the Director, Manned Space Flight, April 28–May 18, 1963," p. 32; "Monthly Progress Report No. 4," LPR-10-7, p. 44; "Activity Report, Apollo Spacecraft Project Office, May 16–June 13, 1963," p. 8.

Early in the Month

Grumman submitted to NASA a Quality Control Program Plan for the LEM, detailing efforts in management, documentation, training, procurement, and fabrication.

    GAEC, "Report No. 1, Grumman Monthly Quality Status Report for Lunar Excursion Module," LPR-50-1, February 14, 1964.

14

Grumman, reporting on the Lunar Landing Research Vehicle's (LLRV) application to the LEM development program, stated the LLRV could be used profitably to test LEM hardware. Also included was a development

15

55

1963
May

schedule indicating the availability of LEM equipment and the desired testing period.

"Monthly Progress Report No. 4," LPR-10-7, p. 39.

15-16

Faith 7, piloted by Astronaut L. Gordon Cooper, Jr., was launched from Cape Canaveral. An Atlas rocket boosted the Mercury spacecraft into a 161.3 by 267 kilometer (100.2 by 165.9 statute mile) orbit. After 22 orbits, Cooper manually fired the retrorockets and the spacecraft reentered the atmosphere, landing safely in the Pacific Ocean 34 hours, 19 minutes, and 49 seconds after liftoff. Astronaut Cooper was reported in good condition. Cooper's one-day flight turned out to be the final Mercury flight. (See June 12.)

James M. Grimwood, *Project Mercury: A Chronology* (NASA SP-4001, 1963), pp. 191-193.

20

In support of NASA's manned space flight programs, Ames Research Center awarded a $150 000 contract to Westinghouse Electric Corporation for a one-year study of potential physiological damage in space caused by cosmic radiation.

NASA News Release 63-107, "NASA Awards Contract for Study of Space Radiation," May 20, 1963.

20-22

At a meeting on mechanical systems at MSC, Grumman presented a status report on the LEM landing gear design and LEM stowage height. (See February and April 17.) On May 9, NASA had directed the contractor to consider a more favorable lunar surface than that described in the original Statement of Work. Accordingly, Grumman recommended an envelope of LEM/S-IVB clearance of 152.4 centimeters (40 inches) for a landing gear radius of 457 centimeters (180 inches). Beyond this radius, a different gear scheme was considered more suitable but would require greater clearances. The landing gear envelope study was extended for one month to establish a stowed height of the LEM above the S-IVB for adapter design. (See June 3 and October 2.)

"Monthly Progress Report No. 4," LPR-10-7, p. 13.

22

Grumman representatives met with the ASPO Electrical Systems Panel (ESP). From ESP, the contractor learned that the communications link would handle voice only. Transmission of physiological and space suit data from the LEM to the CM was no longer required. VHF reception of this data and S-band transmission to ground stations was still necessary. In addition, Grumman was asked to study the feasibility of a backup voice transmitter for communications with crewmen on the lunar surface should the main VHF transmitter fail.

MSC, "Consolidated Activity Report for the Office of the Director, Manned Space Flight, May 19-June 15, 1963," pp. 54-55; "Monthly Progress Report No. 4," LPR-10-7, p. 21.

PART I: DEFINING CONTRACTUAL RELATIONS

A one-man rocket propulsion device, light enough and small enough to be stored in a spacecraft, was designed to give a lunar explorer more range in examining the moon in addition to permitting him to make quick, close examinations. The model was a result of a study made for MSC by Hamilton Standard. —Hamilton Standard photo.

NASA Headquarters, MSC, Jet Propulsion Laboratory, MSFC, North American, and Grumman agreed that the LEM and CSM would incorporate phase-coherent S-band transponders. [The S-band system provides a variety of communications services. Being phase-coherent meant that it could also

1963

May

23

1963

May

provide Mission Control Center with information about the vehicle's velocity and position, and thus was a means of tracking the spacecraft.] Each would have its own allocated frequencies and would be compatible with Deep Space Instrumentation Facilities.

> "Apollo Quarterly Status Report No. 4," p. 22; "Monthly Progress Report No. 4," LPR-10-7, p. 21; MSC, "Consolidated Activity Report for the Office of the Director, Manned Space Flight, May 19-June 15, 1963," p. 62; interview, telephone, Alfred B. Eickmeier, MSC, March 5, 1970.

23

MIT suggested a major redesign of the Apollo guidance computer to make the CM and LEM computers as similar as possible. NASA approved the redesign and the Raytheon Company, subcontractor for the computer, began work.

> Raytheon Company, Space and Information Systems Div., "Quarterly Technical Report No. 4," FR-3-87, April 1-June 30, 1963.

23-24

Meeting in Bethpage, N. Y., officials from MSC, Grumman, Hamilton Standard, International Latex, and North American examined LEM-space suit interface problems. This session resulted in several significant decisions:
- Suit evaluation would include a vehicle mockup in an aircraft flying zero and one-sixth g trajectories
- The suit assembly emergency oxygen supply would serve also as the backup pressurization and oxygen supply during crew transfer from the CM to the LEM
- The four-hour operating requirement for the portable life support system (PLSS) should not be considered for normal operation
- Pending final design of a waste management system, Grumman would retain provisions for stowage of human wastes
- The thermal garment would not normally be worn inside the LEM
- The PLSS battery would be charged before earth launch
- Prototype Apollo space suits were to be delivered to Grumman as soon as possible for evaluation and vehicle design.

> MSC, "Consolidated Activity Report for the Office of the Director, Manned Space Flight, May 19-June 15, 1963," pp. 59-60.

24

North American demonstrated problems with side-arm controller location and armrest design inside the CM. Major difficulties were found when the subject tried to manipulate controls while wearing a pressurized suit. North American had scheduled further study of these design problems.

> "Project Apollo Spacecraft Test Program, Weekly Activity Report (Period 27 May 1963 through 2 June 1963)," p. 5.

28

MSC Director Robert R. Gilruth reported to the MSF Management Council that the lunar landing mission duration profiles, on which North American would base the reliability design objectives for mission success and crew safety and which assumed a 14-day mission, had been documented and approved. The contractor had also been asked to study two other mission profile extremes, a 14-day mission with 110-hour transearth and translunar transfer times and the fastest practicable lunar landing mission.

<small>MSF Management Council Meeting, May 28, 1963, Agenda Item 2, "Technical Highlights," p. 4.</small>

1963
May

Grumman presented its LEM engineering and simulation plans to MSC, stating that their existing facilities and contracted facilities at North American in Columbus, Ohio, and at LTV would be used throughout 1963. Two part-task LEM simulators would be operational at Grumman early in 1964, with a complete mission simulator available in 1965. MSC had approved the contractor's procurement of two visual display systems for use in the simulators.

<small>MSC, "Consolidated Activity Report for the Office of the Director, Manned Space Flight, May 19–June 15, 1963," pp. 62, 63; GAEC, "Monthly Progress Report No. 6," LPR–10–16, August 10, 1963, p. 5.</small>

29

The Operational Evaluation and Test Branch of MSC's Flight Operations Division considered three methods of providing a recovery hoisting loop on the CM: loop separate from the spacecraft and attached after landing, use of the existing parachute bridle, and loop installed as part of the CM equipment similar to Mercury and Gemini. Studies showed that the third method was preferable. (See April 25–26.)

<small>Memorandum, Christopher C. Kraft, Jr., MSC, to Mgr., ASPO, "Command module recovery hoisting loop," May 29, 1963.</small>

29

Rocketdyne reported to Grumman on the LEM descent stage engine development program. Revised measurements for the engine were: diameter, 137 centimeters (54 inches); length, 221 centimeters (87 inches) (30.5 centimeters [twelve inches] more than the original constraint that Grumman had imposed on Rocketdyne).

<small>MSC, "Consolidated Activity Report for the Office of the Director, Manned Space Flight, May 19–June 15, 1963," p. 61; "Apollo Quarterly Status Report No. 4," p. 21.</small>

30

In its first estimates of reliability for the LEM, Grumman reported a .90 probability for mission success and .994 for crew safety. (The probabilities required by NASA were .984 and .9995, respectively.)

<small>"Monthly Progress Report No. 4," LPR–10–7, p. 26.</small>

During the Month

After a detailed comparison of titanium and aluminum propellant tanks for the LEM descent stage, Grumman selected the lighter titanium.

<small>Ibid., p. 7.</small>

During the Month

Grumman studied the possibility of using the portable life support system lithium hydroxide cartridges in the LEM environmental control system, and

During the Month

A mockup of the Rocketdyne descent engine for the LEM spacecraft.

1963

May

determined that such common usage was feasible. This analysis would be verified by tests at Hamilton Standard.

Ibid., p. 12.

During the Month

Grumman completed the LEM M-1 mockup and began installing equipment in the vehicle. Also, the contractor began revising cabin front design to permit comparisons of visibility. (See September 16–18.)

Ibid., p. 8.

During the Month

NASA and General Dynamics/Convair negotiated a major change on the Little Joe II launch vehicle contract. (See February 18.) It provided for two additional launch vehicles which would incorporate the attitude control subsystem (as opposed to the early fixed-fin version). On November 1, MSC announced that the contract amendment was being issued. NASA Headquarters' approval followed a week later.

Little Joe II Test Launch Vehicle, NASA Project Apollo: Final Report, Vol. I, p. 4–3; MSC News Release 63-223, November 1, 1963; MSC, "Consolidated Activity Report for the Office of the Director, Manned Space Flight, October 20–November 16, 1963," p. 57.

PART I: DEFINING CONTRACTUAL RELATIONS

MSC informed MSFC that the length of the spacecraft–Saturn V adapter had been increased from 807.7 centimeters to 889 centimeters (318 inches to 350 inches). The LEM would be supported in the adapter from a fixed structure on the landing gear. (See October 2.)

1963
June
3

"Apollo Quarterly Status Report No. 4," p. 16.

North American announced that it had selected ITT's Industrial Products Division to provide battery chargers for the CSM, designed for an operational lifetime of 40 000 hours.

3

Space Business Daily, June 4, 1963, p. 712.

The $889.3 million definitive Apollo contract with North American was delivered to NASA Headquarters for review and approval. The target date for approval was extended to June 30. (See August 14.)

4

MSC, "Consolidated Activity Report for the Office of the Director, Manned Space Flight, May 19–June 15, 1963," p. 33.

NASA announced that it would select 10 to 15 new astronauts to begin training in October. Civilian applications were due July 1; those from military personnel, prescreened by their services, were due July 15. New selection criteria reduced the maximum age to 35 years and eliminated the requirement for test pilot certifications.

5

NASA News Release 63-122, "NASA to Select New Astronauts," June 5, 1963.

The Operational Evaluation and Test Branch of MSC's Flight Operations Division made the following recommendations on Apollo postlanding water survival equipment:

6

• Development should continue on a three-man life raft for the Apollo mission.
• A 12-hour-duration dye marker packet should be passively deployed on impact. An additional 18 hours of dye marker should be stored in the survival kit.
• Two radio beacons of the type being developed for Gemini should be included in the survival kit.
• Water egress safety features in the Mercury and Gemini space suits should be included in the Apollo space suit.
• All Apollo equipment which might be involved in water egress, survival, and recovery situations should be configured for water landings.

Memorandum, Christopher C. Kraft, Jr., MSC, to Mgr., ASPO, "Apollo postlanding water survival equipment," June 6, 1963.

North American completed a backup testing program (authorized by MSC on November 20, 1962) on a number of ablative materials for the CM heatshield. Only one of the materials (Avcoat 5026–39) performed satisfactorily at low temperatures. During a meeting on June 18 at MSC, company

10

1963

June

representatives discussed the status of the backup heatshield program. This was followed by an Avco Corporation presentation on the primary heatshield development. As a result, MSC directed North American to terminate its backup program. Shortly thereafter, MSC approved the use of an airgun to fill the honeycomb core of the heatshield with ablative material.

> "Apollo Quarterly Status Report No. 4," p. 15; MSC, "Consolidated Activity Report for the Office of the Director, Manned Space Flight, June 16–July 20, 1963," p. 69; MSC, "Weekly Activity Report for the Office of the Director, Manned Space Flight, June 16–22, 1963," p. 8.

10

NASA issued a $1 946 450 definitive contract to Aerojet-General Corporation for Algol solid-propellant motors for GD/C's Little Joe II vehicles.

> MSC, "Consolidated Activity Report for the Office of the Director, Manned Space Flight, May 19–June 15, 1963," p. 33.

10

Christopher C. Kraft, Jr., of the MSC Flight Operations Division, urged that an up-data link (UDL) (see January 17) be included on the LEM. In general, the UDL would function when a great deal of data had to be transmitted during a time-critical phase. It would also permit utilization of the ground operational support system as a relay station for the transmission of data between the CM and LEM. In case of power failure aboard the LEM, the UDL could start the computer faster and more reliably than a manual voice link, and it could be used to resume synchronization in the computer timing system.

> Memorandum, Christopher C. Kraft, Jr., MSC, to Mgr., ASPO, "Up-Digital-Link to the Lunar Excursion Module," June 10, 1963.

A sketch prepared by John Gurley demonstrates the spacecraft's skip when entering the earth's atmosphere.

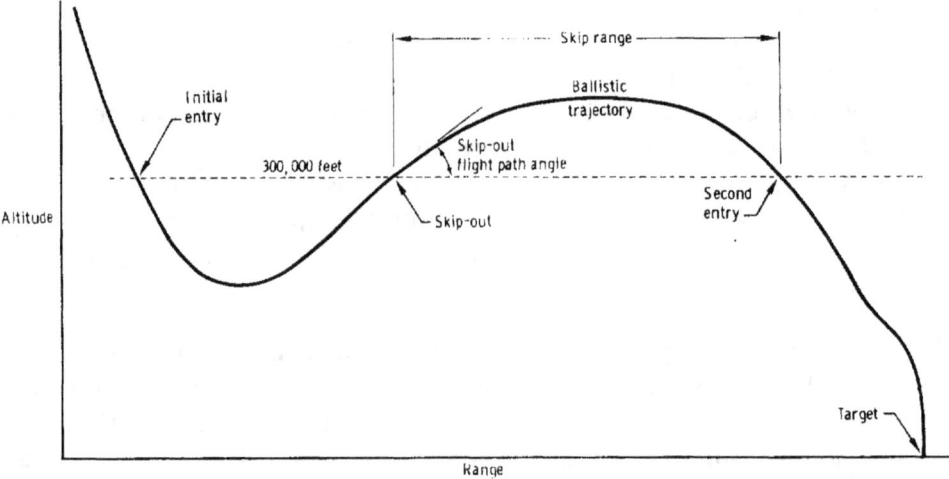

PART I: DEFINING CONTRACTUAL RELATIONS

1963
June
12

The Mission Analysis Branch (MAB) of MSC's Flight Operations Division studied the phenomenon of a spacecraft's "skip" when reentering the earth's atmosphere from lunar trajectories and how that skip relates to landing accuracies. When an Apollo CM encounters the earth's atmosphere (this study used 91 440 meters [300 000 feet] as the practical altitude), the vehicle bounces or "skips" back above the atmosphere. From this point, the spacecraft follows a ballistic trajectory until it re-encounters the atmosphere. During this skip portion of reentry, there is no control of the vehicle's flight trajectory. The length of this skip is, therefore, determined by the angle and speed at the start of this ballistic trajectory. The distance of the skip in turn determines the spacecraft's landing area. Variations in both speed and angle at the start of the skip thus are directly related to landing accuracy, but the effect of these variations is felt much more in shallow than in steep trajectories. In light of these factors, MAB recommended that, for Apollo flights, the skip phase of reentry be made at the steepest practicable angle consistent with maximum allowable acceleration forces.

> Memorandum, John R. Gurley, MSC, to Chief, Flight Operations Div., "A Study of Skip Range Sensitivities and Allowable Errors in Exit Conditions Applicable to the Apollo Missions," June 12, 1963.

NASA Administrator James E. Webb, testifying before the Senate space committee, said that NASA did not plan any further Mercury flights. Project Mercury, America's first manned space flight program, thus was ended.

12

> Loyd S. Swenson, Jr., James M. Grimwood, and Charles C. Alexander, *This New Ocean: A History of Project Mercury* (NASA SP-4201, 1966), p. 503.

D. Brainerd Holmes announced his resignation as NASA's Deputy Associate Administrator and Director of Manned Space Flight, effective sometime in the fall. He had joined NASA in 1961 and was returning to industry.

12

> NASA News Release 63-133, "Holmes Returns to Industry as Mercury Concludes," June 12, 1963.

NASA Headquarters approved a definitive contract for $35 844 550 with AC Spark Plug for the manufacture and testing of navigation and guidance equipment for the CM. This superseded a letter contract of May 30, 1962.

14

> MSC, "Consolidated Activity Report for the Office of the Director, Manned Space Flight, May 19–June 15, 1963," p. 33; NASA News Release 63-136, "Contract Signed with AC Spark Plug for Apollo Guidance System," June 14, 1963; AC Spark Plug, "Apollo Guidance and Navigation System Participating Contractor Quarterly Technical Progress Report," January 1963, p. 2-1.

MSC conducted the final inspection of the Little Joe II launch complex at WSMR.

14

> MSC, "Consolidated Activity Report for the Office of the Director, Manned Space Flight, May 19–June 15, 1963," p. 31.

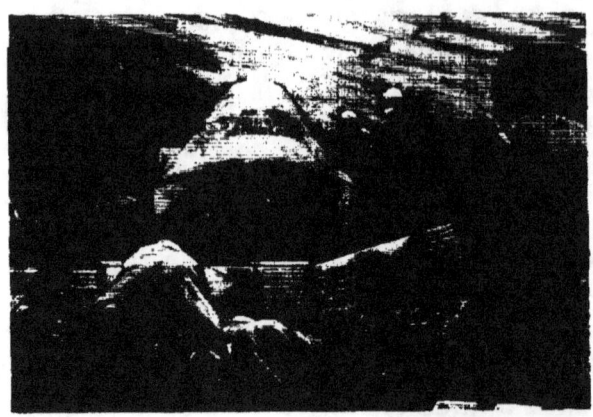

Above is an artist's concept of the instructor-operator station to control the Apollo mission simulator and to monitor crew performance via closed-circuit TV monitors. At right, Lloyd L. Kelly, President of Link Group, General Precision, Inc., inspects a 1/10th-scale model of the simulator for the command module and visual system. This engineering model was used in checking clearances, service access, and overall configuration requirements.
—Link photos.

1963 **June** **14–15**	At its plant in Binghampton, N. Y., Link Division of General Precision, Inc., held a mockup review of the Apollo mission simulator. A number of modifications in the instructor's console were suggested. "Apollo Quarterly Status Report No. 4," p. 40.
14–19	The Soviet Union launched *Vostok V*, piloted by Lt. Col. Valery F. Bykovsky. Two days later Lt. Valentina V. Tereshkova, the first spacewoman, followed in *Vostok VI*. Purposes of the dual mission were to study the medical-biological effects of prolonged space flight upon humans and to perfect spacecraft systems. On its first orbit, *Vostok VI* came within about three miles of *Vostok V*, apparently the closest distance achieved during the flight, and established radio contact. Both cosmonauts landed safely on June 19. The space spectacular featured television coverage of Bykovsky that was viewed in the West as well as in Russia.

PART I: DEFINING CONTRACTUAL RELATIONS

U.S. Congress, Senate, Committee on Aeronautical and Space Sciences, *Soviet Space Programs, 1962–1965; Goals and Purposes, Achievements, Plans, and International Implications*, Staff Report, 89th Cong., 2nd Sess. (December 30, 1966), pp. 180–181.

1963
June

MSC and Grumman assessed crew visibility requirements for the LEM. The study included a series of helicopter flights in which simulated earthshine lighting conditions and LEM window configurations were combined with helicopter landings along representative LEM trajectories. These flights simulated the LEM's attitude, velocity, range, and dive angle in the final approach trajectory.

16–July 20

"Apollo Quarterly Status Report No. 4," p. 18; MSC, "Consolidated Activity Report for the Office of the Director, Manned Space Flight, June 16–July 20, 1963," p. 27.

MSC reported that crew systems engineers at the Center were assessing feasibility of having the LEM crew stand rather than sit. MSC requested Grumman also to look into having the crew fly the vehicle from a standing position. The concept was formally proposed at the August 27 crew systems meeting and was approved at the NASA–Grumman review of the LEM M-1 mockup on September 16–18.

16–July 20

MSC, "Consolidated Activity Report for the Office of the Director, Manned Space Flight, June 16–July 20, 1963," p. 77; "Monthly Progress Report No. 6," LPR-10-16, p. 12; MSC, "Apollo Spacecraft Project Office Activity Report, June 14–July 18, 1963," p. [15].

North American signed (and NASA approved) a definitive contract with Allison Division of General Motors for the service propulsion system propellant tanks.

20

MSC, "Weekly Activity Report for the Office of the Director, Manned Space Flight, June 23–29, 1963," p. 6.

MSC met with those contractors participating in the development of the LEM guidance and navigation system. (See October 18.) Statements of Work for the LEM design concept were agreed upon. (Technical directives covering most of the work had been received earlier by the contractors.)

21–27

MSC, "Activity Report, Apollo Spacecraft Project Office, For Period June 21–27, 1963," p. 2.

North American awarded a contract, valued at $2.8 million, to Avien, Inc., to develop the steerable S-band antenna for the CSM. (See June 11–18, 1964.)

21–27

Ibid.; *Space Business Daily*, July 18, 1963, p. 95.

North American officially froze the design of the CM's stabilization and control system.

22

"Abstract of Proceedings, Command Module Stabilization and Control Systems Meeting No. 16," June 27, 1963, p. 1; MSC, "Activity Report, Apollo Spacecraft Project Office, For Period June 21–27, 1963," p. 2.

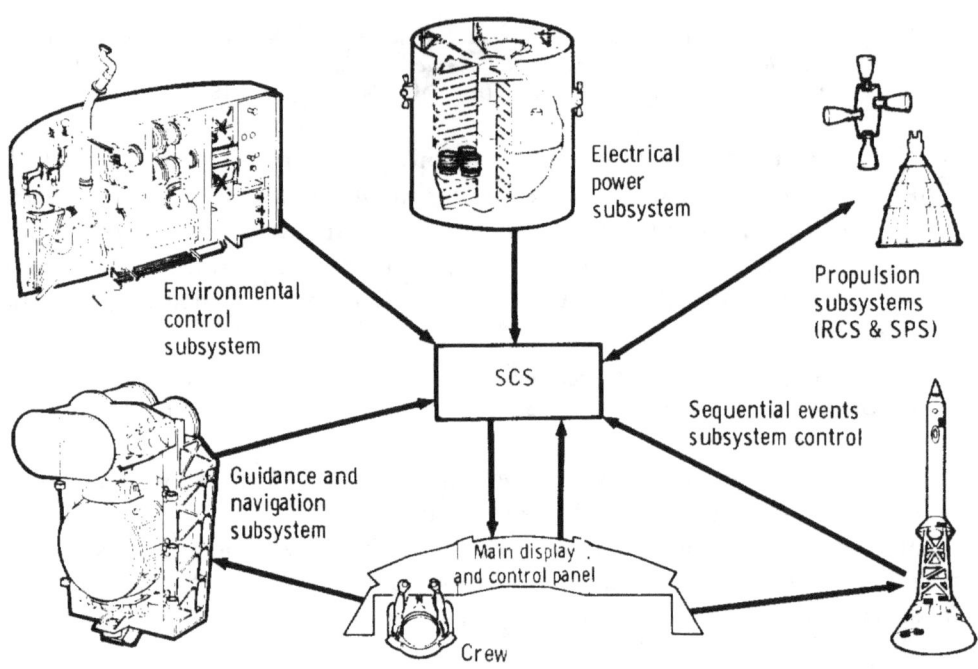

Relationship of SCS to other Apollo subsystems.
—NAA drawing.

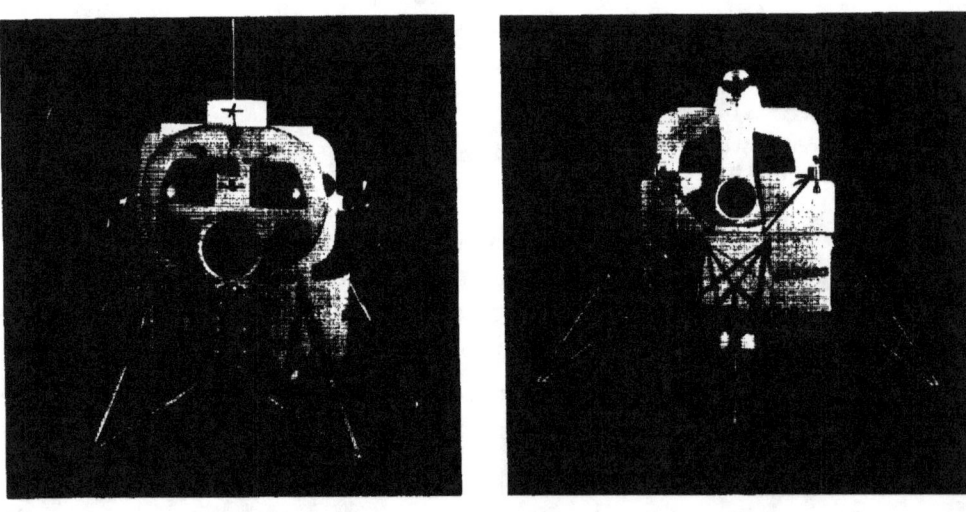

Left, lunar module, 1962; right, lunar module, 1963.
—Grumman photos.

1963

June

25

MSC Director Robert R. Gilruth reported to the MSF Management Council that the LEM landing gear design freeze was now scheduled for August 31. Grumman had originally proposed a LEM configuration with five fixed legs, but LEM changes had made this concept impractical. (See February and

PART I: DEFINING CONTRACTUAL RELATIONS

1963
June

April 17.) The weight and overall height of the LEM had increased, the center of gravity had been moved upward, the LEM stability analysis had expanded to cover a wider range of landing conditions, the cruciform descent stage had been selected, and the interpretation of the lunar model had been revised. These changes necessitated a larger gear diameter than at first proposed. This, in turn, required deployable rather than fixed legs so the larger gear could be stored in the Saturn V adapter. MSC had therefore adopted a four-legged deployable gear, which was lighter and more reliable than the five-legged configuration. (See October 2.)

"Lunar Excursion Module Design Status" (undated), prepared for Gilruth's presentation at the June 25, 1963, meeting of the MSF Management Council, held at the Manned Spacecraft Center.

The first full-scale firing of the SM engine was conducted at the Arnold Engineering Development Center. At the start of the shutdown sequence,

26

Arnold Engineering Development Center altitude rocket test facility.
—AEDC photo.

1963

June

the engine thrust chamber valve remained open because of an electrical wiring error in the test facility. Consequently the engine ran at a reduced chamber pressure while the propellant in the fuel line was exhausted. During this shutdown transient, the engine's nozzle extension collapsed as a result of excessive pressure differential across the nozzle skin.

MSC, "Consolidated Activity Report for the Office of the Director, Manned Space Flight, June 16–July 20, 1963," p. 68.

26

MSC announced that it had contracted with the Martin Company to develop a frictionless platform to simulate the reactions of an extravehicular astronaut in five degrees of freedom—pitch, yaw, roll, forward-backward, and side-to-side. MSC Crew Systems Division would use the simulator to test and evaluate space suits, stabilization devices, tethering lines, and tools.

MSC News Release 63-108, June 26, 1963.

28

A cluster of two Pioneer tri-conical solid parachutes was tested; both parachutes failed. Because of this unsatisfactory performance, the Pioneer solid-parachute program was officially canceled on July 15. (See March 4.)

This Martin Company prototype showed the general configuration of the extravehicular activity simulator developed for MSC. Scheduled for delivery later in 1963 MSC's version would allow the subject to be fully clothed in a pressurized space suit with a portable environmental control system.

PART I: DEFINING CONTRACTUAL RELATIONS

Letter, C. D. Sword, MSC, to NAA, Space and Information Systems Div., "Contract Change Authorization No. Twenty-Seven, Revision 1," July 15, 1963; "Apollo Spacecraft Project Office Activity Report, June 14–July 18, 1963," p. [5].

1963 June

NASA announced its concurrence in Grumman's selection of RCA as subcontractor for the LEM electronics subsystems and for engineering support. Under the $40 million contract, RCA was responsible for five LEM subsystem areas: systems engineering support, communications, radar, inflight testing, and ground support. RCA would also fabricate electronic components of the LEM stabilization and control system. [Engineers and scientists from RCA had been working at Grumman on specific projects since February.]

28

NASA News Release 63–143, "RCA Subcontractor to Grumman for LEM," June 28, 1963; "Monthly Progress Report No. 1," LPR–10–1, p. 2.

The CSM data storage equipment was modified to incorporate a fast-dump capability. Data could thus be recorded at a low speed for later playback at high speed to ground stations.

28

Letter, H. P. Yschek, MSC, to NAA, Space and Information Systems Div., "Contract Change Authorization No. Fifty-Nine," June 28, 1963.

North American reported that mission success predictions continued to be less than the apportioned values. For example, the environmental control subsystem had a predicted mission reliability of .9805, compared to a .997675 apportionment.

During the Month

"Apollo Quarterly Status Report No. 4," pp. 32, 33.

Planning and final details of LTV abort simulation negotiations with Grumman were completed. The abort experiments, to be conducted at LTV's aerospace simulation facility in Dallas, Tex., were scheduled to begin in October. (See April 24, 1964.)

During the Month

GAEC, "Monthly Progress Report No. 5," LPR–10–11, July 10, 1963, p. 19.

MSC reported that two portable life support systems would be stowed in the LEM and one in the CM. Resupplying water, oxygen, and lithium hydroxide could be done in a matter of minutes; however, battery recharging took considerably longer, and detailed design of a charger was continuing.

During the Month

"Apollo Quarterly Status Report No. 4," pp. 24, 25.

Grumman completed the LEM circuit design for suit and cabin pressure control systems. Also the contractor formulated a detailed plan for the evaluation of red and white cockpit lighting; equipment for the test had already been received.

During the Month

"Monthly Progress Report No. 5," LPR–10–11, pp. 13, 20.

BP-6 had arrived and was being off-loaded at WSMR.

1963 **July** 1–2	North American shipped Apollo CM boilerplate 6 and its ground support equipment to WSMR. (See November 7.) "Apollo Quarterly Status Report No. 4," pp. 35, 36; MSC, "Consolidated Activity Report for the Office of the Director, Manned Space Flight, June 16–July 20, 1963," p. 35.
3	Space Technology Laboratories received Grumman's go-ahead to develop the parallel descent engine for the LEM. (See February 27, March 14, and early May.) At the same time, Grumman ordered Bell Aerosystems Company to proceed with the LEM ascent engine. The contracts were estimated at $18 742 820 and $11 205 415, respectively. MSC, "Consolidated Activity Report for the Office of the Director, Manned Space Flight, June 16–July 20, 1963," p. 37; "Monthly Progress Report No. 6," LPR-10-16, p. 50.
9–10	North American held a review of the CM main display console, which would be compatible with the fixed couch and new panel location. The contractor's drawings and comments by the astronauts were then reviewed by MSC. MSC, "Consolidated Activity Report for the Office of the Director, Manned Space Flight, June 16–July 20, 1963," p. 71.
10	As proposed by Joseph F. Shea, Deputy Director (Systems), OMSF, about six weeks earlier, the MSF Management Council established the Panel

PART I: DEFINING CONTRACTUAL RELATIONS

Review Board with broad supervisory and appeal powers over inter-Center panels. (See Volume I, November 8, 1961.) Board members were the Deputy Director (Systems), OMSF, and technical experts from MSC, MSFC, and the Launch Operations Center. OMSF's representative was the chairman.

Recommendations of the board were not binding. If a Center Director decided against a board recommendation, he would, however, discuss and clear the proposed action with the Director of OMSF.

When the Panel Review Board assumed its duties, the Space Vehicle Review Board was abolished. (See Volume I, October 3, 1961.)

>Memorandum, D. Brainerd Holmes, NASA, to Distr., "Panel Review Board," July 10, 1963; MSF Management Council Minutes, May 28, 1963, pp. 3–4.

1963 July

The Marquardt Corporation began testing the prototype engine for the SM reaction control system. Preliminary data showed a specific impulse slightly less than 300 seconds.

>NAA, "Project Apollo Spacecraft Test Program, Weekly Activity Report (Period 8 July 1963 through 14 July 1963)," p. 2.

10

North American reported that it had tried several types of restraint systems for the sleeping area in the equipment bay area of the CM. A "net" arrangement worked fairly well and was adaptable to the constant wear garment worn by the crew. However, North American believed that a simpler restraint system was needed, and was pursuing several other concepts.

>Ibid., p. 4.

10

Aero Spacelines' "Pregnant Guppy," a modified Boeing Stratocruiser, won airworthiness certification by the Federal Aviation Agency. The aircraft would be used to transport major Apollo spacecraft and launch vehicle components.

>Saturn Illustrated Chronology, p. 82; Orlando Sentinel, July 12, 1963.

10

MSC signed a definitive contract, valued at $36.2 million, with International Business Machines (IBM) for the realtime computer complex in the MSC Mission Control Center. IBM was responsible for the design of the computer center, mission and mathematical analyses, programming equipment engineering, computer and program testing, maintenance and operation, and documentation. The complex, consisting of four IBM 7094 computers with their associated equipment, would monitor and analyze data from Gemini and Apollo missions.

>NASA News Release 63-151, "Contract Signed with IBM for Computer Equipment," July 12, 1963; Space Business Daily, July 15, 1963, p. 74.

12

MSC had received 271 applications for the astronaut program. (See June 5.) Seventy-one were military pilots (one from the Army, 34 from the Navy,

15

71

1963

July

26 from the Air Force, and 10 from the Marines). Of the 200 civilians applying, three were women. (See October 18.)

<small>*Astronautics and Aeronautics, 1963* (NASA SP-4004), p. 273; *The Houston Post*, July 17, 1963.</small>

15-16

The Little Joe II qualification test vehicle was shipped from the General Dynamics/Convair plant to WSMR, where the test launch was scheduled for August. (See August 28.)

<small>MSC, "Consolidated Activity Report for the Office of the Director, Manned Space Flight, June 16-July 20, 1963," p. 35; *Little Joe II Test Launch Vehicle, NASA Project Apollo: Final Report*, Vol. I, p. 1-6; TWX, NASA Resident Office, WSMR, to MSC, "Activity Report for MSC-WSMR Office for June 16 through July 20," July 23, 1963.</small>

16

MSC directed North American to concentrate on the extendable boom concept for CSM docking with the LEM. The original impact type of docking had been modified:

(1) The primary mode employed an extendable probe. It would establish initial contact and docking at a separation distance sufficient to prevent dangerous impact as a result of pilot error.

(2) The backup mode consisted of free-flying the two modules together. Mean relative impact velocities established during free-flying docking simulation studies would be used as the design impact velocities.

North American and Grumman began a hardware testing and flight simulation program in late September to evaluate the feasibility of several types of extendable probe/tether systems. The two companies were to determine the stiffness required of the docking structure for compatibility with the stabilization and control system. (See November 19-20.)

<small>"Apollo Monthly Progress Report," SID 62-300-16, pp. 3, 9; MSC, "Weekly Activity Report for the Office of the Director, Manned Space Flight, July 28-August 3, 1963," p. 2; "Monthly Progress Report No. 6," LPR-10-16, p. 3.</small>

Apollo command module probe and drogue assembly. —NAA drawing.

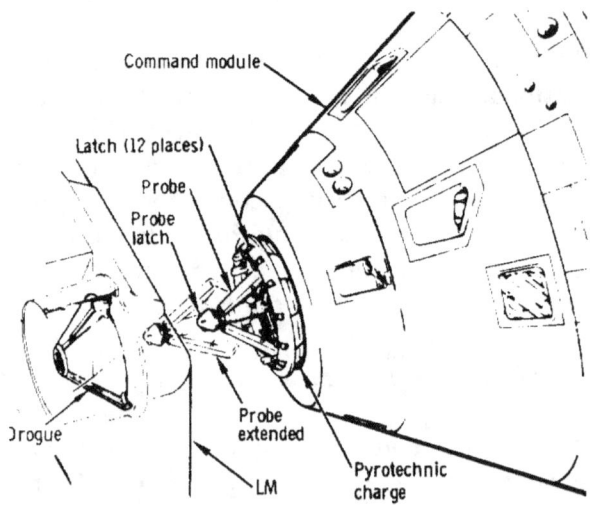

PART I: DEFINING CONTRACTUAL RELATIONS

Grumman presented the results of a study on LEM visibility. A front-face configuration with triangular windows was tentatively accepted by MSC for the ascent stage. Further investigation would be directed toward eliminating the "dead spots" to improve the configuration's visibility.

"Monthly Progress Report No. 6," LPR-10-16, p. 3.

1963
July
16

North American reported that Lockheed Propulsion Company had successfully completed development testing of the launch escape system pitch control motor. (See December 28, 1962.)

"Apollo Monthly Progress Report," SID 62-300-16, p. 18.

16-August 15

MSC authorized North American to fit the launch escape system with a redundant tower separation device. This equipment incorporated an explosive bolt and shaped charge cutter.

Letter, H. P. Yschek, MSC, to NAA, Space and Information Systems Div., "Contract Change Authorization No. Sixty-Two," July 18, 1963.

18

Grumman selected Pratt and Whitney to develop fuel cells for the LEM. Current LEM design called for three cells, supplemented by a battery for power during peak consumption beyond what the cells could deliver. Grumman and Pratt and Whitney completed contract negotiations on August 27, and MSC issued a letter go-ahead on September 5. Including fees and royalties, the contract was worth $9.411 million.

MSC, "Weekly Activity Report for the Office of the Director, Manned Space Flight, July 21-27, 1963," p. 8; MSC, "ASPO Weekly Activity Report, September 5-11, 1963," p. 5; GAEC, "Monthly Progress Report No. 7," LPR-10-22, September 10, 1963, p. 2.

18

North American, Grumman, and Hamilton Standard, meeting at MSC with Crew Systems Division engineers, agreed that the portable life support system (PLSS) would have three attaching points for stowage in the spacecraft. In addition, it was agreed that the PLSS should not be used for shoulder restraint in the LEM.

"Monthly Progress Report No. 6," LPR-10-16, p. 12; MSC, "Apollo Spacecraft Project Office Activity Report, June 14-July 18, 1963," p. [8].

18

Grumman directed the Marquardt Corporation to begin development of the LEM reaction control system thrusters. Negotiations had begun on March 11 on the definitive subcontract, a cost-plus-incentive-fee type with a total estimated cost of $10 871 186.

MSC, "Consolidated Activity Report for the Office of the Director, Manned Space Flight, July 21-August 17, 1963," p. 36; "Monthly Progress Report No. 6," LPR-10-16, p. 50; GAEC, "Monthly Progress Report No. 8," LPR-10-24, October 10, 1963, p. 49.

19

NASA launched a Scout rocket with a nose cone of experimental heatshield material from Wallops Island, Va. The rocket was intentionally destroyed when it deviated from its course a few seconds after liftoff. The nose cone

20

73

1963 July	had been expected to reenter the atmosphere at 27 934 kilometers (18 600 miles) per hour to test the material's thermal performance under heating loads near those of a lunar reentry.

> NASA News Release 63–153, "Reentry Experiment Will Test Ablation Material," July 17, 1963; *The Houston Chronicle*, July 20, 1963.

23 George E. Mueller, Vice President for Research and Development of Space Technology Laboratories, was named NASA Deputy Associate Administrator for Manned Space Flight to succeed D. Brainerd Holmes, effective September 1.

> NASA News Release 63–162, "NASA Names New Head for Manned Space Flight; Succeeds Holmes," July 23, 1963.

23 Grumman authorized Hamilton Standard to begin development of the environmental control system (ECS) for the LEM. The cost-plus-incentive-fee contract was valued at $8 371 465. The parts of the ECS to be supplied by Hamilton Standard were specified by Grumman.

> "Monthly Progress Report No. 6," LPR-10-16, p. 50; MSC, "Consolidated Activity Report for the Office of the Director, Manned Space Flight, July 21–August 17, 1963," p. 36.

A full-scale mockup showed the final configuration of the LEM's ascent rocket engine developed by Bell Aerosystems Company. The entire thrust chamber and nozzle extension were made of an ablative material.

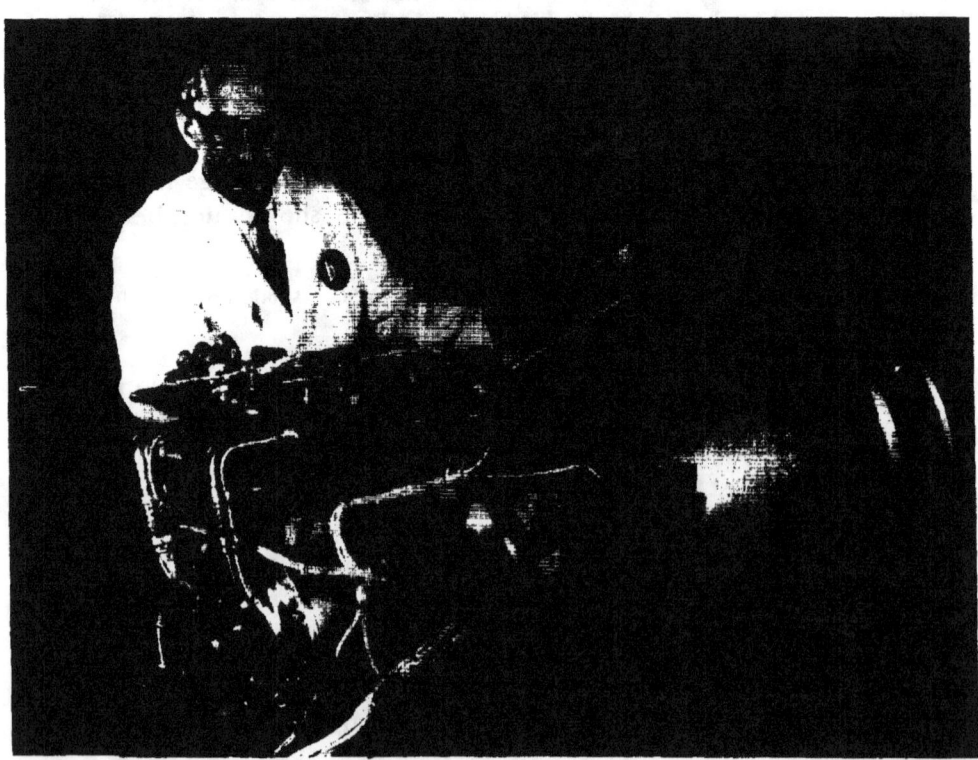

PART I: DEFINING CONTRACTUAL RELATIONS

ASPO reported that a different type of stainless steel would be used for the CM heatshield. The previous type proved too brittle at cryogenic temperatures. Aside from their low temperature properties, the two metals were quite similar and no fabrication problems were anticipated.

1963
July
28–August 3

MSC, "Weekly Activity Report for the Office of the Director, Manned Space Flight, July 28–August 3, 1963," p. 4.

ASPO ordered Grumman to design identical connectors for both ends of the space suit hoses in the LEM. This arrangement, called the "buddy concept," would permit one portable life support system to support two crewmen and thus would eliminate the need for a special suit-to-suit hose. (See August 26, 1964.)

28–August 3

Ibid., p. 6.

MIT and Grumman representatives discussed installing the inertial measurement unit and the optical telescope in the LEM. Of several possible locations, the top centerline of the cabin seemed most promising. Grumman agreed to provide a preliminary structural arrangement of the guidance

30–31

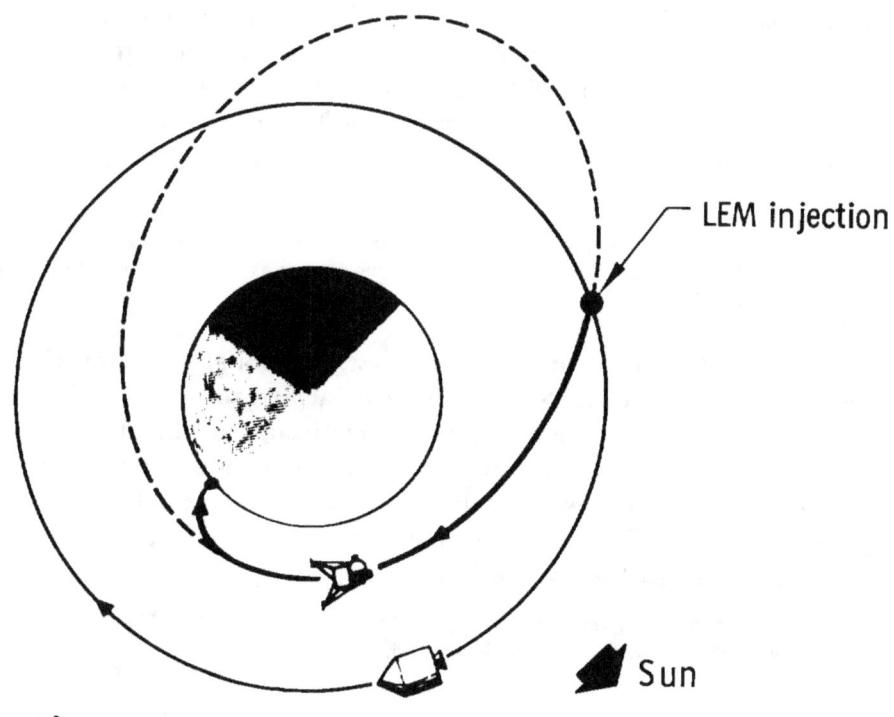

A briefing aid depicted the equal-period orbit method of LEM descent to the lunar surface from lunar orbit.

1963

July

components so that MIT could study problems of installation and integration.

"Monthly Progress Report No. 6," LPR-10-16, pp. 6, 7.

August

1

North American, NASA, and Grumman representatives discussed three methods of descent from lunar parking orbit: (1) descent of the LEM only (the minimum energy Hohmann transfer), (2) the combined descent of both spacecraft, and (3) the synchronous equal period method. While neither contractor felt that weight factors should be of primary concern, Grumman favored the Hohmann transfer and North American the combined descent, which represented the extremes of energy requirements. After considering reliability, fuel consumption, and operational flexibility, NASA chose the synchronous method as the prime mission mode but recommended continued investigation of the other two techniques.

Memorandum, John E. Gerstle, Jr., and Joe D. Payne, MSC, to Chief, Flight Operations Div., "LEM Descent Profile," August 20, 1963.

2

North American asked MSC if Grumman was designing the LEM to have a thrusting capability with the CSM attached and, if not, did NASA intend to require the additional effort by Grumman to provide this capability. North American had been proceeding on the assumption that, should the service propulsion system (SPS) fail during translunar flight, the LEM would make any course corrections needed to ensure a safe return trajectory. [The Guidance and Control Panel, at a meeting on November 29, 1962, had stated that a LEM would be included on all Saturn V flights, thus providing a backup propulsion in case of SPS failure.] On August 6, Robert O. Piland, Acting ASPO Manager, responded by asking North American to investigate the operational and systems aspects of this backup mode before a final decision was made.

TWX, H. G. Osbon, NAA, to MSC, Attn: Robert O. Piland, August 2, 1963; letter, Piland to NAA, Attn: E. E. Sack, "LEM Propulsion System as Backup to SM Propulsion System," August 6, 1963.

5

In what was to have been an acceptance test, the Douglas Aircraft Company static fired the first Saturn S-IV flight stage at Sacramento, Calif. An indication of fire in the engine area forced technicians to shut down the stage after little more than one minute's firing. A week later the acceptance test was repeated, this time without incident, when the vehicle was fired for over seven minutes. [The stage became part of the SA-5 launch vehicle, the first complete Saturn I to fly. See January 29, 1964.]

History of Marshall . . . January 1-June 30, 1963, Vol. I, p. 16; *The Huntsville Times*, August 6, 1963; *The Houston Post*, August 13, 1963.

9-10

The Panel Review Board (see July 10) held its first meeting at the Launch Operations Center (LOC). The board established an Executive Secretariat,

PART I: DEFINING CONTRACTUAL RELATIONS

composed of Bert A. Denicke (OMSF), Joachim P. Kuettner (MSFC), Emil P. Bertram (LOC), and Philip R. Maloney (MSC). Among other actions, the board abolished the GE Policy Review Board (see December 5, 1962).

1963
August

MSC, "Apollo Spacecraft Project Office Activity Report, July 19–August 15, 1963," p. 1.

NASA Administrator James E. Webb signed the definitive contract with North American for the development of the Apollo CSM. This followed by almost two years North American's selection as prime contractor. The $938.4 million cost-plus-fixed-fee agreement was the most valuable single research and development contract in American history. The contract called for the initial production (i.e., through May 15, 1965) of 11 mockups, 15 boilerplate vehicles, and 11 production articles. (See September 1, 1964.)

14

Space News Roundup, August 21, 1963; Oakley, Historical Summary, S&ID Apollo Program, pp. 11, 24–25; Space Business Daily, August 19, 1963, p. 255.

ITT's Kellogg Division delivered to Hamilton Standard the first operational prototype space suit communications system. (See November 27, 1962.)

Mid-month

Aviation Week and Space Technology, 79 (August 19, 1963), p. 29; Space Business Daily, August 20, 1963, p. 263.

At a meeting on the LEM electrical power system, Grumman presented its latest load analysis, which placed the LEM's mission energy requirements at 76.53 kilowatt-hours. (See January 28.) The control energy level for the complete LEM mission had been set at 54 kilowatt-hours and the target energy level at 47.12 kilowatt-hours. Grumman and MSC were jointly establishing ground rules for an electrical power reduction program.

15

MSC, "ASPO Weekly Activity Report, August 15–21, 1963," p. 4.

MSC Crew Systems Division conducted mobility tests of the Apollo prototype space suit inside a mockup of the CM. Technicians also tested the suit on a treadmill. The subjects' carbon dioxide buildup did not exceed two percent; their metabolic rates were about 897 000 joules (850 BTU) per hour at vent pressure, 1 688 000 joules at 2.4 newtons per square centimeter (1600 BTU at 3.5 psi), and 2 320 000 joules at 3.5 newtons per square centimeter (2200 BTU at 5.0 psi).

15–September 21

MSC, "Consolidated Activity Report for the Office of the Director, Manned Space Flight, August 18–September 21, 1963," p. 40.

MSC completed a comparison of 17-volt and 28-volt batteries for the portable life support system. The study showed that a 28-volt battery would provide comparable energy levels without increase in size and weight and would be compatible with the spacecraft electrical system.

Week of August 18

MSC, "Weekly Activity Report for the Office of the Director, Manned Space Flight, August 18–24, 1963," p. 6.

1963
August
21

John P. Bryant, of the Flight Operations Division's (FOD) Mission Analysis Branch (MAB), reported to FOD that the branch had conducted a rough analysis of the effects of some mission constraints upon the flexibility possible with lunar launch operations. (As a base, MAB used April and May 1968, called "a typical two-month period.") First, Bryant said, MAB used the mission rules demanded for the Apollo lunar landing (e.g., free-return trajectory; predetermined lunar landing sites; and lighting conditions on the moon—"by far the most restrictive of the lot"). Next, MAB included a number of operational constraints, ones "reasonably representative of those expected for a typical flight," but by no means an "exhaustive" list:

- A minimum daily launch window of three hours
- A 26-degree maximum azimuth variation
- An earth landing within 40 degrees of the equator
- A minimum of three successive daily launch windows
- A daylight launch with at least three hours of daylight following liftoff
- Transposition and docking in sunlight
- Use of but one of the two daily windows available for translunar injection.

Bryant advised that, taken just by themselves, these various constraints, both mission and operational, had a "restrictive effect" and that operational flexibility was thereby "dramatically curtailed." Moreover, "there are still a number of possible constraints which have not been considered which could still further affect the size of the ultimate launch window" (and the list was "increasing almost daily"): requirements for tracking coverage and for lighting during rendezvous and reentry; and restrictions imposed by solar activity, launch environment, and—no small matter—weather conditions at the launch site.

"The consequences," Bryant concluded, "of imposing an ever-increasing number of these flight restrictions is obvious—the eventual loss of almost all operational flexibility. The only solution is . . . [a] meticulous examination of every constraint which tends to reduce the number of available launch opportunities," looking toward eliminating "as many as possible."

Memorandum, John Bryant, MSC, to Chief, Flight Operations Div., "Planning Apollo missions with imposed operational contraints," September 5, 1963.

Week of August 22–29

An Ad Hoc Rendezvous Working Group was formed at MSC to study the possibility of substituting a unified S-band system for the rendezvous X-band radar on the LEM and CSM.

"ASPO Weekly Activity Report, August 22–29, 1963," p. 7; MSC, "Weekly Activity Report for the Office of the Director, Manned Space Flight, September 1–7, 1963," p. 11.

26

MSC received proposals for the visual displays for the LEM simulator. Because of the changed shape of that vehicle's windows, however, Grumman had to return those proposals to the original bidders, sending revised proposals to MSC in December. Farrand Optical Company was selected to develop the display, and the Center approved Grumman's choice. Negotiations between Grumman and Farrand were completed during March 1964.

> "Apollo Quarterly Status Report No. 5," pp. 55–56; MSC, "Consolidated Activity Report for the Office of the Director, Manned Space Flight, August 18–September 21, 1963," p. 28; "Consolidated Activity Report for the Office of the Associate Administrator, Manned Space Flight, December 22, 1963–January 18, 1964," p. 39; GAEC, "Monthly Progress Report No. 14," LPR-10-30, April 10, 1964, p. 35.

1963
August

The MSF Management Council decided that, as part of the proposed reorganization of NASA Headquarters (see October 9), a Deputy Associate Administrator for Manned Space Flight would become responsible for all manned space flight activities within NASA.

> MSF Management Council Meeting, August 27, 1963, Agenda Item 10, "Responsibility of the Deputy Associate Administrator for Manned Space Flight For Technical Matters," p. 1.

27

At left is the scene at White Sands as NASA and contractor engineers and technicians worked through the night to assure that everything was in readiness for the launch of the Little Joe II qualification test vehicle. At right is the scene the next morning just after all seven motors of the vehicle ignited simultaneously, providing a thrust of about 141 000 kilograms (310 000 pounds).

1963

August

27 A LEM crew systems meeting was held at Grumman. The standing arrangement proposed for the crew (see June 16–July 20) promised to reduce the weight of the LEM by as much as 27.2 kilograms (60 pounds), and would improve crew mobility, visibility, control accessibility, and ingress-egress. Pending more comprehensive analysis, crew systems designers also favored the revised front-face configuration (see July 16).

MSC, "ASPO Weekly Activity Report, August 22–29, 1963," p. 7.

28 The Little Joe II qualification test vehicle was launched from WSMR. Its objectives were to prove the Little Joe's capability as an Apollo spacecraft test vehicle and to determine base pressures and heating on the missile. These aims were achieved. The lone failure was a malfunction in the destruct system.

Little Joe II Test Launch Vehicle, NASA Project Apollo: Final Report, Vol. I, pp. 1–11, 1–13, 1–17.

PART II

Developing Hardware Distinctions

August 30, 1963, through April 28–30, 1964

PART II

The Key Events

1963

August 30: Lunar Orbiter program officially approved.
September 16–18: Grumman Aircraft Engineering Corporation (GAEC) held inspection and review of first lunar excursion module (LEM) ascent stage mockup M–1.
October 8: Joseph F. Shea named manager of Apollo Spacecraft Project Office at Manned Spacecraft Center (MSC).
October 18: Third "class" of astronauts introduced.
October 24: George E. Mueller, the new NASA Associate Administrator for Manned Space Flight, held first meeting of NASA–Industry Apollo Executives.
November 1: Major reorganization of NASA Headquarters and Office of Manned Space Flight (OMSF) took effect; Mueller directed the revision of Saturn–Apollo flight schedules.
November 7: Apollo Pad Abort Mission 1, using command module (CM) boilerplate 6 was conducted at White Sands Missile Range, N. Mex.
November 22: Preliminary ground rules for the Spacecraft Development Test Program and gross lunar landing sites selected.
December 31: Samuel C. Phillips (Brig. Gen., USAF) announced as new NASA Deputy Director for Apollo Program.

1964

January 3: Apollo prime contractors issued joint report on spacecraft development test plan.
January 19: George M. Low assigned to MSC as Deputy Director.
January 21: North American Aviation, Inc. (NAA), presented a design concept for the Block II command and service module (CSM), designed for lunar missions.
January 29: Saturn-Apollo 5 flight marked first mission of Block II Saturn with two live stages.
March 9: MSC assigned funds and responsibility for developing scientific instruments for lunar exploration.
March 23: OMSF outlined Saturn-Apollo mission plans.
March 24–26: GAEC held first complete LEM mockup TM–1 inspection and review.
April 8: First Gemini mission performed.
April 14: Project Fire tested heat transfer concepts for Apollo at 40 230 kilometers (25 000 miles) per hour lunar return velocity.
April 21: Basic rules for Apollo space suit operation established.
April 28–30: NAA held basic mockup inspection and review for Block II CSM.

PART II

Developing Hardware Distinctions

August 30, 1963, through April 28–30, 1964

NASA Associate Administrator Robert C. Seamans, Jr., approved the Lunar Orbiter program. Objectives of the program were reconnaissance of the moon's topography, investigation of its environment, and collection of selenodetic information. (See May 12, 1964.)

1963
August
30

The document called for five flight and three test articles. The Lunar Orbiter spacecraft would be capable of photographing the moon from a distance of 22 miles above the surface. Overall cost of the program was estimated at between $150 and $200 million.

<small>NASA Office of Space Sciences (OSS) Review, "Lunar Orbiter Program Status Report," September 4, 1963; *Space Business Daily*, September 3, 1963, p. 327; NASA Project Approval Document, "Research and Development Project: Lunar Orbiter," Cost No.</small>

A scale model of the Lunar Orbiter spacecraft.

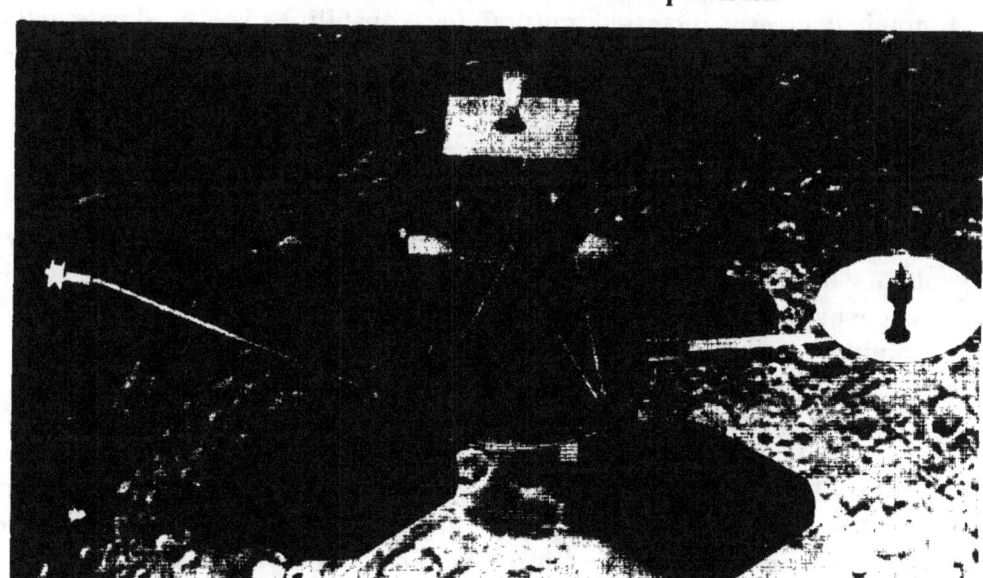

1963

August

84-800-804, undated; memorandum, Dir., OSS, to Langley Research Center, Attn: Floyd L. Thompson, "Implementation of the Lunar Orbiter Project," August 30, 1963. See also Bruce K. Byers, "Lunar Orbiter: A Preliminary History" (NASA HHN-71, Comment Edition), p. 30.

Week of August 30

North American defined the maximum, nominal, and minimum CM ablation heatshield thicknesses for lunar reentry. The maximum and minimum limits represented variations that might arise as studies progressed.

MSC, "ASPO Weekly Activity Report, August 30-September 4, 1963," p. 3.

During the Month

Grumman built a full-scale cardboard model of the LEM to aid in studying problems of cockpit geometry, specifically the arrangement of display panels. This mockup was reviewed by MSC astronauts and the layout of the cockpit was revised according to some of their suggestions.

Also Grumman reported that a preliminary analysis showed the reaction control system plume heating of the LEM landing gear was not a severe problem. [This difficulty had been greatly alleviated by the change from five to four landing legs on the vehicle. (See April 17 and May 20-22.)]

"Monthly Progress Report No. 7," LPR-10-22, pp. 7, 25.

During the Month

At a meeting at MSC, Grumman representatives submitted the cost proposal for LEM test articles LTA-8 and LTA-9, and suggested a testing program for the two vehicles: LTA-8 should be used for restrained integrated systems testing in the altitude propulsion test facilities at the Atlantic Missile Range; LTA-9 should be used for manned atmospheric tethered operation tests. The contractor also recommended an early flight demonstration program to verify the helicopter tether operation potential, which promised greatly increased mission test capability over fixed-base tether facilities. The tether method (helicopter or fixed-base) should be determined after the verification. LTA-8 should be considered as a constraint to LEM-5, and LTA-9 as a constraint to the lunar landing mission.

Ibid., pp. 45, 46.

September 1-7

MSC reported that design of the control and displays panel for the CM was about 90 percent complete. North American was expected to release the design by September 20. Qualification testing of the panels would begin around December 1.

MSC, "Weekly Activity Report for the Office of the Director, Manned Space Flight, September 1-7, 1963," p. 3.

4

Director Robert R. Gilruth established the MSC Manned Spacecraft Criteria Board to set up engineering, design, and procedural standards for manned spacecraft and associated systems. The board was composed of Maxime A. Faget, Chairman; James A. Chamberlin; Kenneth S. Kleinknecht; F. John Bailey, Jr.; G. Barry Graves; Jacob C. Moser; and Norman F. Smith,

PART II: DEVELOPING HARDWARE DISTINCTIONS

Secretary. Board criteria would become MSC policy; and unless specific waivers were obtained, compliance by project offices was mandatory.

> MSC Circular No. 85, "MSC Manned Spacecraft Criteria Board," September 17, 1963.

MSC Flight Operations Division (FOD) recommended a series of water impact tests to establish confidence in the CM's recovery systems under a variety of operating conditions. FOD suggested several air drops with water landings under various test conditions. Among these were release of the main parachutes at impact, deployment of the postlanding antennas, actuation of the mechanical location aids, and activation of the recovery radio equipment.

> Memorandum, Christopher C. Kraft, Jr., MSC, to Mgr., ASPO, "Recommendation for a water landing operational qualification test series using AFRM 005," September 4, 1963.

MSC began a study to define the stability limits of a 457-centimeter (180-inch) radius LEM gear configuration. The study, in two phases, sought to examine factors affecting stability (such as lunar slope, touchdown velocity and direction, and the effects of soil mechanics) in direct support of the one-sixth model and full-scale drop test programs and to complete definition of landing capabilities of the LEM. (See October 2.)

> MSC, "ASPO Weekly Activity Report, September 5–11, 1963," pp. 7–8.

MSC announced a $7.658 million definitive contract with Kollsman Instrument Corporation for the CM guidance and navigation optical equipment, including a scanning telescope, sextant, map and data viewer, and related ground support equipment. MSC had awarded Kollsman a letter contract on May 28, 1962, and had completed negotiations for the definitive contract on March 29, 1963. "The newly signed contract calls for delivery of all hardware to AC Spark Plug by August 1, 1964."

> MSC News Release 63–147, September 6, 1963; MSC, "Weekly Activity Report for the Office of the Director, Manned Space Flight, May 27–June 3, 1962," p. 12; Kollsman Instrument Corporation, "Apollo Program Quarterly Progress Report No. 3," March 31, 1963, p. 2; *ibid.*, "Apollo Program Quarterly Progress Report No. 6," December 31, 1963, pp. 10–11.

MSC Flight Operations Division (FOD) established a 72-hour lifetime for Apollo recovery aids. This limitation was derived from considerations of possible landing footprints, staging bases, and aircraft range and flying time to the landing areas. Primary location aids were the spacecraft equipment (VHF AM transceiver, VHF recovery beacon, and HF transceiver) and the VHF survival radio. Because of battery limitations, current planning called for only a 24-hour usage of the VHF recovery beacon. If electronic aids were needed beyond this time, the VHF survival radio would be used. If the spacecraft were damaged or lost, the VHF survival radio would be the only electronic location aid available. MSC had recently selected

1963
September

4

5–11

6

6

1963

September

the Sperry Phoenix Company to produce the Gemini VHF survival radio, which was expected to meet the Apollo requirements. FOD recommended that the current contract with Sperry Phoenix be extended to provide the units needed for Apollo missions.

<small>Memorandum, Christopher C. Kraft, Jr., MSC, to ASPO, Attn: L. N. McMillion, "Apollo VHF survival radio," undated (ca. September 1963).</small>

6

At El Centro, Calif., CM boilerplate (BP) 3, a parachute test vehicle, was destroyed during tests simulating the new BP-6 configuration (without strakes or apex cover). Drogue parachute descent, disconnect, and pilot mortar fire appeared normal. However, one pilot parachute was cut by contact with the vehicle and its main parachute did not deploy. Because of harness damage, the remaining two main parachutes failed while reefed. Investigation of the BP-3 failure resulting in rigging and design changes on BP-6 and BP-19.

<small>"Apollo Monthly Progress Report," SID 62-300-17, p. 11; ibid., SID 62-300-18, pp. 15-16.</small>

9

MSC ordered North American to make provisions in the CM to permit charging the 28-volt portable life support system battery from the spacecraft battery charger.

On the following day, the Center informed North American also that a new mechanical clock timer system would be provided in the CM for indicating elapsed time from liftoff and predicting time to and duration of various events during the mission.

<small>Letter, H. P. Yschek, MSC, to NAA, Space and Information Systems Div., "Contract Change Authorization No. Eighty-Two," September 9, 1963; ibid., "Contract Change Authorization No. Eighty-Four," September 10, 1963.</small>

12

NASA announced that, in the future, unmanned lunar landing spacecraft (e.g., Rangers and Surveyors) will be assembled in "clean rooms" and treated with germ-killing substances to reduce the number of microbes on exposed surfaces. These sterilization procedures, less stringent than earlier methods, were intended to prevent contamination of the lunar surface and, at the same time, avoid damage to sensitive electronic components. Heat sterilization was suspected as one of the reasons for the failure of Ranger spacecraft.

<small>*The Washington Post*, September 13, 1963.</small>

16

A tone warning signal was added to the CM instrumentation system. If a system malfunctioned, this warning would be heard through both the master caution and warning subsystem and the astronauts' earphones.

<small>Letter, H. P. Yschek, MSC, to NAA, Space and Information Systems Div., "Contract Change Authorization No. Eighty-Nine," September 16, 1963.</small>

16

The launch escape system was modified so that, under normal flight conditions, the crew could jettison the tower. On unmanned Saturn I flights,

PART II: DEVELOPING HARDWARE DISTINCTIONS

A design engineering inspection (DEI) and Apollo program design review were held at NAA's El Segundo, Calif., facilities September 10-12, 1963. About 70 NASA personnel members participated in the DEI of boilerplate 12 before it was shipped to WSMR to test the launch escape system. The following two days approximately 100 NASA officials including personnel from most NASA Centers and Headquarters attended the program design review. Topics included structural design, the propulsion, power, and electrical systems, guidance and navigation, simulation and trainers, ground support equipment, and a program hardware summary.

tower jettison was initiated by a signal from the instrument unit of the S-IV (second) stage.

> Letter, H. P. Yschek, MSC, to NAA, Space and Information Systems Div., "Contract Change Authorization No. Ninety-One," September 16, 1963.

NASA representatives held a formal review of Grumman's LEM M-1 mockup, a full-scale representation of the LEM's crew compartment. MSC decided that (1) the window shape (triangular) and visibility were satisfactory; (2) a standing position for the crew was approved, although, in general, it was believed that restraints restricted crew mobility; (3) the controllers were positioned too low and lacked suitable arm support for fine control; and (4) crew station arrangement was generally acceptable, although specific details required further study. (See June 16-July 20 and August 27.)

> MSC, "ASPO Weekly Status Report, September 19-25, 1963."

1963

September

16–18

1963

September

17 — LTV presented the preliminary results of a manual rendezvous simulation study. Their studies indicated that a pilot trained in the technique could accomplish lunar launch and rendezvous while using only two to three percent more fuel than the automatic system. (See May 6 and October 10, 1963, and April 24, 1964.)

> MSC, "Consolidated Activity Report for the Office of the Director, Manned Space Flight, September 22–October 19, 1963," p. 31.

18 — The AiResearch Manufacturing Company announced that it had been awarded a $20 million definitive contract for the CM environmental system. [AiResearch had been developing the system under a letter contract since 1961. See Volume I, December 21, 1961.]

> *The Houston Post*, September 19, 1963.

19 — MSC made several changes in the CM's landing requirements. Impact attenuation would be passive, except for that afforded by the crew couches and the suspension system. The spacecraft would be suspended from the landing parachutes in a pitch attitude that imposed minimum accelerations on the crew. A crushable structure to absorb landing shock was required in the aft equipment bay area.

> Letter, H. P. Yschek, MSC, to NAA, Space and Information Systems Div., "Contract Change Authorization No. Ninety-Three," September 19, 1963.

19–25 — The space suit umbilical disconnects were being redesigned to the "buddy concept" and for interchangeability between the CM and the LEM. (See September 29, 1964.) MSC was reviewing methods for a crewman to return to the LEM following space suit failure on the lunar surface. (See July 28–August 3.)

> MSC, "ASPO Weekly Activity Report, September 19–25, 1963," p. 4.

19–25 — North American incorporated an automatic radiator control into the CM's environmental control system to eliminate the need for crew attention during lunar orbit.

Recent load analysis at North American placed the power required for a 14-day mission at 577 kilowatt-hours, a decrease of about 80 kilowatt-hours from earlier estimates.

> *Ibid.*, pp. 2, 3.

19–October 16 — Grumman directed Bell Aerosystems Company to establish the ablative nozzle extension as the primary design for the LEM's ascent stage engine. The radiation-cooled nozzle design, a weight-saving alternative, must be approved by NASA. (See March; also January and May 4–11, 1964.)

> MSC, "ASPO Monthly Activity Report, September 19–October 16, 1963," p. 18.

PART II: DEVELOPING HARDWARE DISTINCTIONS

President John F. Kennedy, during an address before the United Nations General Assembly, suggested the possibility of Russian-American "cooperation" in space. Though not proposing any specific program, Kennedy stated that, "in a field where the United States and the Soviet Union have a special capacity—the field of space—there is room for new cooperation, for further joint efforts in the regulation and exploration of space. I include among these possibilities," he said, "a joint expedition to the moon. . . . Surely we should explore whether the scientists and astronauts of our two countries—indeed, of all the world—cannot work together in the conquest of space, sending some day in this decade to the moon, not the representatives of a single nation, but the representatives of all humanity."

During a news conference in Houston that same day, several NASA officials commented on the President's address. Associate Administrator Robert C. Seamans, Jr., stated that Kennedy's proposals came as no great surprise. He said that many "large areas" for cooperation exist, such as exchanges of scientific information and in space tracking, but emphasized that no cosmonauts would be flying in Apollo spacecraft. Deputy Associate Administrator George E. Mueller shared Seamans' views, comparing future U.S.–U.S.S.R. cooperation in space to joint explorations in Antarctica. Scientists from both nations work together, but "they get there in different ships." Just three days earlier, MSC Director Robert R. Gilruth had told the National Rocket Club that a joint American–Russian space flight—especially one to the moon—would present almost insuperable technological difficulties. "I tremble at the thought of the integration problems . . . ," he said. Gilruth cautioned his audience that he was speaking "not as an international politician," but as an engineer. The task of mating American and Russian spacecraft and launch vehicles would make such international cooperation "hard to do in a practical sort of way." And at the September 20 MSC news conference he added that such problems "are very difficult even when they [hardware components] are built by American contractors."

1963
September
20

Robert L. Rosholt, *An Administrative History of NASA, 1958–1963* (NASA SP-4101), p. 288; *Astronautics and Aeronautics, 1963*, pp. 343, 347; *The Houston Chronicle*, September 19, 20, 21, 1963.

North American checked out the test fixture that was slated for the astronaut centrifuge training program, resolving interfaces between test fixture, centrifuge, and the test conductor's console, and familiarizing astronauts with controls and displays inside the spacecraft.

22–29

On October 1, North American delivered the test fixture to the U.S. Navy Aviation Medical Acceleration Laboratory, where the first phase of the manned centrifuge program was scheduled to begin that month.

"Apollo Monthly Progress Report," SID 62-300-18, pp. 4–5; MSC, "ASPO Weekly Activity Report, October 3–9, 1963," p. 3.

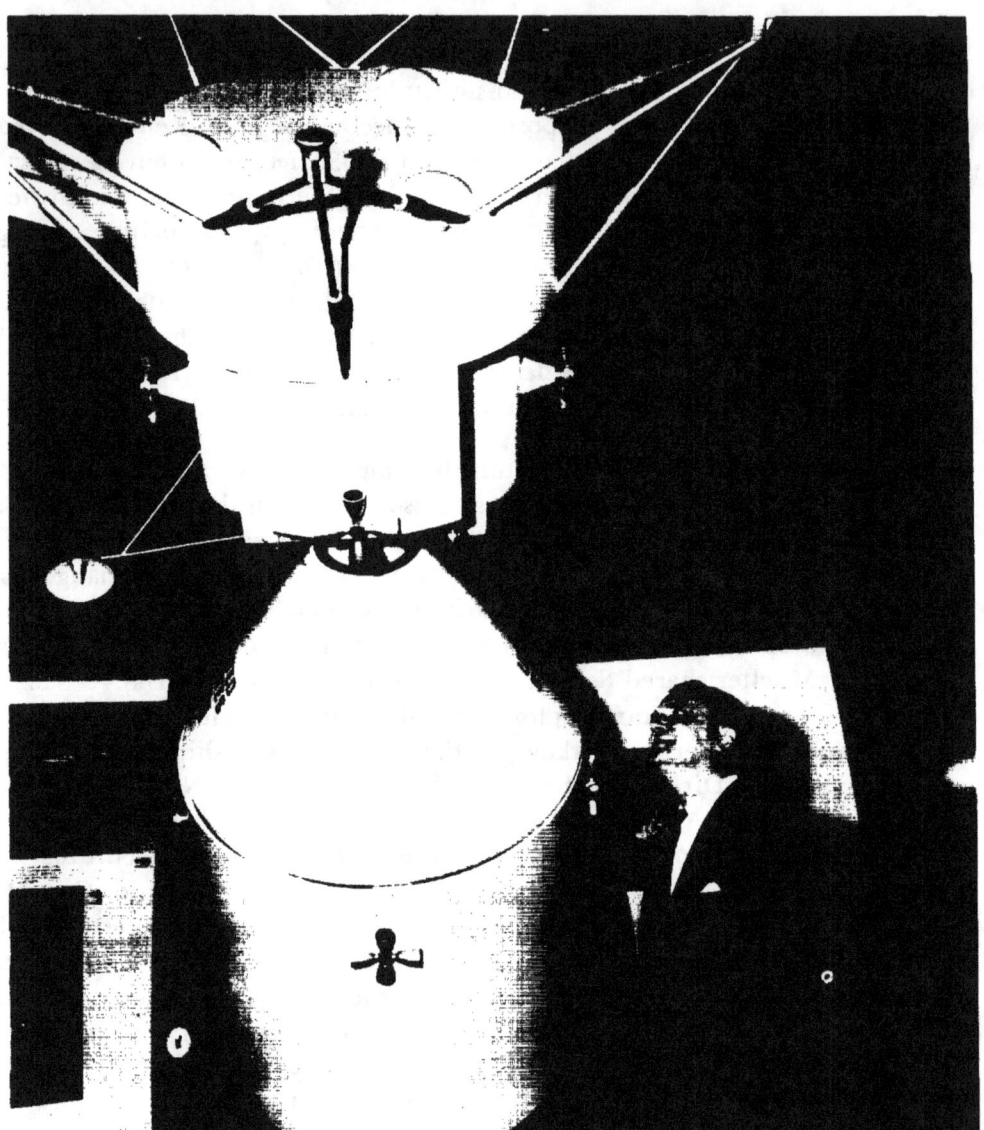

NASA Administrator James E. Webb examined a docked configuration of the Apollo spacecraft model during a visit to Houston September 24, 1963.

1963

September

24

MSC advised North American that the television camera in the CM was being modified so that ground personnel could observe the astronauts and flight operations. Television images would be transmitted directly to earth via the Deep Space Instrumentation Facility.

>Letter, H. P. Yschek, MSC, to NAA, Space and Information Systems Div., "Contract Change Authorization No. Ninety-Five," September 24, 1963.

25–26

MSC representatives reviewed Grumman's program for thermal testing for the LEM, to be conducted with the test model 2 (TM-2) vehicle. Because the vehicle's configuration had changed so extensively, the Center canceled

PART II: DEVELOPING HARDWARE DISTINCTIONS

the currently planned TM-2 ascent stage and ordered another stage to be substituted. TM-2's descent stage needed only small design changes to make it suitable for the program.

>MSC, "ASPO Weekly Activity Report, September 26–October 2, 1963," p. 12.

At a meeting at MSC, Grumman representatives presented 18 configurations of the LEM electrical power system, recommending a change from three to two fuel cells, still supplemented by an auxiliary battery system, with continued study on tankage design. On December 10, ASPO authorized the contractor to proceed with this configuration.

>Letter, Owen E. Maynard, MSC, to GAEC, Attn: R. S. Mullaney, "Contract NAS 9-1100, Electrical Power Subsystem Configuration Recommendation," December 10, 1963; MSC, "ASPO Weekly Activity Report, September 26–October 2, 1963," p. 11.

OMSF, MSC, and Bellcomm representatives, meeting in Washington, D.C., discussed Apollo mission plans: OMSF introduced a requirement that the first manned flight in the Saturn IB program include a LEM. ASPO had planned this flight as a CSM maximum duration mission only.

• Bellcomm was asked to develop an Apollo mission assignment program without a Saturn I.

• MSFC had been asking OMSF concurrence in including a restart capability in the S-IVB (second) stage during the Saturn IB program. ASPO would agree to this, but only if the H-1 engine were uprated from 85 275 to 90 718 kilograms (188 000 to 200 000 pounds) of thrust, resulting in a 907-kilogram (2000-pound) payload gain.

>MSC, "ASPO Weekly Activity Report, September 26–October 2, 1963."

MSC representatives visited Grumman for a preliminary evaluation of the Apollo space suit integration into the LEM. A suit failure ended the exercise prematurely. Nonetheless, leg and foot mobility was good, but the upper torso and shoulder needed improvement.

On October 11, MSC Crew Systems Division (CSD) tested the suit's mobility with the portable life support system (PLSS). CSD researchers found that the PLSS did not restrict the wearer's movement because the suit supported the weight of the PLSS. Shifts in the center of gravity appeared insignificant. The PLSS controls, because of their location, were difficult to operate, which demanded further investigation.

>Ibid.; MSC, "Consolidated Activity Report for the Office of the Director, Manned Space Flight, September 22–October 19, 1963," p. 48.

North American recommended that the portable life support system in the CM be deleted. Current planning placed two units in the LEM and one in the CM.

>MSC, "ASPO Weekly Activity Report, September 26–October 2, 1963," p. 3.

1963
September

26

26

26–27

26–October 2

THE APOLLO SPACECRAFT: A CHRONOLOGY

1963
September
30

MSC awarded Texas Instruments, Inc., a $194 000 contract to study experiments and equipment needed for scientific exploration of the lunar surface. The analysis was to be completed by the end of May 1964. (See March 17, 1964.)

> MSC, "Consolidated Activity Report for the Office of the Director, Manned Space Flight, September 22–October 19, 1963," p. 41; "Apollo Quarterly Status Report No. 6," p. 34; MSC News Release 63–171, October 16, 1963.

30

Qualification testing began on fuel tanks for the service propulsion system (SPS). The first article tested developed a small crack below the bottom weld, which was being investigated, but pressurization caused no expansion of the tank. During mid-October, several tanks underwent proof testing. And, on November 1, the first SPS helium tank was burst-tested.

> MSC, "ASPO Status Report for Period Ending October 16, 1963"; "ASPO Status Report for Period October 16–November 12, 1963"; "ASPO Status Report for Period Ending October 23, 1963."

During the Month

The interrelationships between all major LEM test vehicles, including all test constraints and documentation requirements, were developed. This logic study, prepared by Grumman and forwarded to MSC, stressed the feasibility of alterations in the LEM test program as needed.

> "Monthly Progress Report No. 8," LPR–10–24, p. 45.

A stack of logic and rope memory modules for the Apollo onboard computer was checked by Ralph R. Ragan, left, operations manager of Space and Information Systems Division of Raytheon's Sudbury (Mass.) Laboratory, and Eldon Hall, Director of Apollo Computer Division at MIT's Instrumentation Laboratory.

PART II: DEVELOPING HARDWARE DISTINCTIONS

At right, David G. Hoag, technical director of the Apollo guidance and navigation system design program at MIT's Instrumentation Laboratory, inspected a mockup of the inertial measurement unit in the system. Below left, director of the Laboratory Dr. C. Stark Draper posed beside a mockup of the guidance and navigation system. Below right, the mockup was checked by Milton B. Trageser, director of the Apollo program at the Laboratory, and David W. Gilbert, right, head of the Guidance and Control Division, Apollo Project Office, MSC.

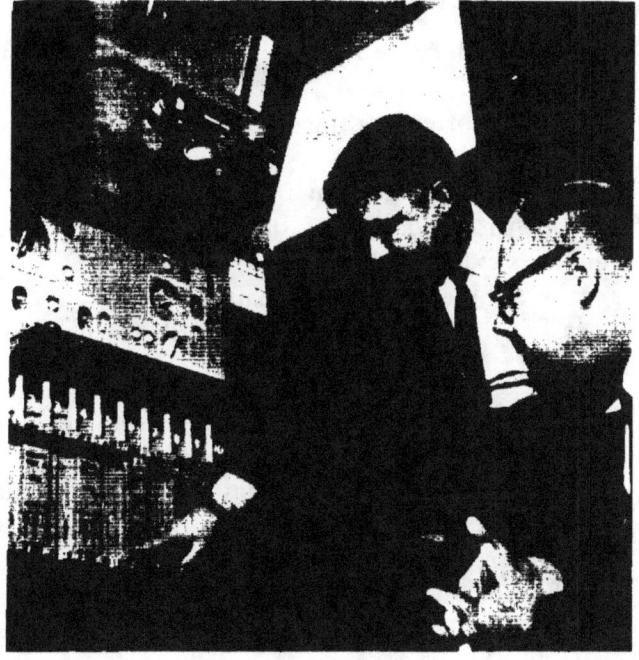

1963

October

2

At a LEM Mechanical Systems Meeting in Houston, Grumman and MSC agreed upon a preliminary configuration freeze for the LEM-adapter arrangement. The adapter would be a truncated cone, 876 centimeters (345 inches) long. The LEM would be mounted inside the adapter by means of the outrigger trusses on the spacecraft's landing gear. This configuration provided ample clearance for the spacecraft, both top and bottom (i.e., between the service propulsion engine bell and the instrument unit of the S-IVB). (See June 3 and December 5.)

At this same meeting, Grumman presented a comparison of radially and laterally folded landing gears (both of 457-centimeter [180-inch] radius). The radial-fold configuration, MSC reported, promised a weight savings of 22.2 kilograms (49 pounds). MSC approved the concept, with an 876-centimeter (345-inch) adapter. Further, an adapter of that length would accommodate a larger, lateral fold gear (508 centimeters [200 inches]), if necessary. During the next several weeks, Grumman studied a variety of gear arrangements (sizes, means of deployment, stability, and even a "bending" gear). At a subsequent LEM Mechanical Systems Meeting, on November 10, Grumman presented data (design, performance, and weight) on several other four-legged gear arrangements—a 457-centimeter (180-inch), radial fold "tripod" gear (i.e., attached to the vehicle by three struts), and 406.4-centimeter (160-inch) and 457-centimeter (180-inch) cantilevered gears. As it turned out, the 406.4-centimeter (160-inch) cantilevered gear, while still meeting requirements demanded in the work statement, in several respects was more stable than the larger tripod gear. In addition to being considerably lighter, the cantilevered design offered several added advantages:

• A reduced stowed height for the LEM from 336.5 to 313.7 centimeters (132.5 to 123.5 inches)
• A shorter landing stroke (50.8 instead of 101.6 centimeters) (20 instead of 40 inches)
• Better protection from irregularities (protuberances) on the surface
• An alleviation of the gear heating problem (caused by the descent engine's exhaust plume)
• Simpler locking mechanisms
• A better capability to handle various load patterns on the landing pads.

Because of these significant (and persuasive) factors, MSC approved Grumman's change to the 406.4-centimeter (160-inch) cantilevered arrangement as the design for the LEM's landing gear. By mid-November, MSC reported to OMSF that Grumman was pursuing the 406.4-centimeter (160-inch) cantilevered gear. Although analyses would not be completed for some weeks, the design was "shown . . . to be the lightest gear available to date. . . . Tentative estimates indicate a gear stowed height reduction of about 9" [22.9 centimeters], which will still accommodate the 180" [45.7 centimeter] cantilever or 200" [508-centimeter] lateral fold gear as growth potential." Grumman's effort continued at "firming up" the design, including folding and docking mechanisms.

1963
October

> GAEC, "Monthly Progress Report No. 9," LPR-10-25, November 10, 1963, pp. 3, 12; MSC, "ASPO Weekly Activity Report, September 26–October 2, 1963," p. 15; "ASPO Monthly Activity Report, September 19–October 16, 1963," p. 5; MSC, "Weekly Activity Report for the Office of the Director, Manned Space Flight, September 8–14, 1963," pp. 10–11; "Weekly Activity Report for the Office of the Director, Manned Space Flight, November 17–23, 1963," pp. 9–10; MSC, "Consolidated Activity Report for the Office of the Director, Manned Space Flight, October 20–November 16, 1963," p. 36; "Apollo Quarterly Status Report No. 6," p. 27; "ASPO Status Report for Period Ending October 16, 1963"; "ASPO Weekly Status Report, November 12–19, 1963"; "Monthly Progress Report No. 7," LPR-10-22, p. 10; "Monthly Progress Report No. 8," LPR-10-24, p. 11; GAEC, "Monthly Progress Report No. 10," LPR-10-26, December 10, 1963, p. 10; GAEC, "Monthly Progress Report No. 11," LPR-10-27, January 10, 1964, p. 11.

NASA announced the appointment of Joseph F. Shea as ASPO Manager effective October 22. He had been Deputy Director (Systems) in OMSF. George M. Low, OMSF Deputy Director (Programs), would direct the Systems office as well as his own. Robert O. Piland, Acting Manager of ASPO since April 3, resumed his former duties as Deputy Manager.

8

> NASA News Release 63-226, "Shea to Head Apollo Spacecraft Development at Manned Spacecraft Center," October 8, 1963; MSC News Release 63-163, October 8, 1963; MSC Announcement No. 263, "Manager, Apollo Spacecraft Program Office," October 22, 1963.

Verne C. Fryklund, Jr., of NASA's Office of Space Sciences (OSS), in a memorandum to MSC Director Robert R. Gilruth, recommended some general guidelines for Apollo scientific investigations of the moon (which OSS already was using). "These guidelines," Fryklund told Gilruth, ". . . should be followed in the preparation of your plans," and thus were "intended to place some specific constraints on studies. . . . The primary scientific objective of the Apollo project," Fryklund said, was, of course, the "acquisition of comprehensive data about the moon." With this as a starting point, he went on, ". . . it follows that the structure of the moon's surface, gross body properties and large-scale measurements of physical and chemical characteristics, and observation of whatever phenomena may occur at the actual surface will be the prime scientific objectives." Basically, OSS's guidelines spelled out what types of activity were and were not part of Apollo's immediate goals. These activities were presumed to be mostly reconnaissance, "to acquire knowledge of as large an area as possible, and by as simple a means as possible, in the limited time available." The three principal scientific activities "listed in order of decreasing importance" were: (1) "comprehensive observation of lunar phenomena," (2) "collection of representative samples," and (3) "emplacement of monitoring equipment."

8

1963

October

These guidelines had been arrived at after extensive consultation within NASA as a whole as well as with the scientific community.

> Memorandum, Verne C. Fryklund, Jr., NASA Office of Space Sciences (OSS), to Director, MSC, "Scientific Guidelines for the Apollo Project," October 6, 1963; OSS, "NASA Program Planning in Space Sciences," September 1963, pp. VI-3 through VI-8.

8

At MSC, the Spacecraft Technology Division reported to ASPO the results of a study on tethered docking of the LEM and CSM. The technology people found that a cable did not reduce the impact velocities below those that a pilot could achieve during free flyaround, nor was fuel consumption reduced. In fact, when direct control of the spacecraft was attempted, the tether proved a hindrance and actually increased the amount of fuel required.

> MSC, "Flight Crew Operations Division, Activity Report, September 16–October 21, 1963," pp. 2–3.

9

NASA Administrator James E. Webb announced a major reorganization of NASA Headquarters, effective November 1, to consolidate management of major programs and direction of research and development centers and to realign Headquarters management of agency-wide support functions. On October 28, NASA Headquarters announced a similar reorganization within OMSF, also to take effect on November 1, to strengthen NASA Headquarters' control of the agency's manned space flight programs. In effect, these administrative adjustments "recombined program and institutional management by placing the field centers under the Headquarters program directors instead of under general management (i.e., the Associate Administrator)."

> NASA News Release 63–225, "NASA Announces Reorganization," October 9, 1963; NASA News Release 63–241, "NASA Realigns Office of Manned Space Flight," October 28, 1963; Rosholt, *Administrative History of NASA, 1958–1963*, pp. 289–96.

10

LTV announced the results of tests performed by astronauts in the Manned Space Flight Mission Simulator in Dallas, Tex. (See May 6 and September 17, 1963, and April 24, 1964.) These indicated that, should the primary guidance and navigation system fail, LEM pilots could rendezvous with the CM by using a circular slide rule to process LEM radar data.

> *Tulsa Daily World*, October 11, 1963; *The Houston Post*, October 11, 1963.

14

Langley Research Center's Lunar Landing Research Facility was nearing completion. A gantry structure 121.9 meters (400 feet) long and 76.2 meters (250 feet) high would suspend a model of the LEM. It would sustain five-sixths of the model's weight, simulating lunar gravity, and thus would enable astronauts to practice lunar landings. (See Volume I, Summer 1961.)

> *Aviation Week and Space Technology*, 79 (October 14, 1963), pp. 83, 86; MSC, *Space News Roundup*, November 27, 1963, p. 8.

PART II: DEVELOPING HARDWARE DISTINCTIONS

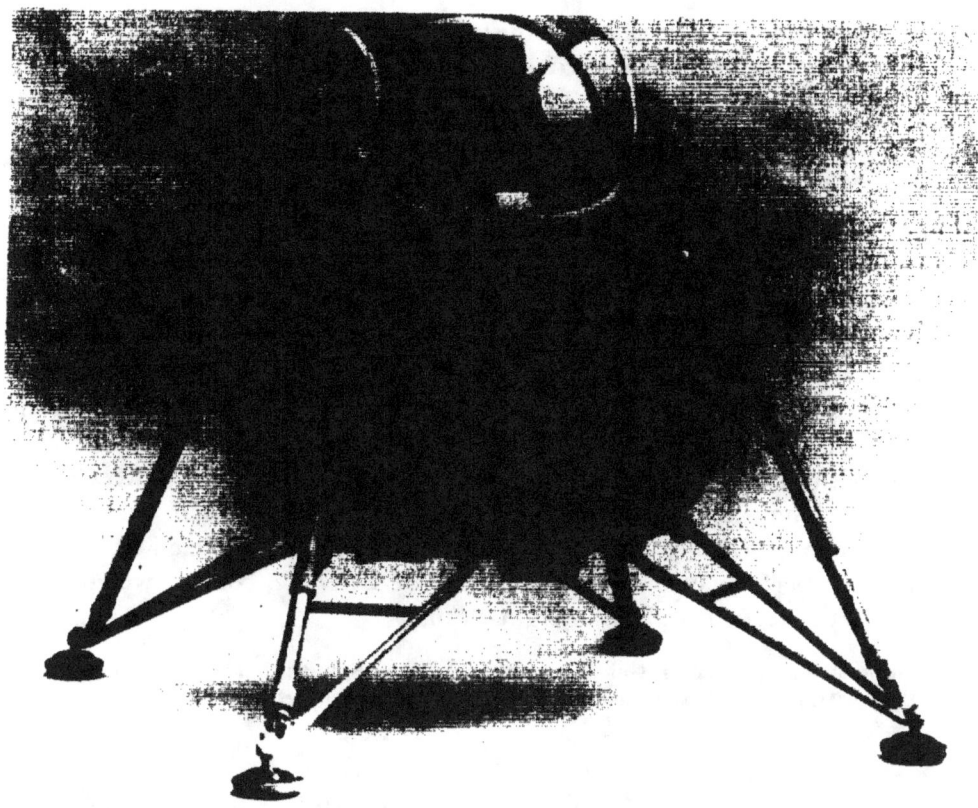

Above is a model of the prototype rocket-powered vehicle to be used in the lunar landing test facility at NASA's Langley Research Center. At the right is an engineering sketch of the way the facility would look when completed and in use.

1963

October

14 — ASPO established criteria for combustion stability in the service propulsion engine. The engine had to recover from any instability, whether induced or spontaneous, within 20 milliseconds during qualification testing.

> MSC, "ASPO Monthly Activity Report, September 19–October 16, 1963," p. 3.

15 — The Guidance and Performance Sub-Panel, at its first meeting, began coordinating work at MSC and MSFC. The sub-panel outlined tasks for each Center: MSFC would define the dispersions comprising the launch vehicle performance reserves, prepare a set of typical translunar injection errors for the Saturn V launch vehicle, and give MSC a typical Saturn V guidance computation for injection into an earth parking orbit. MSC would identify the constraints required for free-return trajectories and provide MSFC with details of the MIT guidance method. Further, the two Centers would exchange data each month showing current launch vehicle and spacecraft performance capability. (For operational vehicles, studies of other than performance capability would be based on control weights and would not reflect the current weight status.)

> Memorandum, Secretaries, Guidance and Performance Sub-Panel, MSFC and MSC, to Distr., "Minutes of First Guidance and Performance Sub-Panel Meeting," October 16, 1963.

16–17 — MSC discussed commonality of displays and controls with its two principal spacecraft contractors. A review of panel components suggested that Grumman might use the same vendors as North American for such items as switches, potentiometers, and indicators.

> MSC, "ASPO Activity Report, October 16–22, 1963," pp. 1–2.

16–23 — An MSC Spacecraft Technology Division Working Group reexamined Apollo mission requirements and suggested a number of ways to reduce spacecraft weight: eliminate the free-return trajectory; design for slower return times; use the Hohmann descent technique, rather than the equal period orbit method, yet size the tanks for the equal period mode; eliminate the CSM/LEM dual rendezvous capability; reduce the orbital contingency time for the LEM (the period of time during which the LEM could remain in orbit before rendezvousing with the CSM); reduce the LEM lifetime.

> MSC, "ASPO Status Report for Period Ending October 23, 1963."

16–November 15 — Because of an electrical equipment failure on Mercury MA-9, North American began a CM humidity study. The company found in the crew compartment major spacecraft systems which were not designed for operation in the presence of corrosive moisture. (The environmental control system did not guarantee complete humidity control.) Investigators also examined in minute detail all electrical/electronic components. North

PART II: DEVELOPING HARDWARE DISTINCTIONS

Above, the MSC "Navy" was arriving at its Seabrook docking facility and, at right, Skipper Frank Gammon of Flight Operations Division took command. The modified Army LCU (landing craft, utility), painted "NASA blue and white," was named *Retriever* to indicate its function in recovering spacecraft in drop and flotation tests in Galveston and Trinity Bays and in the Gulf of Mexico. The 115-foot *Retriever* could recover heavy spacecraft and could spend five days in the open sea. It had a permanent crew of three.

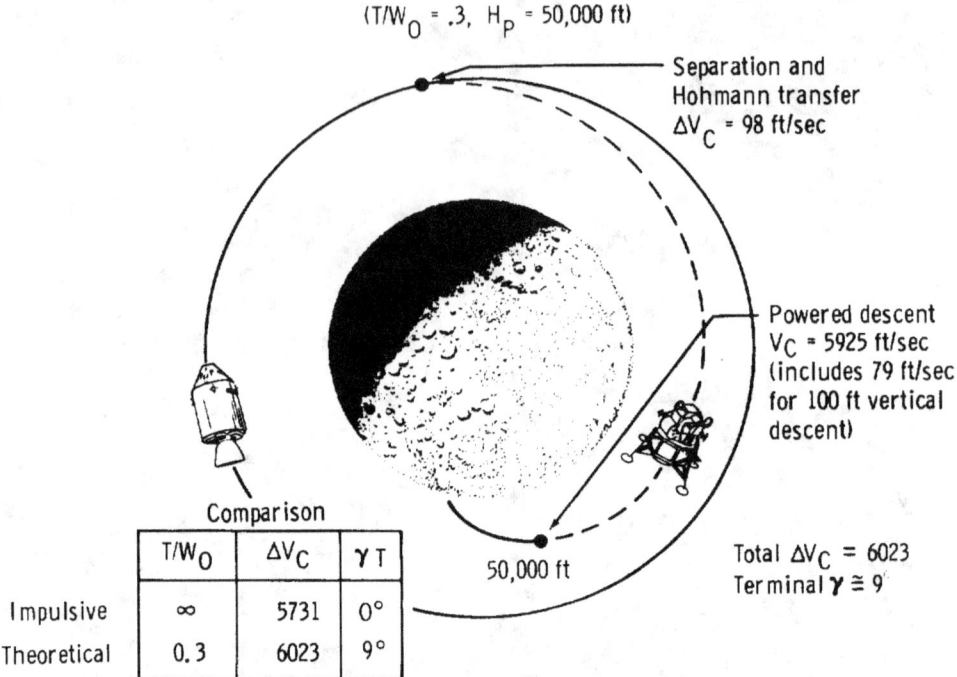

Theoretical optimum lunar model descent with thrust-to-weight ratio (initial value in lunar orbit) at 0.3, height at perilune of the transfer orbit at 15 200 meters (50 000 feet), and using the Hohmann transfer technique. The diagram showed the velocity change (ΔV_c, in feet per second) and approach flight-path angle (γ) close to those for an impulsive orbital change (an instantaneous change, without time value, taken as the ultimate though unachievable ideal for comparison). —NASA drawing.

1963
October

American was considering design changes that would protect all components from moisture.

"Apollo Monthly Progress Report," SID 62-300-19, p. 25.

18

NASA and GD/C negotiated amendments totaling $354 737 to Little Joe II contract. This sum covered study activity and several relatively small changes that came out of a Design Engineering Inspection on May 3. More ground support equipment was authorized, as was fabrication of an additional breadboard autopilot system for use at MSC. The dummy payload was deleted and the instrumentation was limited to a control system on the vehicle to be used for Mission A-002 (BP-23).

Little Joe II Test Launch Vehicle, NASA Project Apollo: Final Report, Vol. I, p. 4-3.

18

NASA Headquarters announced the selection of five organizations for contract negotiations totaling $60 million for the development, fabrication, and

PART II: DEVELOPING HARDWARE DISTINCTIONS

testing of LEM guidance and navigation equipment: (1) MIT, overall direction; (2) Raytheon, LEM guidance computer; (3) AC Spark Plug, inertial measurement unit, gyroscopes, navigation base, power and servo assembly, coupling display unit, and assembly and testing of the complete guidance and navigation system; (4) Kollsman Instrument Corporation, scanning telescope, sextant, and map and data viewer; and (5) Sperry Gyroscope Company, accelerometers. (All five had responsibility for similar equipment for the CSM as well. See Vol. I, August 9, 1961, and May 8, 1962.)

1963
October

MSC News Release 63-17, October 18, 1963.

NASA announced the selection of 14 astronauts for Projects Gemini and Apollo, bringing to 30 the total number of American spacemen. They were Maj. Edwin E. Aldrin, Jr., Capt. William A. Anders, Capt. Charles A. Bassett II, Capt. Michael Collins, Capt. Donn F. Eisele, Capt. Theodore C. Freeman, and Capt. David R. Scott of the Air Force; Lt. Cdr. Richard F. Gordon, Jr., Lt. Alan L. Bean, Lt. Eugene A. Cernan, and Lt. Roger B. Chaffee of the Navy; Capt. Clifton C. Williams, Jr., of the Marine Corps; R. Walter Cunningham, research scientist for the Rand Corporation; and Russell L. Schweickart, research scientist for MIT.

18

MSC News Release 63-180, October 18, 1963; *Space News Roundup*, October 30, 1963.

MSC reported that preliminary testing had begun on the first prototype extravehicular suit telemetry and communications system and on the portable life support system of which it was an integral part. The hardware had recently been received from the prime contractor, Hamilton Standard.

20–November 16

MSC, "Consolidated Activity Report for the Office of the Director, Manned Space Flight, October 20–November 16, 1963," p. 67.

The second prototype space suit was received by MSC's Crew Systems Division. (See August 15–September 21.) Preliminary tests showed little improvement in mobility over the first suit. On October 24–25, a space suit mobility demonstration was held at North American. The results showed that the suit had less shoulder mobility than the earlier version, but more lower limb mobility. (See September 26–27.) Astronaut John W. Young, wearing the pressurized suit and a mockup portable life support system (PLSS), attempted an egress through the CM hatch but encountered considerable difficulty. At the same time, tests of the suit-couch-restraint system interfaces and control display layout were begun at the Navy's Aviation Medical Acceleration Laboratory centrifuge in Johnsville, Pa. Major problems were restriction of downward vision by the helmet, extension of the suit elbow/arm beyond the couch, and awkward reach patterns to the lower part of the control panel. On October 30–November 1, lunar task studies with the suit were carried out at Wright-Patterson Air Force Base in a KC–

21

Apollo prototype space suit with reflective coverall and life-support-equipment back pack.

1963
October

135 aircraft at simulated lunar gravity. Mobility tests were made with the suit pressurized and a PLSS attached.

"Apollo Quarterly Status Report No. 6," p. 25; MSC, "Weekly Activity Report for the Office of the Director, Manned Space Flight, October 27–November 2, 1963," p. 6; MSC, "ASPO Status Report for Week Ending November 6, 1963;" "ASPO Status Report for Period Ending October 23, 1963;" "ASPO Status Report for Period October 16–November 12, 1963."

22

George E. Mueller, NASA Associate Administrator for Manned Space Flight, appointed Walter C. Williams Deputy Associate Administrator for Manned Space Flight in OMSF. Williams would direct operations at MSC, MSFC, and LOC for all manned space flight missions.

MSC News Release 63–179, October 22, 1963.

23

MSC Flight Operations Division defined systems and outlined ground rules for the lunar landing mission. System definitions were: (1) primary, most efficient or economic; (2) alternate, either redundant (identical to but independent of the primary) or backup (not identical but would perform the same function); (3) critical (failure would jeopardize crew safety); (4) repairable (for which tools and spares were carried and which the crew

PART II: DEVELOPING HARDWARE DISTINCTIONS

could service in flight); and (5) operational, which must be working to carry out a mission.

Mission rules established crew safety as the major consideration in all mission decisions and detailed actions to be taken in the event of a failure in any system or subsystem.

> Memorandum, Eugene L. Duret, MSC, to Chief, Flight Operations Div., "Project Apollo, operational ground rules for the Lunar Landing Mission," October 23, 1963, with enclosure.

MSC Instrumentation and Electronic Systems Division awarded a $50 000 contract to the Hughes Aircraft Company for a study of backup high gain directable antennas for the LEM lunar surface equipment.

> MSC, "ASPO Status Report for Week Ending October 30, 1963."

Because OMSF had requested OSSA to provide lunar surface microrelief and bearing strength data to support LEM landing site selection and to permit LEM landing-gear design validation, the Ad Hoc Working Group on Follow-On Surveyor Instrumentation met at NASA Headquarters. Attending were Chairman Verne C. Fryklund, Clark Goodman, Martin Swetnick, and Paul Brockman of the NASA Office of Space Sciences and Applications; Harry Hess and George Derbyshire of the National Academy of Sciences; Dennis James of Bellcomm (for OMSF); and Milton Beilock of the Jet Propulsion Laboratory (JPL). The group proposed "a fresh look at the problem of instrumenting payloads of Surveyor spacecraft that may follow the currently approved developmental and operational flights, so that these spacecraft will be able to determine that a particular lunar site is suitable for an Apollo landing." The study was assigned to JPL.

> Summary Minutes, "Ad Hoc Working Group on Follow-On Surveyor Instrumentation, October 24, 1963," October 28, 1963, pp. 1–2.

The NASA–Industry Apollo Executives Group, composed of top managers in OMSF and executives of the major Apollo contractors, met for the first time. The group met with George E. Mueller, NASA Associate Administrator for Manned Space Flight, for status briefings and problem discussions. In this manner, NASA sought to make executives personally aware of major problems in the program.

> *Tenth Semiannual Report to Congress of the National Aeronautics and Space Administration, July 1–December 31, 1963* (1964), p. 43.

MSC directed Grumman to schedule manned environmental control system (ECS) development tests, using a welded-shell cabin boilerplate and air lock. At about the same time, the company was also requested to quote cost and delivery schedule for a second boilerplate vessel, complete with prototype ECS. Although this vessel would be used by the MSC Crew Systems Division for in-house investigation and evaluation of ECS development problems, its

1963
October

23–30

24

24

25

1963

October

major purpose was to serve as a tool for trouble-shooting during the operational phase.

> MSC, "Weekly Activity Report for the Office of the Director, Manned Space Flight, October 27–November 2, 1963," p. 11; MSC, "ASPO Status Report for Period October 16–November 12, 1963."

29

After a program review at an MSF Management Council meeting, George E. Mueller, head of OMSF, suggested several testing procedures. To meet schedules, "dead-end" testing, that is, "tests involving components or systems that [would] not fly operationally without major modification," should be minimized. Henceforth, Mueller said, NASA would concentrate on "all-up" testing. [In "all-up" testing, the complete spacecraft and launch vehicle configuration would be used on each flight. Previously, NASA plans had called for a gradual buildup of subsystems, systems stages, and modules in successive flight tests.] To simplify both testing and checkout at Cape Canaveral, complete systems should be delivered. An instrumentation task force with senior representatives from each Center, one outside member, and Walter C. Williams of OMSF should be set up immediately; a second task force, to study storable fuels and small motors, would include members from Lewis Research Center, MSC, MSFC, as well as representatives from outside the government.

> Memorandum, Clyde Bothmer, MSF Management Council, for Distribution, "Management Council Meeting, October 29, 1963, in Washington, D.C.," October 31, 1963.

30

NASA canceled four manned earth orbital flights with the Saturn I launch vehicle. Six of a series of 10 unmanned Saturn I development flights were still scheduled. Development of the Saturn IB for manned flight would be accelerated and "all-up" testing would be started. (See November 1.) This action followed Bellcomm's recommendation of a number of changes in the Apollo spacecraft flight test program. The program should be transferred from Saturn I to Saturn IB launch vehicles; the Saturn I program should end with flight SA-10. All Saturn IB flights, beginning with SA-201, should carry operational spacecraft, including equipment for extensive testing of the spacecraft systems in earth orbit.

Associate Administrator for Manned Space Flight George E. Mueller had recommended the changeover from the Saturn I to the Saturn IB to NASA Administrator James E. Webb on October 26. Webb's concurrence came two days later.

> Memoranda: Mueller to Robert F. Freitag, "Replacement of Scheduled Manned Flights on Saturn I," October 18, 1963; Mueller to Webb, "Reorientation of Apollo Plans," October 26, 1963, with handwritten notation signed by Webb, undated; OMSF, *Recommended Changes in the Use of Space Vehicles in the Apollo Test Program*, Technical Memorandum, MD(S) 3100.180 (October 29, 1963), pp. 1–4; NASA News Release 63-246, "NASA Announces Changes in Saturn Missions," October 30, 1963.

PART II: DEVELOPING HARDWARE DISTINCTIONS

The Marquardt Corporation received a definitive $9 353 200 contract from North American for development and production of reaction control engines for the SM. Marquardt, working under a letter contract since April 1962, had delivered the first engine to North American that November.

1963
October
31

> MSC News Release 63–22, October 31, 1963; MSC, *Space News Roundup*, November 13, 1963, p. 8.

The first production F–1 engine was flown from Rocketdyne's Canoga Park, Calif., facility, where it was manufactured, to MSFC aboard Aero Spacelines' "Pregnant Guppy."

31

> David S. Akens, A. Ruth Jarrell, and Leo L. Jones, *History of the George C. Marshall Space Flight Center From July 1 Through December 31, 1963* (MHM–8, July 1964), Vol. I, p. 129.

NASA tentatively approved Project Luster, a program designed to capture lunar dust deflected from the moon by meteorites and spun into orbit around the earth. An Aerobee 150 sounding rocket containing scientific equipment built by Electro-Optical Systems, Inc., was scheduled for launch in late 1964.

During the Month

> *Missiles and Rockets*, 13 (October 14, 1963), p. 9.

NASA Associate Administrator for Manned Space Flight George E. Mueller notified the Directors of MSC, MSFC, and LOC that he intended to plan

November
1

A drawing of the 445-newton-thrust (100-lb-thrust) reaction control rocket, left, shows the major components of the vital rocket engine used to maintain attitude and perform maneuvers in space. At right is a photo of a production model. The engine could be commanded to fire for periods of time ranging from milliseconds to long continuous operations.
—Marquardt drawing and photo.

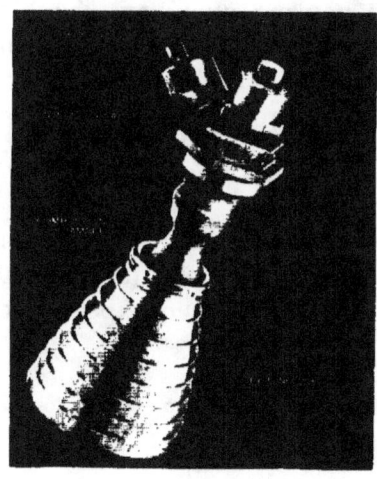

1963

November

a flight schedule which would have a good chance of being met or exceeded. To this end, he directed that "all-up" spacecraft and launch vehicle tests be started as soon as possible; all Saturn IB flights would carry CSM and CSM/LEM configurations; and two successful unmanned flights would be flown before a manned mission on either the Saturn IB or Saturn V.

On November 18, Mueller further defined the flight schedule planning. Early Saturn IB flights might not be able to include the LEM, but every effort must be made to phase the LEM into the picture as early as possible. Launch vehicle payload capability must be reached as quickly as practicable. Subsystems for the early flights should be the same as those intended for lunar missions. To conserve funds, the first Saturn V vehicle would be used to obtain reentry data early in the Saturn test program.

By December 31 the official schedule showed:
 Final Saturn I flight (SA–10): June 1965
 First Saturn IB flight (SA–201): first quarter, 1966
 First manned Saturn IB flight: either SA–203, third quarter of 1966, or SA–207, third quarter of 1967
 First Saturn V flight (SA–501): first quarter, 1967
 First manned Saturn V flight: either SA–503, third quarter of 1967, or SA–507, second quarter of 1968.

> TWX, Mueller to Dir., MSC, MSFC, and LOC, "Revised Manned Space Flight Schedule," November 1, 1963; memorandum, Mueller to Dir., MSC, MSFC, and LOC, "Manned Space Flight Schedule," November 18, 1963; "Apollo Quarterly Status Report No. 6," fig. 9, 10, 11.

1

MSC Flight Operations Division outlined the advantages inherent in the CSM's capability to use the HF transceiver during earth orbit. The HF transceiver would allow the CSM to communicate with any one tracking station at any time during earth orbit, even when the spacecraft had line-of-sight (LOS) contact with only one or two ground stations in some orbits. It would give the astronauts an additional communications circuit. Most important, this HF capability could alert the network about any trouble in the spacecraft and give the Flight Director more time to make a decision while the spacecraft was out of LOS communication with the ground stations.

> Memorandum, Christopher C. Kraft, Jr., MSC, to Mgr., ASPO, "Apollo HF communications during earth orbit," November 1, 1963.

3

MSC Crew Systems Division, conducting flammability tests on the constant wear garment material in a 3.5 newtons per square centimeter (5 psi), 100 percent oxygen atmosphere, reported that no fires had been experienced thus far.

> MSC, "Weekly Activity Report for the Office of the Director, Manned Space Flight, November 3–9, 1963," p. 7.

PART II: DEVELOPING HARDWARE DISTINCTIONS

1963
November
5

MSC Director Robert R. Gilruth announced a reorganization of MSC to strengthen the management of the Apollo and Gemini programs. Under Gilruth and Deputy Director James C. Elms, there were now four Assistant Directors, Managers for both the Gemini and Apollo programs, and a Manager for MSC's Florida Operations. Assigned to these positions were:

- Maxime A. Faget, Assistant Director for Engineering and Development
- Christopher C. Kraft, Jr., Assistant Director for Flight Operations
- Donald K. Slayton, Assistant Director for Flight Crew Operations
- Wesley L. Hjornevik, Assistant Director for Administration
- Joseph F. Shea, Manager, Apollo Spacecraft Program Office
- Charles W. Mathews, Manager, Gemini Program Office and
- G. Merritt Preston, Manager, MSC Florida Operations.

MSC News Release 63–277, November 5, 1963; *The Houston Post*, November 6, 1963.

5

MSC accepted the final items of a $237 000 vibration test system from the LTV Electronics Division to be used in testing spacecraft parts.

On this same day, MSC awarded a $183 152 contract to Wyle Laboratories to construct a high-intensity acoustic facility, also for testing spacecraft parts. The facility would generate noise that might be encountered in space flight.

MSC News Release 63–224, November 5, 1963; MSC News Release 63–225, November 5, 1963.

7

North American presented to MSC the results of a three-month study on radiation instrumentation. Three general areas were covered: radio-frequency (RF) warning systems, directional instrumentation, and external environment instrumentation. The company concluded that, with the use of an RF system, astronauts would receive about two hours' notice of any impending solar proton event and could take appropriate action. Proper orientation of the spacecraft could reduce doses by 17 percent, but this could be accomplished only by using a directional detection instrument. There was a 70 percent chance that dosages would exceed safe limits unless such an instrument was used. Consequently North American recommended prompt development.

Despite the contractor's findings, MSC concluded that there was no need for an RF warning system aboard the spacecraft, believing that radiation warning could be handled more effectively by ground systems. But MSC did concur in the recommendation for a combined proton direction and external environment detection system and authorized North American to proceed with its design and development.

MSC, "ASPO Status Report for Period October 16–November 12, 1963"; memorandum, David M. Hammock and Lee N. McMillion, MSC, to E. E. Sack, NAA, "Contract NAS 9–150, Radiation Instrumentation for Apollo," November 27, 1963; "Apollo Monthly Progress Report," SID 62–300–20, pp. 12–13.

Key sequences of the first pad abort test of the Apollo program: At left, the liftoff, just after the escape tower ignition. At center, top, the drogue chutes deployed and, below, the pilot chutes pulling the main chutes out. At right, the three main chutes fully open as they lowered boilerplate 6 to earth.

1963

November

7

Apollo Pad Abort Mission I (PA-1), the first off-the-pad abort test of the launch escape system (LES), was conducted at WSMR. PA-1 used CM boilerplate 6 and an LES for this test.

All sequencing was normal. The tower-jettison motor sent the escape tower into a proper ballistic trajectory. The drogue parachute deployed as programmed, followed by the pilot parachute and main parachutes. The test lasted 165.1 seconds. The postflight investigation disclosed only one significant problem: exhaust impingement that resulted in soot deposits on the CM.

"Postlaunch Memorandum Report for Apollo Pad Abort I," November 13, 1963, pp. 1-1, 1-2, 3-1.

8

Grumman issued a go-ahead to RCA to develop the LEM radar. Negotiations on the $23.461 million cost-plus-fixed-fee contract were completed on December 10. Areas yet to be negotiated between the two companies were LEM communications, inflight test, ground support, and parts of the stabilization and control systems. (See June 28.)

MSC, "Consolidated Activity Report for the Office of the Director, Manned Space Flight, October 20-November 16, 1963," p. 57; *Wall Street Journal*, December 10, 1963.

8

MSFC directed Rocketdyne to develop an uprated H-1 engine to be used in the first stage of the Saturn IB. In August, Rocketdyne had proposed that the H-1 be uprated from 85 275 to 90 718 kilograms (188 000 to 200 000 pounds) of thrust. The uprated engine promised a 907-kilogram (2000-pound) increase in the Saturn IB's orbital payload, yet required no major systems changes and only minor structural modifications.

1963
November

Akens *et al.*, *History of Marshall . . . July 1–December 31, 1963*, Vol. I, pp. 65, 66.

At El Centro, Calif., a drop test was conducted to evaluate a dual drogue parachute arrangement for the CM. The two drogues functioned satisfactorily. The cargo parachute used for recovery, however, failed to fully inflate, and the vehicle was damaged at impact. This failure was unrelated to the test objectives.

8

MSC, "ASPO Status Report for Period October 16–November 12, 1963."

A joint North American–MSC meeting reviewed the tower flap versus canard concept for the earth landing system (ELS). (See January 18.) During a low-altitude abort, MSC thought, the ELS could be deployed apex forward with a very high probability of mission success by using the tower flap configuration. The parachute system proposed for this mode would be very reliable, even though this was not the most desirable position for deploying parachutes. Dynamic stability of the tower flap configuration during high-altitude aborts required further wind tunnel testing at Ames Research Center. Two basic unknowns in the canard system were deployment reliability, and the probability of the crew's being able to establish

12

One of the functions of Ames Research Center was supporting research for NASA's manned space flight projects. In the photo, the launch escape system of the Apollo command module was readied for aerodynamic testing in the Ames Unitary Wind Tunnel.

1963
November

the flight direction and trim the CM within its stability limits for a safe reentry. Design areas to be resolved were a simple deployment scheme and a spacecraft system that would give the crew a direction reference.

MSC directed North American to proceed with the tower flap as its prime effort, and attempt to solve the stability problem at the earliest possible date. MSC's Engineering and Development Directorate resumed its study of both configurations, with an in-depth analysis of the canard system, in case the stability problem on the tower flap could not be solved by the end of the year. (See February 7 and 25, 1964.)

> Memorandum, David M. Hammock, MSC, to Asst. Dir. for Engineering and Development, "Analysis of the abort and earth landing systems if implemented by a tower flap versus a canard mode," November 18, 1963.

12

The Boeing Company and NASA signed a $27.4 million supplemental agreement to the contract for development, fabrication, and test of the S-IC (first) stage of the Saturn V launch vehicle.

> *Aviation Week and Space Technology*, 79 (November 25, 1963), p. 67; Akens et al., *History of Marshall . . . July 1-December 31, 1963*, Vol. I, p. 97.

12

NASA awarded a $19.2 million contract to Blount Brothers Corporation and M. M. Sundt Construction Company for the construction of Pad A, part of the Saturn V Launch Complex 39 at LOC.

> Akens et al., *History of Marshall . . . July 1-December 31, 1963*, Vol. I, p. 169.

12-15

North American representatives reviewed Farrand Optical Company's subcontract with Link for visual displays in the Apollo Mission Simulator. MSC officials attended the technical portion of the meeting, which was held at Link. Farrand and Link had established window fields of view and optical axis orientations. Designs were to be reviewed to verify accuracy and currency of window locations and crew eye position parameters.

> MSC, "ASPO Status Report for Week Ending November 19, 1963."

12-19

ASPO reviewed Grumman's evaluation of series and parallel propellant feed systems for the LEM ascent stage. Because of the complications involved in minimizing propellant residuals in a parallel system, a series feed appeared preferable, despite an increase in LEM structural weight. Further study of the vehicle showed the feasibility of a two-tank configuration which would be lighter and have about the same propellant residual as the four-tank series-feed arrangement. (See December 17.)

> "Monthly Progress Report No. 10," LPR-10-26, p. 16; MSC, "ASPO Status Report for Week Ending November 19, 1963"; "Apollo Quarterly Status Report No. 6," p. 33.

13-14

After careful study, Grumman proposed to MSC 15 possible means for reducing the weight of the LEM. These involved eliminating a number of

PART II: DEVELOPING HARDWARE DISTINCTIONS

1963
November

hardware items in the spacecraft; two propellant tanks in the vehicle's ascent stage and consequent changes in the feed system; two rather than three fuel cells; and reducing reaction control system propellants and, consequently, velocity budgets for the spacecraft. If all these proposed changes were made, Grumman advised, the LEM could be lightened significantly, perhaps by as much as 454 kilograms (1000 pounds).

MSC, "ASPO Status Report for Week Ending November 19, 1963."

14

ASPO revised the normal and emergency impact limits (20 and 40 g, respectively) to be used as human tolerance criteria for spacecraft design. [These limits superseded those established in the August 14, 1963, North American contract and subsequent correspondence.]

Memorandum, David M. Hammock, MSC, to NAA, Attn: E. E. Sack, "Contract 9-150, Impact Acceleration Limits," November 14, 1963.

15

NASA and contractor studies showed that, in the event of an engine hard-over failure during maximum q, a manual abort was impractical for the Saturn I and IB, and must be carried out by automatic devices. Studies were continuing to determine whether, in a similar situation, a manual abort was possible from a Saturn V.

Memorandum, Maxime A. Faget, MSC, to ASPO, Attn: Calvin H. Perrine, "Apollo abort mode in event of maximum 'q' engine hard-over malfunction," November 15, 1963.

16–December 15

All production drawings for the CM environmental control system were released. AiResearch Manufacturing Company reported the most critical pacing items were the suit heat exchanger, cyclic accumulator selector valve, and the potable and waste water tanks.

The Garrett Corporation, AiResearch Manufacturing Division, "Monthly Progress Report, Environmental Control System, NAA/S&ID, Project Apollo, 16 November 1963–15 December 1963," SS-1013-R(19) January 2, 1964, p. 4.

16–December 15

North American conducted an eight-day trial of the prototype Apollo diet. Three test subjects, who continued their normal activities rather than being confined, were given performance and oxygen consumption tests and lean body mass and body compartment water evaluations. The results showed insignificant changes in weight and physiology.

"Apollo Monthly Progress Report," SID 62-300-20, p. 6.

17–December 21

As a result of an MSC Crew Systems Division–Hamilton Standard meeting on the space suit, MSC directed the company to develop a micrometeoroid protective garment to be worn over the suit. (See August 13–20, 1964.)

MSC, "Consolidated Activity Report for the Office of the Associate Administrator, Manned Space Flight, November 17–December 21, 1963," p. 54.

1963

November 19–20

At a meeting of the Apollo Docking Interface Panel, North American recommended and Grumman concurred that the center probe and drogue docking concept be adopted. (See July 16.) MSC emphasized that docking systems must not compromise any other subsystem operations nor increase the complexity of emergency operations. In mid-December, MSC/ASPO notified Grumman and North American of its agreement. At the same time, ASPO laid down docking interface ground rules and performance criteria which must be incorporated into the spacecraft specifications.

There would be two ways for the astronauts to get from one spacecraft to the other. The primary mode involved docking and passage through the transfer tunnel. An emergency method entailed crew and payload transfer through free space. The CSM would take an active part in translunar docking, but both spacecraft must be able to take the primary role in the lunar orbit docking maneuver. A single crewman must be able to carry out the docking maneuver and crew transfer.

> MSC, "ASPO Status Report for Week Ending December 4, 1963"; "ASPO Status Report for Week Ending December 17, 1963"; "Apollo Monthly Progress Report," SID 62-300-20, pp. 7, 8, 18; "Apollo Quarterly Status Report No. 6," pp. 3–4.

21

MSC approved Grumman's $19 383 822 cost-plus-fixed-fee subcontract with Rocketdyne for the LEM descent engine development program. (See January 30, February 13, and May 1.)

> MSC, "Consolidated Activity Report for the Office of the Associate Administrator, Manned Space Flight, November 17–December 21, 1963," p. 42.

22

MSC's Space Environment Division (SED) recommended (subject to reconnaissance verification) 10 lunar landing areas for the Apollo program:

(1)	36°55' E.	1°45' N.
(2)	31° E.	0° N.
(3)	28°22' E.	1°10' N.
(4)	24°10' E.	0°10' N.
(5)	12°50' E.	0°20' N.
(6)	1°28' W.	0°30' S.
(7)	13°15' W.	2°45' N.
(8)	28°15' W.	2°45' N.
(9)	31°30' W.	1°05' S.
(10)	41°30' W.	1°10' S.

SED chose these sites on the basis of regional slopes, surface texture and strength, landmarks, isolated features, and the size, shape, and position of the various areas. The list included several sites that the Division had designated earlier in the year.

> NASA Project Apollo Working Paper No. 1100, "Environmental Factors Involved in the Choice of Lunar Operational Dates and the Choice of Lunar Landing Sites" (November 22, 1963), pp. 30–33.

PART II: DEVELOPING HARDWARE DISTINCTIONS

1963

November

22

ASPO developed ground rules and guidelines for the Spacecraft Development Test Program being conducted by Grumman, North American, and MIT Instrumentation Laboratory. (See January 3, 1964.)

<small>NAA, "Apollo Spacecraft Development Test Plan," Study Report, SID 64-66-1, February 3, 1964, Vol. I, pp. v, 26, 53-57.</small>

27

At its Santa Susana facility, Rocketdyne conducted the first long-duration (508 seconds) test firing of a J-2 engine. In May 1962 the J-2's required firing time was increased from 250 to 500 seconds.

<small>Akens et al., History of Marshall . . . July 1–December 31, 1963, Vol. I, p. 242; Missiles and Rockets, 13 (December 9, 1963), p. 10; interview, telephone, Erika Fry, Rocketdyne, February 24, 1969.</small>

27

ASPO Manager Joseph F. Shea asked NASA Headquarters to revise velocity budgets for the Apollo spacecraft. (Studies had indicated that those budgets could be reduced without degrading performance.) He proposed that the 10 percent safety margin applied to the original budget be eliminated in favor of specific allowances for each identifiable uncertainty and contingency; but, to provide for maneuvers which might be desired on later Apollo missions, the LEM's propellant tanks should be oversized. (See December 1963.)

The ASPO Manager's proposal resulted from experience that had arisen because of unfortunate terminology used to designate the extra fuel. Originally the fuel budget for various phases of the mission had been analyzed and a 10 percent allowance had been made to cover—at that time, unspecified—contingencies, dispersions, and uncertainties. Mistakenly this fuel addition became known as a "*10% reserve*"! John P. Mayer and his men in the Mission Planning and Analysis Division worried because engineers at North American, Grumman, and NASA had "been freely 'eating' off the so-called 'reserve'" before studies had been completed to define what some of the contingencies might be and to apportion some fuel for that specific situation. Mayer wanted the item labeled a "10% uncertainty."

Shea recommended also that the capacity of the LEM descent tanks be sufficient to achieve an equiperiod orbit, should this become desirable. However, the spacecraft should carry only enough propellant for a Hohmann transfer. This was believed adequate, because the ascent engine was available for abort maneuvers if the descent engine failed and because a low-altitude pass over the landing site was no longer considered necessary. By restricting lunar landing sites to the area between ±5° latitude and by limiting the lunar stay time to less than 48 hours, a one-half-degree, rather than two-degree, plane change was sufficient.

In the meantime, Shea reported, his office was investigating how much weight could be saved by these propellant reductions.

<small>Memorandum, Shea to NASA Headquarters, Attn: Mgr., Apollo Program Office, "Revised Apollo Spacecraft ΔV Budget," November 27, 1963; memorandum, Christopher C.</small>

113

THE APOLLO SPACECRAFT: A CHRONOLOGY

1963
November

Kraft, Jr., MSC, to Mgr., ASPO, "Use of 10% 'reserve' ΔV in CSM and LEM ΔV Budgets," October 21, 1963.

28

In honor of the late President John F. Kennedy, who was assassinated six days earlier, President Lyndon B. Johnson announced that LOC and Station No. 1 of the Atlantic Missile Range would be designated the John F. Kennedy Space Center (KSC), ". . . to honor his memory, and the future of the works he started . . . ," Johnson said. On the following day, he signed an executive order making this change official. With the concurrence of Florida Governor Farris Bryant, he also changed the name of Cape Canaveral to Cape Kennedy.

Angela C. Gresser, "Historical Aspects Concerning the Redesignation of Facilities at Cape Canaveral," KHN-1, April 1964, p. 15; *The New York Times*, November 29, 1963; *The Houston Chronicle*, November 30, 1963.

28–December 4

MSC reviewed a North American proposal for adding an active thermal control system to the SM to maintain satisfactory temperatures in the propulsion and reaction control engines. The company's scheme involved two water-glycol heat transport loops with appropriate nuclear heaters and radiators. During December, MSC directed North American to begin preliminary design of a system for earth orbit only. Approval for spacecraft intended for lunar missions was deferred pending a comprehensive review of requirements.

MSC, "ASPO Status Report for Week Ending December 4, 1963"; "Apollo Quarterly Status Report No. 6," p. 15.

29

After a meeting with Grumman officials on November 27, ASPO directed the contractor to begin a Grumman-directed Apollo mission plan development study. (See January 16, 1964.)

TWX, Owen E. Maynard, MSC, to GAEC, Attn: R. S. Mullaney, November 29, 1963.

During the Month

MSC directed Grumman to halt work on LEM test article 9, pending determination of its status as a tethered flight vehicle. (See August 1963.) As a result, the proposed flight demonstration of the tether coupler, using an S-64A Skycrane helicopter, was canceled.

"Monthly Progress Report No. 10," LPR-10-26, p. 37.

During the Month

Ames Research Center performed simulated meteoroid impact tests on the Avco Corporation heatshield structure. Four targets of ablator bonded to a stainless steel backup structure were tested. The ablator, in a Fiberglas honeycomb matrix, was 4.369 millimeters (0.172 inch) thick in two targets and 17.424 millimeters (0.686 inch) thick in the other two. Each ablator was tested at 116.48 K (−250 degrees F) and at room temperature, with no apparent difference in damage.

PART II: DEVELOPING HARDWARE DISTINCTIONS

Penetration of the thicker targets was about 13.970 millimeters (0.55 inch). In the thinner targets, the ablator was pierced. Debris tore through the steel honeycomb and produced pinholes on the rear steel sheet. Damage to the ablator was confined to two or three honeycomb cells and there was no cracking or spalling on the surface.

Tests at Ames of thermal performance of the ablation material under high shear stress yielded favorable preliminary results.

> MSC, "ASPO Status Report for Week Ending December 4, 1963."

1963
November

Verne C. Fryklund of NASA's Manned Space Sciences Division advised Bellcomm of the procedure for determining Apollo landing sites on the moon. The Manned Space Sciences chief outlined an elimination for the site selection process. For the first step, extant selenographic material would be used to pick targets of interest for Lunar Orbiter spacecraft photography. After study of the Lunar Orbiter photography, a narrower choice of targets then became the object of Surveyor spacecraft lunar missions, with final choice of potential landing sites to be made after the Surveyor program. (See December 20.)

The selection criteria at all stages were determined by lunar surface requirements prepared by OMSF. Fryklund emphasized that a landing at the least hazardous spot, rather than in the area with the most scientific interest, was the chief aim of the site selection process.

> Memorandum, Verne C. Fryklund, NASA Manned Space Sciences Division, to B. T. Howard, Bellcomm, "Your memorandum of October 31, 1963 about Apollo Landing Sites," November 4, 1963.

During the Month

Grumman selected AiResearch Manufacturing Company to supply cryogenic storage tanks for the LEM electrical power system. Final negotiations on the cost-plus-incentive-fee contract were held in June 1964.

On this same date, Grumman concluded negotiations with Allison Division of General Motors Corporation for design and fabrication of the LEM descent engine propellant storage tanks (at a cost of $5 479 560).

> "Apollo Quarterly Status Report No. 6," pp. 30, 32; MSC, "Project Apollo Quarterly Status Report No. 8 for Period Ending June 30, 1964," p. 38; MSC, "Consolidated Activity Report for the Office of the Associate Administrator, Manned Space Flight, November 17–December 21, 1963," p. 42.

December 2

A design review of the CSM part-task trainer was held at North American. Briefings included general design criteria and requirements, physical configuration, simulation models, and scheduling. The trainer was expected to be operational in December 1964.

> "Apollo Monthly Progress Report," SID 62-300-20, pp. 20-21; MSC, "ASPO Status Report for Week Ending December 10, 1963."

3–5

1963
December

5

Primarily to save weight, the length of the adapter was shortened to 853 centimeters (336 inches), as recommended by Grumman. (See October 2.)

> Letter, Owen E. Maynard, MSC, to GAEC, Attn: R. S. Mullaney, "Contract NAS 9–1100, Line Items 1 and 6, Implementation of Actions Recommended in Apollo Program Systems Meetings," December 5, 1963; TWX, David M. Hammock and Maynard, MSC, to GAEC, Attn: Mullaney, and NAA, Attn: E. E. Sack, December 5, 1963.

9

ASPO requested that Grumman make a layout for transmittal to MSFC showing space required in the S–IVB instrument unit for 406.4- and 457-centimeter (160- and 180-inch) cantilevered gears and for 508-centimeter (200-inch)-radius lateral fold gears. (See October 2.)

> Letter, Owen E. Maynard, MSC, to GAEC, Attn: R. S. Mullaney, "Contract NAS 9–1100, Implementation of Actions in MSC–MSFC Mechanical Integration Panel," December 9, 1963.

10–17

As a result of wind tunnel tests, Langley Research Center researchers found the LEM/Little Joe II configuration to be aerodynamically unstable. To achieve stability, larger booster fins were needed. However, bigger fins caused more drag, shortening the length of the flight. MSC was investigating

> To define the aerodynamic forces and moments on the Apollo launch escape system during the most critical period of flight, NASA scientists investigated a 0.085 scale Apollo launch escape vehicle model. Mounted in Langley Research Center's 16-Foot Transonic Wind Tunnel, the model was studied at transonic speeds to determine aerodynamic characteristics during separation from the service module. The decomposition products of hydrogen peroxide were used to simulate the rocket exhaust. The experiments were designed to help ensure that the command module and crew could be safely recovered if a launch vehicle should malfunction.

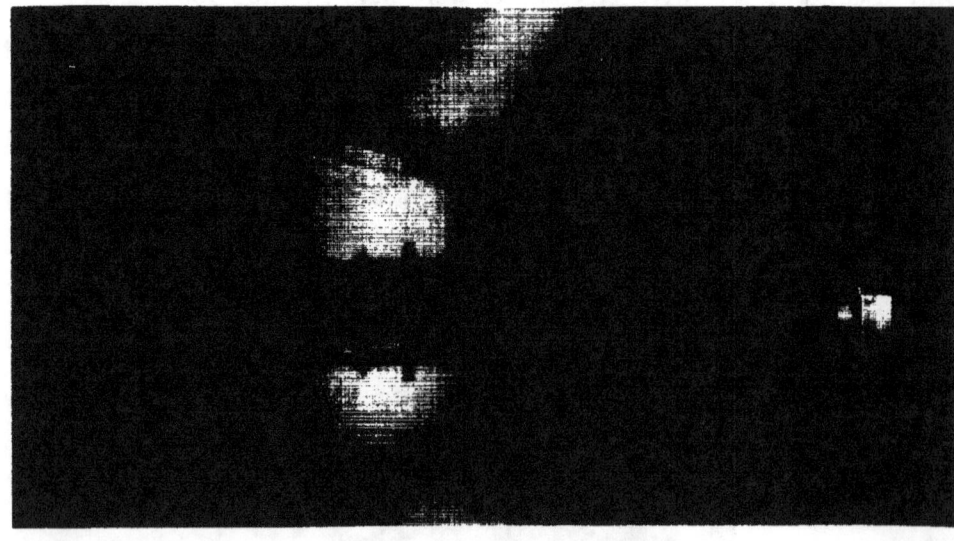

PART II: DEVELOPING HARDWARE DISTINCTIONS

the possibility of using more powerful rocket engines to overcome this performance degradation. (See February 10, 1964.)

1963

December

"Monthly Progress Report No. 11," LPR-10-27, p. 42; MSC, "ASPO Status Report for Week Ending December 17, 1963."

The MSC Operations Planning Division (OPD) reviewed the operational demands upon the CM from the time of CM–SM separation until splashdown. OPD concluded that the CM should be designed to operate for 45 minutes during this phase of the mission.

10–17

MSC, "ASPO Status Report for Week Ending December 17, 1963."

NASA Headquarters approved a $48 064 658 supplement to the Douglas Aircraft Company, Inc., contract for 10 additional S–IVB stages, four for the Saturn IB and six for the Saturn V missions.

11

Akens et al., *History of Marshall . . . July 1–December 31, 1963*, Vol. I, p. 69.

NASA canceled five Ranger flights (numbers 10 through 14) designed to take high-resolution photographs of the lunar surface before impact. [Five Rangers had thus far been launched.] OSS Associate Administrator Homer E. Newell stated that NASA would depend on the remaining four Rangers, the Lunar Orbiters, and the Surveyors for information about the lunar surface. Cancellation of the flights promised to save $90 million.

13

NASA News Release 63–276, "NASA Cancels Five Follow-On Rangers," December 13, 1963.

The Ad Hoc Working Group on Apollo Experiments submitted its final recommendations on what should be Apollo's principal scientific objectives:

15

    (1) Examination of the physical and geological properties of the moon in the area surrounding the spacecraft
    (2) Geological mapping
    (3) Investigations of the moon's interior
    (4) Studies of the lunar atmosphere
    (5) Radio astronomy from the surface.

This group, which had as its chairman Charles P. Sonett of NASA's Ames Research Center and thus was known as the Sonett Committee, had been formed wholly within NASA for just this purpose. Much of the Sonett Committee's report already was contained in the Office of Space Sciences' guidelines transmitted earlier to MSC (see October 8); their reception was not what one could call enthusiastic.

"Final Report of the Ad Hoc Working Group on Apollo Experiments and Training on the Scientific Aspects of the Apollo Program," December 15, 1963, p. 4; letter, Willis B. Foster, to Associate Administrator for Manned Space Flight, "Apollo Scientific Guidelines," December 19, 1963.

1963
December
16

MSC and the U.S. Air Force Aerospace Medical Division completed a joint manned environmental experiment at Brooks Air Force Base, Tex. After spending a week in a sea-level atmospheric environment, the test subjects breathed 100 percent oxygen at 3.5 newtons per square centimeter (5 psi) at a simulated altitude of 8230 meters (27 000 feet) for 30 days. They then reentered the test capsule for observation in a sea-level environment for the next five days. This experiment demonstrated that men could live in a 100 percent oxygen environment under these conditions with no apparent ill effects.

> MSC, "Consolidated Activity Report for the Office of the Director, Manned Space Flight, October 20–November 16, 1963," p. 63; *The Houston Chronicle,* November 4, 1963; *Missiles and Rockets,* 13 (November 11, 1963), p. 31; *The Evening Star,* Washington, December 17, 1963.

16

To ensure MSC's use of its manpower resources to the fullest extent possible, the Engineering and Development Directorate (EDD) assigned a subsystem manager to each of the major subsystems in the Apollo program. EDD provided such support as was needed for him to carry out his assignment effectively. These subsystem managers were responsible to ASPO for the development of systems within the cost and schedule constraints of the program. Primary duties were management of contractor efforts and testing.

> MSC, "Apollo Subsystem Management Plan," December 16, 1963.

16

General Dynamics Corporation announced the receipt of a contract (worth about $4 million) from the Philco Corporation for fabrication of the computer display equipment for the Integrated Mission Control Center at MSC.

> *Wall Street Journal,* December 16, 1963.

16

ASPO concurred in Grumman's recommendation to delete the redundant gimbal actuation system in the LEM's descent engine. A nonredundant configuration would normally require mission abort in case of actuator failure. Consequently, in making this change, Grumman must ensure that mission abort and the associated staging operation would not compromise crew survival and mission reliability.

> Letter, Owen E. Maynard, MSC, to GAEC, Attn: R. S. Mullaney, "Contract NAS 9–1100, Item 2, Descent Engine Gimbal Drive Actuator," December 16, 1963.

16–January 15

Phase I of the Apollo manned centrifuge program was completed at the U.S. Navy Aerospace Medical Acceleration Laboratory, Philadelphia, Pa. The tests pointed up interface problems between couch, suit, and astronaut. For example, pressurizing the suit increased the difficulty of seeing the lower part of the instrument panel. The test fixture was disassembled and the couch, framework, and empty instrument panel were shipped to International Latex Corporation to serve as a mockup for further study.

PART II: DEVELOPING HARDWARE DISTINCTIONS

"Apollo Monthly Progress Report," SID 62-300-21, p. 6.

1963

December

16–January 15

North American completed a study to determine, for automatic modes of reentry, adequacy of the current CM reaction control system (RCS) and compatibility of the RCS with other reentry subsystems.

Ibid., p. 8.

MSC directed North American to redesign the CM environmental control system compressor to provide .283 cubic meters (10 cubic feet) of air per minute to each space suit at 1.8 newtons per square centimeter (3.5 psi), 16.78 kilograms (37 pounds) per hour total.

16–January 15

Ibid., p. 10.

Grumman proposed a two-tank ascent stage configuration for the LEM. (See November 12–19.) On January 17, 1964, ASPO formally concurred and authorized Grumman to go ahead with the design. The change was expected to reduce spacecraft weight by about 45 kilograms (100 pounds) and would make for a simpler, more reliable ascent propulsion system. ASPO also concurred in the selection of titanium for the two propellant tanks.

17

"Monthly Progress Report No. 11," LPR-10-27, p. 1; letter, William F. Rector III, MSC, to GAEC, Attn: R. S. Mullaney, "Contract NAS 9-1100, LEM Program Review," January 17, 1964.

MSC directed North American to assign bioinstrumentation channels to the CM for early manned flights for monitoring the crew's pulse rate, blood pressure, respiration, and temperature. These readings could be obtained simultaneously on any one crew member and by switching from man to man for monitoring the entire crew.

18–January 14

MSC, "ASPO Status Report for Period December 18–January 14, 1964."

The System Engineering Division (SED) examined the feasibility of performing an unmanned earth orbital mission without the guidance and navigation system. SED concluded that the stabilization and control system could be used as an attitude reference for one to two orbits and would have accuracies at retrofire suitable for recovery. The number of orbits depended upon the number of maneuvers performed by the vehicle, since the gyros tended to drift.

18–January 14

Ibid.

Pratt and Whitney Aircraft delivered the first three prototype-A fuel cells to North American.

19

"Apollo Monthly Progress Report," SID 62-300-21, p. 11.

119

1963

December 20

MSC announced that Grumman and Hamilton Standard had signed an $8 371 465 definitive contract for the LEM environmental control system. A go-ahead had been issued to Hamilton Standard on July 23.

MSC News Release 63-257, December 20, 1963; *The Houston Post*, December 22, 1963.

20

NASA selected The Boeing Company to build five Lunar Orbiter spacecraft. (See August 30.) Beginning in 1966, Lunar Orbiters would take close-range photographs of the moon and transmit them by telemetry back to earth. The spacecraft would also detect radiation and micrometeoroid density and supply tracking data on the gravitational field of the moon. Information derived from the project (managed by Langley Research Center) would aid in the selection of lunar landing sites. (See November 1963 and May 8, 1964.)

NASA News Release 63-280, "NASA to Negotiate with Boeing for Lunar Orbiter," December 20, 1963.

20–January 18

MSC awarded the U.S. Army Corps of Engineers contracts valued at $4 211 377 (to be subcontracted to W. S. Bellows Construction Corporation and Peter Kiewit and Sons, Inc.) for the construction of the MSC Mission and Training Facility and for additions to several existing facilities at the Center.

MSC, "Consolidated Activity Report for the Office of the Associate Administrator, Manned Space Flight, December 22, 1963–January 18, 1964," p. 38; MSC News Release 64-46, March 5, 1964; *The Houston Post*, January 9, 1964.

21

MSC defined the LEM terminal rendezvous maneuvers. That phase of the mission would begin at a range of 9.3 kilometers (five nautical miles) from the CSM and terminate at a range of 152.4 meters (500 feet). Before rendezvous initiation, closing velocity should be reduced to 61 meters (200 feet) per second by use of the ascent engine. The reaction control system should be used exclusively thereafter.

Letter, Owen E. Maynard, MSC, to GAEC, Attn: R. S. Mullaney, "Contract NAS 9-1100, Definition of LEM Terminal Rendezvous Model," December 21, 1963.

23

Motorola, Inc., received a follow-on contract from the Jet Propulsion Laboratory for the manufacture and integration of at least three S-band receiving subsystems for NASA's Deep Space Network and Manned Space Flight Network ground stations. Within the unified S-band system adopted by NASA, receiving equipment of the two networks would be identical except for a slight difference in operating frequency. This enabled all communications between ground stations and spacecraft to be on a single frequency. It also allowed more efficient power transfer between the directive antennas and the spacecraft and would greatly reduce galactic noise encountered with UHF frequencies.

NASA News Release 63-284, "Motorola to Make S-Band Radio Receiving Equipment for NASA Ground Stations," December 23, 1963.

PART II: DEVELOPING HARDWARE DISTINCTIONS

Based upon centrifuge test results, MSC directed Hamilton Standard to modify the space suit helmet. The vomitus port and other obstructions to the line of sight in the downward direction were deleted.

> MSC, "Weekly Activity Report for the Office of the Director, Manned Space Flight, December 29, 1963–January 4, 1964," p. 4.

1963

December

29–January 4

NASA announced the appointment of Air Force Brig. Gen. Samuel C. Phillips as Deputy Director of the NASA Headquarters Apollo Program Office. General Phillips assumed management of the manned lunar landing program, working under George E. Mueller, Associate Administrator of Manned Space Flight and Director of the Apollo Program Office.

> NASA News Release 63–287, "NASA Appoints General Phillips to Assist in Apollo Program Management," December 31, 1963.

31

MSC decided to supply television cameras for the LEM as government-furnished items. Grumman was ordered to cease its effort on this component.

Resizing of the LEM propulsion tanks was completed by Grumman. The cylindrical section of the descent tank was extended 34.04 millimeters (1.34 inches), for a total of 36.27 centimeters (14.28 inches) between the spherical end bells. The ascent tanks (two-tank series) were 1240.54 centimeters (48.84 inches) in diameter.

> "Monthly Progress Report No. 11," LPR–10–27, pp. 18, 30.

During the Month

RCA, contractor to Grumman for the LEM rendezvous and landing radars, chose Ryan Aeronautical Company as vendor for the landing radar. The contract was signed March 16, 1964.

> "Apollo Quarterly Status Report No. 6," p. 34.

During the Month

North American, Grumman, and MIT Instrumentation Laboratory summarized results of a six-week study, conducted at ASPO's request, on requirements for a Spacecraft Development Program. Purpose of the study was to define joint contractor recommendations for an overall development test plan within resource constraints set down by NASA. ASPO required that the plan define individual ground test and mission objectives, mission descriptions, hardware requirements (including ground support equipment), test milestones, and individual subsystem test histories.

Intermediate objectives for the Apollo program were outlined: the qualification of a manned CSM capable of earth reentry at parabolic velocities after an extended space mission; qualification of a manned LEM both physically and functionally compatible with the CSM; and demonstration of

1964

January

3

1964

January

manned operations in deep space, including lunar orbit. The most significant basic test plan objective formulated during the study was the need for flexibility to capitalize on unusual success or to compensate for unexpected difficulties with minimum impact on the program.

Only one major issue in the test plan remained unresolved—lunar descent radar performance and actual lunar touchdown. Two possible solutions were suggested:

(1) Landing of an unmanned spacecraft. If this failed, however, there would be little or no gain, since there was not yet a satisfactory method for instrumenting the unmanned vehicle for necessary failure data. If the landing were successful, it would prove only that the LEM was capable of landing at that particular location.

(2) Designing the LEM for a reasonably smooth surface. This would avoid placing too stringent a requirement on the landing criteria to accommodate all lunar surface unknowns. A block change to the LEM design could then be planned for about mid-1966. By that time, additional lunar data from Ranger, Surveyor, and Lunar Orbiter flights would be available. The group agreed the second solution was more desirable.

The contractors recommended: (1) ASPO concur with the proposed plan as a planning basis for implementation; (2) ASPO issue a Development Test Plan to all three contractors (preferably within 30 to 60 days); (3) each contractor analyze the effect of the plan upon spacecraft, facility, and equipment contracts; and (4) ASPO and the contractors conduct periodic reviews of the plan once it was formalized.

In addition, the test plan should be coordinated with the lunar landing mission study, as well as development testing and systems engineering for the complete Apollo program.

The complete findings of this joint study were contained in a five-volume report issued by North American and submitted to MSC early in February 1964. [This document became known informally as the "Project Christmas Present Report."]

"Apollo Spacecraft Development Test Plan," SID 64-66-1, Vol. I, pp. v, 1, 3-5, 195-197.

3

MSC forwarded a $1.4 million contract to Control Data Corporation for two computer systems and peripheral equipment which would be supplied to GE as part of the preflight acceptance checkout equipment.

MSC, "Consolidated Activity Report for the Office of the Associate Administrator, Manned Space Flight, December 22, 1963–January 18, 1964," p. 39.

7

ASPO directed Grumman to implement a number of recommendations on space suit oxygen umbilical hoses discussed at a joint Grumman/North American meeting and forwarded to ASPO on December 4, 1963: (1) adopt a design that would permit use of CM hose sets in the LEM after crew

PART II: DEVELOPING HARDWARE DISTINCTIONS

transfer; (2) place connectors on short hoses permanently attached to the suit, because suit vision and arm mobility did not permit use of on-suit connectors; (3) determine exact placement and hose angles to route the suit/portable life support system umbilicals between the legs of the suit; (4) build the "buddy concept" into the umbilical design by ensuring that one of the LEM hoses had valve and safety provisions; and (5) design the CM and LEM oxygen hose umbilicals to be interchangeable. (MSC would select a contractor for the connectors.)

1964
January

> MSC "ASPO Status Report for Week Ending December 10, 1963"; TWX, William F. Rector III, MSC, to GAEC, Attn: R. S. Mullaney, "Space Suit Oxygen Umbilical Hoses," January 7, 1964.

MSC directed Grumman to integrate LEM translation and descent engine thrust controllers. The integrated controller would be lighter and easier to install; also it would permit simultaneous reaction control system translation and descent engine control. Grumman had predicted that such a capability might be required for touchdown.

8

> MSC, "ASPO Status Report for Week Ending January 7, 1963."

The Flight Data Systems Branch of the Engineering and Development Directorate provided ASPO's Lunar Mission Planning Branch with information about the LEM extravehicular suit telemetry and communications system. No line of sight (LOS) communications were possible, and there would be no ground wave propagation and no atmospheric reflection. The link between astronaut and LEM would be limited to LOS of the two antennas, and surface activities by an extravehicular astronaut must be planned accordingly.

10

> Memorandum, Ragan Edmiston, MSC, to Richard H. Kohrs, "Lunar transmission range for Astro/LEM communications link," January 10, 1964.

Three U. S. Air Force test pilots began a five-week training period at the Martin Company leading to their participation in a simulated seven-day lunar landing mission. This was part of Martin's year-long study of crew performance during simulated Apollo missions (under a $771 000 contract from NASA).

11

> *The Houston Post,* January 13, 1964; *The Houston Chronicle,* January 13, 1964.

Based on the LEM mockup review of September 16–18, 1963, MSC established criteria for redundancy of controls and displays in the LEM crew station. Within the framework of apportioned reliability requirements for mission success and crew safety, these guidelines applied: (1) the LEM must be provisioned so that hover to touchdown could be flown manually by the crew; (2) no single failure in the controls and displays should cause an abort; and (3) the unknowns associated with lighting conditions or dust caused by rocket exhaust impingement on the lunar surface might require

14

1964

January

a joint effort by the crew. Although duplication of all equipment was not required, dual flight controls and windows, as well as gross attitude, attitude error, and vehicle rates information, were necessary. Other flight displays should be dual or be readable from either station.

>Letter, William F. Rector III, MSC, to GAEC, Attn: R. S. Mullaney, "Contract NAS 9-1100, Requirements for Dual Flight Controls and Displays in the LEM," January 14, 1964.

14

At an MSC–North American meeting, spacecraft communications problems were reviewed. Testing had indicated that considerable redesign was essential to ensure equipment operation in a high-humidity environment. Also antenna designs had created several problem areas, such as the scimitar antenna's causing the CM to roll during reentry. The amount of propellant consumed in counteracting this roll exceeded reentry allowances. Further, because the CM could float upside down, the recovery antenna might be pointed at the ocean floor. In fact, many at this meeting doubted whether the overall communications concept was satisfactory "without having detailed ground receiver characteristics." The situation derived from "one of the primary problems in the area of communications system design . . . the lack of functional requirements specifications."

>"Minutes of NASA–NAA Technical Management Meeting, January 14–15, 1964," p. 4.

15

MSC and Bellcomm agreed upon a plan for testing the Apollo heatshield under reentry conditions. Following Project Fire and Scout tests, the Saturn IB would be used to launch standard "all-up" spacecraft into an elliptical orbit; the SM engine would boost the spacecraft's velocity to 8839 meters (29 000 feet) per second. Two flights were scheduled, one a test of ablator performance and the other a long-range flight to achieve a high total heat load and assess the interaction of the ablator, its backup structure, and other related structural members. This degree of heat rate and loading would permit "demonstration" rather than "development" tests on the Saturn V.

>Memorandum, Robert O. Piland, MSC, to Joseph F. Shea, "Apollo Reentry Testing," January 16, 1964.

15

The first fuel cell module delivered by Pratt and Whitney Aircraft to North American was started and put on load. The module operated normally and all test objectives were accomplished. Total operating time was four hours six minutes, with one hour at each of four loads—20, 30, 40, and 50 amperes. The fuel cell was shut down without incident and approximately 1500 cubic centimeters (1.6 quarts) of water were collected.

>"Apollo Monthly Progress Report," SID 62-300-21, p. 11.

PART II: DEVELOPING HARDWARE DISTINCTIONS

Bendix Products Aerospace Division was awarded a $99 973 contract by MSC to study crushable aluminum honeycomb, a lightweight, almost non-elastic, shock-absorbing material for LEM landing gears. Bendix would test the honeycomb structures in a simulated lunar environment.

MSC News Release 64-9, January 15, 1964.

1964
January
15

MSC's Systems Engineering Division met with a number of astronauts to get their comments on the feasibility of the manual reorientation maneuver required by the canard abort system concept. (See November 12, 1963.) The astronauts affirmed that they could accomplish the maneuver and that manual control during high-altitude aborts was an acceptable part of a launch escape system design. They pointed out the need to eliminate any possibility of sooting of the windows during normal and abort flight. Although the current design did not preclude such sooting, a contemplated boost protective cover might satisfy this requirement.

MSC, "ASPO Status Report for Week Ending January 23, 1964."

15-23

ASPO asked the Flight Crew Operations Directorate to study whatever was necessary to ensure that the LEM crew could reorient their spacecraft manually in an abort 36 600 meters (120 000 feet) above the moon.

Ibid.

15-23

MSC's Center Medical Office was reevaluating recommendations for LEM bioinstrumentation. The original request was for three high-frequency channels (two electrocardiogram and one respiration) that could be switched to monitor all crew members. Grumman wanted to provide one channel for each astronaut with no switching.

Ibid.

15-23

ASPO and the Astronaut Office agreed to provide the crew with food that could be eaten in a liquid or semi-liquid form during emergency pressurized operation. This would permit considerable reduction in the diameter of the emergency feeding port in the helmet visor.

Ibid.

15-23

Representatives of Grumman, MSC's Instrumentation and Electronics Systems Division, ASPO, and Resident Apollo Spacecraft Program Office (RASPO) at Bethpage met at Grumman to plan the LEM's electrical power system. The current configuration was composed of three fuel cell generators with a maximum power output of 900 watts each, spiking stabilizing batteries, one primary general-purpose AC inverter, and a conventional bus

16

1964
January

arrangement. To establish general design criteria, the primary lunar mission of the LEM-10 vehicle was analyzed. This "critical" mission appeared to be the "worst case" for the electrical power system and established maximum power and usage rate requirements.

Those attending the meeting foresaw a number of problems:
- Grumman allowed only 10 percent margin for all contingencies and errors in energy requirements
- Fuel cells and cryogenic fuels needed testing in a simulated space environment
- Grumman depended upon its subcontractors to develop component testing procedures
- Optimum power supply modes and motors for the environmental control system were still to be selected
- "Essential loads" needed standardizing to allow the proper bus loading structure
- Proper charging rates and equipment for the portable life support system extravehicular suit batteries needed to be selected.

Memorandum, Donald G. Wiseman, MSC, to Deputy Asst. Dir. for Engineering and Development, "Meetings attended by Instrumentation and Electronics Systems Division personnel at the Grumman Aircraft Engineering Corporation," January 24, 1964.

16

Grumman presented to MSC the first monthly progress report on the Lunar Mission Planning Study. (See November 29, 1963.) The planning group, designated the Apollo Mission Planning Task Force (AMPTF), established ground rules and constraints to serve as a base line around which mission flexibilities and contingency analyses could be built. Main topics of discussion at the meeting were the reference mission, study ground rules, task assignments, and future plans. The following week, MSC Flight Operations Directorate provided a reference trajectory for the AMPTF's use. Major constraints were daylight launch, translunar injection during the second earth parking orbit, free-return trajectory, daylight landing near the lunar equator, 24-hour lunar surface staytime, and a water landing on earth. (See May 4.)

MSC, "ASPO Status Report for Week Ending January 23, 1964"; "ASPO Status Report for Period December 18–January 14, 1964."

16–February 12

The first full-throttle firing of Space Technology Laboratories' LEM descent engine (being developed as a parallel effort to the Rocketdyne engine) was carried out. The test lasted 214 seconds, with chamber pressures from 66.2 to 6.9 newtons per square centimeter (96 to 10 psi). Engine performance was about five percent below the required level.

MSC, "Monthly ASPO Status Report for Period January 16–February 12, 1964."

16–February 15

Two astronauts took part in tests conducted by North American to evaluate equipment stowage locations in CM mockup 2. Working as a team, the

PART II: DEVELOPING HARDWARE DISTINCTIONS

astronauts simulated the removal and storage of docking mechanisms. Preliminary results indicated this equipment could be stowed in the sleeping station. When his suit was deflated, the subject in the left couch could reach, remove, and install the backup controllers if they were stowed in the bulkhead, couch side, or headrest areas. When his suit was pressurized, he had difficulty with the bulkhead and couch side locations. The subject in the center couch, whose suit was pressurized, was unable to be of assistance.

1964
January

> NAA, "Apollo Monthly Progress Report," SID 62-300-22, March 1, 1964, p. 6.

AiResearch Manufacturing Company reported that it had completed design effort on all components of the CM environmental control system. (See January 23-29.)

16-February 15

> The Garrett Corporation, AiResearch Manufacturing Division, "Monthly Progress Report, Environmental Control System, NAA/S&ID, Project Apollo, 16 January 1964-15 February 1964," SS-1013-R(21), February 29, 1964.

Grumman was studying problems of transmitting data if the LEM missed rendezvous with the CSM after lunar launch. This meant that the LEM had to orbit the moon and a data transmission blackout would occur while the LEM was on the far side of the moon. There were two possible solutions, an onboard data recorder or dual transmission to the CSM and the earth. This redundancy had not previously been planned upon, however.

17

> Memorandum, Donald G. Wiseman, MSC, to Deputy Asst. Dir. for Engineering and Development, "Meetings attended by Instrumentation and Electronics Systems Division personnel at the Grumman Aircraft Engineering Corporation," January 24, 1964.

A design review of the CM reaction control system (RCS) was held. Included was a discussion of possible exposure of the crew to hazardous fumes from propellants if the RCS ruptured at earth impact. For the time being, the RCS design would not be changed, but no manned flights would be conducted until the matter had been satisfactorily resolved. A detailed study would be made on whether to eliminate, reduce, or accept this crew safety hazard.

17

> "Apollo Monthly Progress Report," SID 62-300-22, p. 22.

NASA assigned George M. Low to the position of Deputy Director of MSC. He would replace James C. Elms, who had resigned on January 17 to return to private industry. Although Low continued as Deputy Associate Administrator for Manned Space Flight at NASA Headquarters until May 1, he assumed his new duties at MSC the first part of February.

19

> MSC News Release 64-13, January 17, 1964; NASA News Release 64-13, "NASA Names Low Deputy Director of Manned Spacecraft Center," January 19, 1964.

North American gave a presentation at MSC on the block change concept with emphasis on Block II CSM changes. These were defined as modifications necessary for compatibility with the LEM, structural changes to reduce

21

1964
January

weight or improve CSM center of gravity, and critical systems changes. [Block I spacecraft would carry no rendezvous and docking equipment and would be earth-orbital only. Block II spacecraft would be flight-ready vehicles with the final design configuration for the lunar missions.] (See February 13–20 and April 16, 1964.)

"Apollo Monthly Progress Report," SID 62-300-22, pp. 1-2.

22

Representatives of MSC, North American, Collins Radio Company, and Motorola, Inc., met in Scottsdale, Ariz., to discuss a proposed redesign of

The test firing of a launch escape motor for the Apollo spacecraft's launch escape system made a spectacular flame pattern during static firing at Lockheed Propulsion Company's Potrero facility near Beaumont, Calif. Four nozzles, canted outboard, split the flame of the solid-fuel rocket motor into four equal tails.

PART II: DEVELOPING HARDWARE DISTINCTIONS

the unified S-band to make it compatible with the Manned Space Flight Network. To ensure that there would be no schedule impact, North American proposed only a limited capability on the Block I vehicles. MSC deferred a decision on the redesign pending equipment compatibility tests at Motorola; spacecraft/network compatibility tests by MSC, North American, and the Jet Propulsion Laboratory; and cost analyses.

1964
January

MSC, "ASPO Status Report for Period January 23-29, 1964;" "ASPO Status Report for Period January 30-February 5, 1964;" "Apollo Monthly Progress Report," SID 62-300-22, p. 10.

NASA and North American discussed visibility requirements on the CM and came to the following conclusions: the contractor would provide four portholes in the protective shroud so the astronauts could see through both side and forward viewing windows, and ensure that all windows were clean after launch escape tower separation. North American proposed the addition to Block II CM of a collimated optical device for orientation and alignment during docking. MSC Flight Crew Operations Directorate recommended that mirrors be added to increase external and internal field of vision.

23

MSC, "Minutes, Project Apollo Window and Vision Requirements Meeting, January 23, 1964," January 24, 1964; MSC, "Consolidated Activity Report for the Office of the Associate Administrator, Manned Space Flight, January 19-February 15, 1964," pp. 29-30; MSC, "ASPO Status Report for Period January 23-29, 1964."

MSC issued a $9.2 million contract amendment to North American for the construction and modification of buildings at Downey, Calif., and for research and development work on the CM.

23

MSC News Release 64-17, January 23, 1964.

The AiResearch Manufacturing Company began qualification testing of the first group of components of the CM environmental control system.

23-29

MSC, "ASPO Status Report for Period January 30-February 5, 1964"; "Monthly Progress Report, Environmental Control System," SS-1013-R(21), p. 2.

The second phase of docking simulation studies ended at North American-Columbus (Ohio). Tests included 170 runs simulating transposition and lunar orbital docking with stable and unstable targets, and two extendible probe concepts: cable and rigid boom.

24

"Apollo Monthly Progress Report," SID 62-300-22, p. 2.

A design review of crew systems checkout for the CM waste management system was held at North American. As a result, MSC established specific requirements for leakage flow measurement and for checkout at North American and Cape Kennedy. The current capability of the checkout unit restricted it to measuring only gross leakage of segments of the system.

24

1964

January

Further analysis of the management system was necessary to determine changes needed in the checkout unit.

> *Ibid.,* p. 22.

26–February 1

MSC authorized AiResearch Manufacturing Company and the Linde Company to manufacture high-pressure insulated tanks. This hardware, to be available about May 15, would be used in a study of the feasibility of a supercritical helium pressurization system for the LEM.

> MSC, "Weekly Activity Report for the Office of the Associate Administrator, Manned Space Flight, January 26–February 1, 1964," p. 11.

27

ASPO asked Grumman to study whether attitude control of the docked vehicles was practicable using the LEM's stabilization and control system (RCS). Grumman also was to evaluate the RCS fuel requirements for a five-minute alignment period to permit two star sightings. ASPO further directed the contractor to determine RCS fuel requirements for a second alignment of the LEM's inertial measurement unit during descent coast. This second alignment was needed for the required landing accuracy from a Hohmann descent.

> Letter, W. F. Rector III, MSC, to GAEC, Attn: R. S. Mullaney, "Contract NAS 9–1100, Request for Study of LEM Capability to Stabilize the Command and Service Modules in Lunar Orbit," January 27, 1964.

27

Studies on the LEM's capability to serve as the active vehicle for lunar orbit docking showed the forward docking tunnel to be the best means of accomplishing this. ASPO requested Grumman to investigate the possibility of this docking approach and the effect it might have on the spacecraft's configuration.

> Letter, W. F. Rector III, MSC, to GAEC, Attn: R. S. Mullaney, "Contract NAS 9–1100, Effects of Docking Requirements on the LEM Configuration," January 27, 1964.

28

The United States and Spain agreed to the construction and operation of a $1.5 million space tracking and data acquisition station about 48 kilometers (30 miles) west of Madrid, Spain. Spanish firms would construct the storage and other support structures, and Spanish technicians would participate in operating the station. Linked with the NASA Deep Space Instrumentation Facility, the station included a 26-meter (85-foot)-diameter parabolic antenna and equipment for transmitting, receiving, recording, data handling, and communications with the spacecraft. Later, unified S-band equipment was added to join the facility with the Manned Space Flight Network to support the Apollo program.

> NASA News Release 64–22, "Spain Becomes Site of Major U.S. Space Tracking Station," January 28, 1964; U.S. Congress, *Eleventh Semiannual Report to Congress,* House Doc. No. 63, 98th Cong., 1st Sess. (January 26, 1965), p. 146.

PART II: DEVELOPING HARDWARE DISTINCTIONS

1964
January
29

SA-5, a vehicle development flight, was launched from Cape Kennedy Complex 37B at 11:25:01.41, e.s.t. This was the first flight of the Saturn I Block II configuration (i.e., lengthened fuel tanks in the S-I and stabilizing tail fins), as well as the first flight of a live (powered) S-IV upper stage. The S-I, powered by eight H-1 engines, reached a full thrust of over 680 400 kilograms (1.5 million pounds) the first time in flight. The S-IV's 41 000-kilogram (90 000-pound)-thrust cluster of six liquid-hydrogen RL-10 engines performed as expected. The Block II SA-5 was also the first flight test of the Saturn I guidance system.

> MSFC, *Results of the Fifth Saturn I Launch Vehicle Test Flight, SA-5* (MPR-SAT-FE-64-17, September 22, 1964), pp. 1-5, 8, 82, 85; *Missiles and Rockets*, 14 (February 3, 1964), pp. 17-18.

29

NASA announced the award of a $1.356 million contract to the Blaw-Knox Company for design and construction of three parabolic antennas, each 26 meters (85 feet) in diameter, for the Manned Space Flight Network stations at Goldstone, Calif.; Canberra, Australia; and near Madrid, Spain.

> *Missiles and Rockets*, 14 (February 10, 1964), p. 42; *Astronautics and Aeronautics, 1964* (NASA SP-4005, 1965), p. 33.

30

NASA launched *Ranger VI* from Cape Kennedy. (See December 19, 1962.) The probe, which sought to obtain television pictures of the lunar surface, landed in the moon's Sea of Tranquility on February 2. Despite being the subject of an intensive quality and reliability testing program, *Ranger VI* was a failure—no pictures were obtained. The cause was believed to exist in the power system for the spacecraft's television cameras.

> *Astronautics and Aeronautics, 1964*, pp. 34-35, 41; Henry L. Richter, Jr., (ed.), *Space Measurements Survey: Instruments and Spacecraft, October 1957-March 1965* (NASA SP-3028), p. 468.

30-February 5

MSC and North American representatives discussed preliminary analysis of the probabilities of mission success if the spacecraft were hit by meteoroids. The contractor believed that pressurized tankage in the SM must be penetrated before a failure was assumed. To MSC, this view appeared overly optimistic. MSC held that, as the failure criterion, no debris should result from meteoroid impact of the SM outer structure. [This change in criteria would cost several hundred pounds in meteoroid protection weight in the SM and LEM.] North American thought that penetration of one half the depth of the heatshield on the conical surface of the CM was a failure. Here, MSC thought the contractor too conservative; full penetration could probably be allowed.

> MSC, "ASPO Status Report for Period January 30-February 5, 1964."

During the Month

Grumman began initial talks with Bell Aerosystems Company looking toward concentrating on the all-ablative concept for the LEM's ascent engine, thus abandoning the hope of using the lighter, radiatively cooled

1964

January

nozzle extension. (See September 19–October 16, 1963; also May 4–11.) These talks culminated in July, when Bell submitted to Grumman a revised development and test plan for the engine, now an all-ablative design.

> GAEC, "Monthly Progress Report No. 12," LPR-10-28, February 10, 1964, p. 16; GAEC, "Monthly Progress Report No. 18," LPR-10-34, August 10, 1964, p. 5.

February

1

At an Apollo Program Review held at MSC, Maxime A. Faget reported that Crew Systems Division had learned that the metabolic rate of a man walking in an unpressurized suit was twice that of a man in everyday clothes. When the suit was pressurized to 1.8 newtons per square centimeter (3.5 psi), the rate was about four times as much. To counteract this, a water-cooled undergarment developed by the British Ministry of Aviation's Royal Aircraft Establishment was being tested at Hamilton Standard. These "space-age long johns" had a network of small tubes through which water circulated and absorbed body heat. Advantages of the system were improved heat transfer, low circulating noise levels, and relatively moderate flow rates required. An MSC study on integration of the suit with the LEM environmental control system showed a possible weight savings of 9 kilograms (20 pounds).

> NASA, "Apollo, Program Review Document, February 1, 1964," p. 109; MSC, "Monthly ASPO Status Report for Period January 16–February 12, 1964"; *Space Business Daily*, February 3, 1964; MSC, "ASPO Status Report for Period February 13–20, 1964"; *Aviation Week and Space Technology*, 80 (February 17, 1964), p. 29; MSC, "ASPO Status Report for Period Ending February 27–March 4, 1964"; TWX, W. F. Rector III, MSC, to GAEC, Attn: R. S. Mullaney, March 2, 1964.

3

Fourteen new astronauts, chosen in October 1963, reported at MSC for training for the Gemini and Apollo programs. (See October 18, 1963.)

> MSC News Release 64-24, February 3, 1964.

4

MSC and MSFC officials discussed development flight tests for Apollo heatshield qualification. Engineers from the Houston group outlined desired mission profiles and the number of missions needed to qualify the component. MSFC needed this information to judge its launch vehicle development test requirements against those of MSC to qualify the heatshield. By the middle of the month, Richard D. Nelson of the Mission Planning and Analysis Division (MPAD) had summarized the profiles to be flown with the Saturn V that satisfied MSC's needs. Nelson compiled data for three trajectories that could provide reentry speeds of around 11 000 meters (36 000 feet) per second, simulating lunar return. As an example, "Trajectory 1" would use two of the booster's stages to fire into a suborbital ballistic path, and then use a third stage to accelerate to the desired reentry speed.

Flight profiles for Saturn IB missions for heatshield qualification purposes proved to be a little more difficult because "nobody would or could define the requirements or constraints, or test objectives." In other words, MSFC requirements for booster development test objectives and those of MSC for

PART II: DEVELOPING HARDWARE DISTINCTIONS

the spacecraft heatshield conflicted. So compromises had to be forged. Finally Ted H. Skopinski and other members of MPAD bundled up all of ASPO's correspondence on the subject generated from the various pertinent sources: MSFC, MSC, and contractors. From this, the Skopinski group drafted "broad term test objectives and constraints" for the first two Saturn IB flights (missions 201 and 202). Generally, these were to man-rate the launch vehicle and the CSM and to "conduct entry tests at superorbital entry velocities" (8500 to 8800 meters per second) (28 000 to 29 000 feet per second). Skopinski also enumerated specific test objectives covering the whole spacecraft-launch vehicle development test program. These were first distributed on March 27, and adjustments were made several times later in the year.

1964
February

> MSC, "ASPO Status Report for Period January 30–February 5, 1964"; memorandum, Carl R. Huss, MSC, to BE4/Historical Office, "Comments on Volume II of *The Apollo Spacecraft: A Chronology*," March 30, 1970; memorandum, Richard D. Nelson, MSC, to Chief, Mission Planning and Analysis Division, "Mission profiles for Saturn V superorbital heat shield qualification test," February 13, 1964; memorandum, Ted H. Skopinski, MSC, to Distr., "Summary of broad term test objectives and constraints for Saturn IB development missions 201 and 202," March 27, 1964; memorandum, E. D. Murrah and R. E. McAdams, MSC, to Distr., "Possible change in trajectory profile for Apollo mission SA-201," September 29, 1964; memorandum, McAdams, to Distr., "Revised preliminary trajectory profile for Apollo Mission SA-201," October 19, 1964; memorandum, McAdams, to Distr., "Preliminary Reference Trajectory for Apollo Mission SA-201," October 26, 1964.

Minneapolis-Honeywell Regulator Company reported it had developed an all-attitude display unit for the CM to monitor the guidance and navigation system and provide backup through the stabilization and control system. The Flight Director Attitude Indicator (or "eight-ball") would give enough information for all spacecraft attitude maneuvers during the entire mission to be executed manually, if necessary.

6

> Honeywell News Release, "All-Attitude Display Produced By Honeywell For Apollo Spacecraft," February 6, 1964; *Space Business Daily*, February 24, 1964, p. 290.

Grumman received MSC's response to the "Project Christmas Present Report" (see January 3), and accordingly reevaluated its testing concept for the LEM. On February 19, the contractor proposed to ASPO Manager Joseph F. Shea a flight program schedule, which was tentatively approved. ASPO's forthcoming proposal was identical to Grumman's proposal. It called for 11 LEMs (which were now renumbered consecutively) and two flight test articles. All LEMs were to have full mission capability, but numbers one through three had to be capable of either manned or unmanned flight.

7

> GAEC, "Monthly Progress Report No. 13," LPR-10-29, March 10, 1964, p. 35; "Monthly Progress Report No. 14," LPR-10-30, p. 36.

Engineers from ASPO and Engineering and Development Directorate (EDD) discussed the current status of the tower flap versus the canard launch escape vehicle (LEV) configurations. (See November 12, 1963.) Their aim was to

7

THE APOLLO SPACECRAFT: A CHRONOLOGY

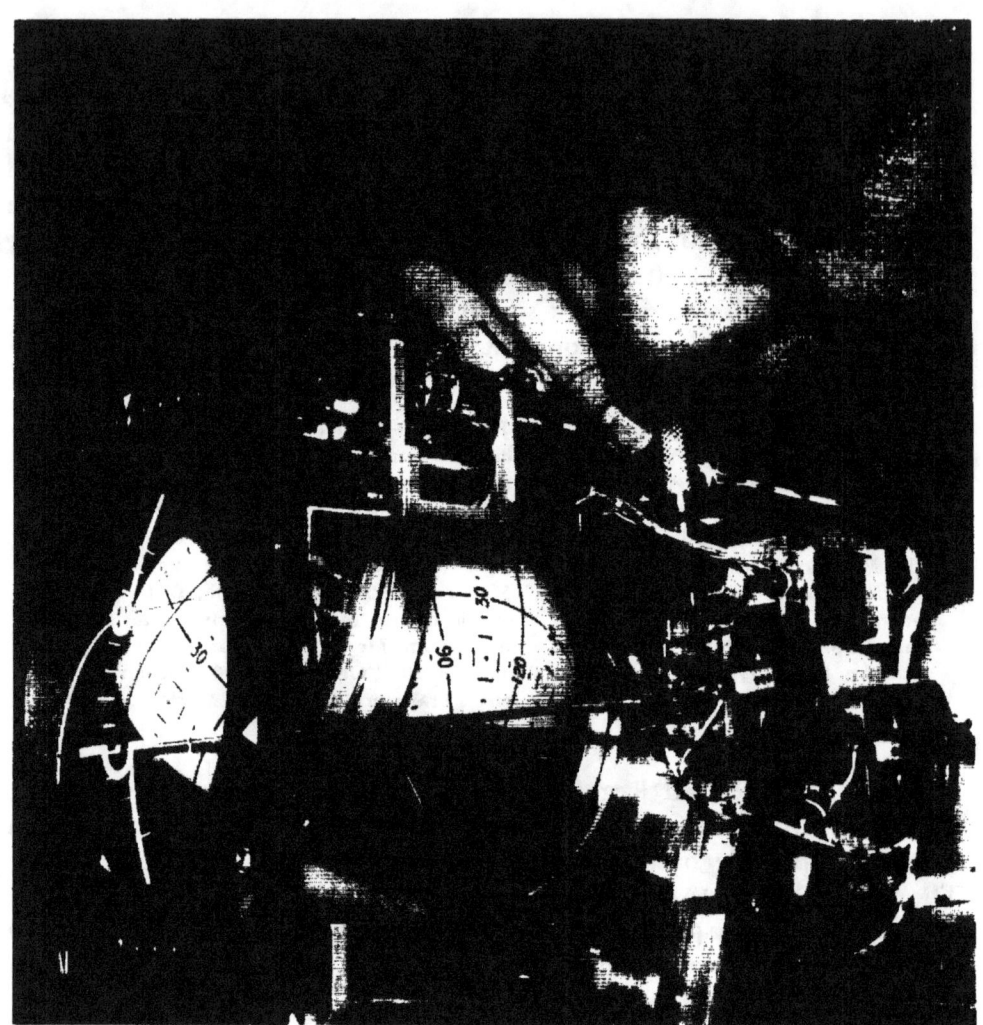

Space globe built by Honeywell would help guide Project Apollo astronauts on their 800 000-kilometer (500 000-mile) flight to the moon and back. Figures on the globe and pointers on the instrument face (left) would tell astronauts at a glance which way their spacecraft faced and how fast it was moving in any direction. The new device, produced with watchmaker precision at Honeywell's Aeronautical Division in Minneapolis, was called a flight director attitude indicator. In the photo, engineer Bill Coleman made final adjustments to a unit ready for shipment to North American Aviation.
—Minneapolis-Honeywell photo.

| 1964 February | select one of the two LEV configurations for Block I spacecraft. (See February 25.) ASPO and EDD concluded that the canard was aerodynamically superior; that arguments against the canard, based on sequencing, mechanical complexity, or schedule effect, were not sufficient to override this aerodynamic advantage; and that this configuration should be adopted for |

PART II: DEVELOPING HARDWARE DISTINCTIONS

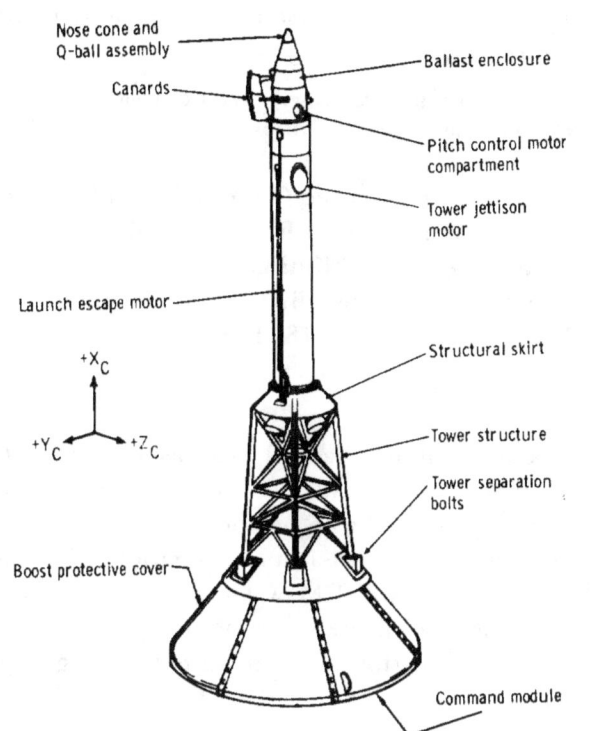

Launch escape vehicle configuration.

Block I spacecraft. However, further analysis was needed to choose the design for the Block II LEV.

<blockquote>Memorandum, Calvin H. Perrine, Jr., MSC, to Distribution, "Minutes of meeting on tower flap and canards, February 7, 1964," February 12, 1964.</blockquote>

During a meeting at MSC, North American and MSC Crew Systems Division agreed that there should be a central authority with total cognizance over Gemini and Apollo food and survival equipment, and that all this equipment should be government furnished.

<blockquote>MSC, "Monthly ASPO Status Report for Period January 16–February 12, 1964."</blockquote>

MSC directed Grumman to stop all work on the LEM/Little Joe II program. This action followed the ASPO Manager's decision against a testing program for the LEM comparable to that for the CSM. (See December 10–17, 1963.)

<blockquote>Ibid.; memorandum, Joseph F. Shea, MSC, to Distr., "Cancellation of LEM/LJ II Program," February 10, 1964.</blockquote>

ASPO directed Grumman to provide an abort guidance system (AGS) in the LEM using an inertial reference system attached to the structure of the vehicle. Should the spacecraft's navigation and guidance system fail, the crew could use the AGS to effect an abort. Such a device eliminated the

1964
February

7

10

11

1964

February

need for redundancy in the primary guidance system (and proved to be a lighter and simpler arrangement).

> Letter, Joseph F. Shea, MSC, to GAEC, Attn: R. S. Mullaney, "Abort Guidance System," February 11, 1964; interview, telephone, Enoch M. Jones, Houst., February 27, 1970.

12

NASA gave credit to two MSC engineers, George C. Franklin and Louie G. Richard, for designing a harness system for the LEM that enabled the crew to fly the vehicle from a standing position. Eliminating the seats reduced the LEM's weight and gave the crew better visibility and closer observation of controls and intruments. (See September 16–18, 1963.)

> MSC News Release 64-27, February 12, 1964.

13

MSC issued Requests for Proposals to more than 50 firms asking for studies and recommendations on how the lunar surface should be explored. Studies should show how lunar surveys could be performed and how points on the lunar surface might be located for future lunar navigation. Maximum use of equipment planned for the LEM and CM was expected. Part of the scientific apparatus aboard the LEM would be selenodetic equipment. The study would not include actual fabrication of hardware but might give estimates of cost and development times.

> *Space Business Daily*, February 13, 1964, p. 238; *ibid.*, March 2, 1964, p. 329.

13–19

Boilerplate (BP) 13 spacecraft was flown from North American, Downey, Calif., to MSC's Florida Operations facility at Cape Kennedy, where the vehicle was inspected and checked out. On April 2, the spacecraft and launch escape system were moved to the pad and mated to the launch vehicle, SA–6. After exhaustive testing, a Flight Readiness Review on May 19 established that BP–13 was ready for launch. (See May 28.)

> MSC, "Postlaunch Report for Apollo Mission A–101 (BP–13)," MSC–R–A–64–2 (June 18, 1964), pp. 6–1 through 6–4.

13–20

The Block II CSM configuration (see January 21) was based on three classes of changes: mandatory changes necessary to meet the

(1) Functional requirements of the lunar mission
(2) Manufacturing or fabrication changes (identified only with improved fabrication techniques)
(3) Technically desirable and weight reduction changes.

> MSC, "ASPO Status Report for Period February 13–20, 1964."

14

MSC ordered North American to design the SM's reaction control system with the capability for emergency retrograde from earth orbit.

> Letter, H. P. Yschek, MSC, to NAA, Space and Information Systems Div., "Contract Change Authorization No. One-Hundred, Forty-Seven," February 14, 1964.

PART II: DEVELOPING HARDWARE DISTINCTIONS

North American completed its initial phase of crew transfer tests using a mockup of the CM/LEM transfer tunnel. Subjects wearing pressure suits were suspended and counterbalanced in a special torso harness to simulate weightlessness; hatches and docking mechanisms were supported by counterweight devices. The entire tunnel mockup was mounted on an air-bearing, frictionless table. Preliminary results showed that the crew could remove and install the hatches and docking mechanisms fairly easily.

1964

February

16–March 15

"Apollo Monthly Progress Report," SID 62-300-23, p. 5.

The potable water system was changed to meter both hot and cold water in one-ounce increments to provide accurate measurements for food rehydration. The previous water valve was a full-flow tap.

16–March 15

Ibid., p. 10.

MSC gave its formal consent to two of Grumman's subcontracts for engines for the LEM:

(1) With Bell Aerosystems for the ascent engine ($11 205 416 incentive-fee contract)

(2) With Space Technology Laboratories for a descent engine to parallel that being developed by Rocketdyne ($18 742 820 fixed-fee contract). (See May 1963.)

16–March 21

MSC, "Consolidated Activity Report for the Office of the Associate Administrator, Manned Space Flight, February 16–March 21, 1964," p. 45.

MSC completed and forwarded to NASA Headquarters a plan for changing the relationship of the navigation and guidance contractors. AC Spark Plug would become the principal contractor, with the Raytheon Company and Kollsman Instrument Corporation as subcontractors. MIT would still have primary responsibility for system design and analysis. (See June 20.)

16–March 21

Ibid.

MSC announced that, during a 14-day lunar mission, fuel cells in the Apollo CSM would produce about 16 liters (60 gallons) of potable water while furnishing power to operate the electronic equipment.

17

MSC News Release 64-32, February 17, 1964.

General Dynamics/Convair delivered to White Sands Missile Range (WSMR) the second Little Joe II launch vehicle, the first Little Joe II scheduled to fly with a production Apollo spacecraft. (See May 13.)

17

MSC, "Postlaunch Report for Apollo Mission A-001 (BP-12)," MSC-R-A-64-1, May 28, 1964, p. 2-1.

Motorola, Inc., submitted a proposal to NASA for the Apollo Unified S-Band Test Program, a series of tests on the unified S-band transponder and

17

1964
February

premodulation processor. Motorola had already begun test plans, analytical studies, and fabrication of special test equipment. (See December 23, 1963.)

> MSC, "ASPO Status Report for Period February 20–26, 1964"; "ASPO Status Report for Period Ending February 27–March 4, 1964."

19–20   MSC officials conducted acceptance testing of the 024 prototype space suit at the International Latex Corporation. [Reviewers identified several faults, but they were minor and the suit was accepted.]

> MSC, "ASPO Status Report for Period February 20–26, 1964."

20–26   Trajectory analyses by North American indicated that, with the tower flap configuration, it was highly probable that crew acceleration limits would be exceeded during high-altitude abort.

> MSC, "ASPO Status Report for Period Ending February 27–March 4, 1964."

20–26   North American submitted to ASPO a proposal for dynamic testing of the docking subsystem, which called for a full-scale air-supported test vehicle. The contractor estimated the program cost at $2.7 million for facilities, vehicle design, construction, and operation.

> MSC, "ASPO Status Report for Period February 20–26, 1964."

20–26   ASPO decided upon transfer through free space as the backup mode for the crew's getting from the LEM back to the CM if the two spacecraft could not be pressurized. North American had not designed the CM for extravehicular activity nor for passage through the docking tunnel in a pressurized suit. Thus there was no way for the LEM crew to transfer to the CM unless docking was successfully accomplished. ASPO considered crew transfer in a pressurized suit both through the docking tunnel and through space to be a double redundancy that could not be afforded.

> Ibid.

20–26   North American conducted three tests (4, 20, and 88 hours) on the CSM fuel cell. The third ended prematurely because of a sudden drop in output. (Specification life on the modules was 100 hours.)

During this same week, Pratt and Whitney Aircraft tested a LEM-type fuel cell for 400 hours without shutdown and reported no leaks.

> Ibid.

20–26   Grumman completed negotiations with Bell Aerosystems Company for the LEM's reaction control system propellant tanks.

> Ibid.

22   George E. Mueller, NASA Associate Administrator for Manned Space Flight, summarized recent studies of the dangers of meteoroids and radiation in the Apollo program. Data from the *Explorer XVI* satellite and ground observations indicated that meteoroids would not be a major hazard. Clouds of protons ejected by solar flares would present a risk to astronauts, but studies of the largest solar flares recorded since 1959 showed that maximum radiation dosages in the CM and the Apollo space suit would have been far below acceptable limits (set in July 1962 by the Space Science Board of the National Academy of Sciences). Cosmic rays would not be a hazard because of their rarity. Radiation in the Van Allen belts was not dangerous because the spacecraft would fly through the belts at high speeds.

1964
February

NASA News Release 64–43, "Radiation, Technical Problems Won't Bar Moon Landing in This Decade, Mueller Says," February 22, 1964.

RCA presented results of a weight and power tradeoff study on the LEM's radar systems, which were over Grumman's specification in varying amounts from 100 to 300 percent. RCA proposed that the accuracy requirements be relaxed to cope with this problem. MSC requested Grumman, on the basis of this report, to estimate a slippage in the schedule and the effects of additional weight and power. (See February 27–March 4.)

24

MSC, "ASPO Status Report for Period Ending February 27–March 4, 1964."

At a NASA–North American Technical Management Meeting at Downey, Calif., North American recommended that Apollo earth landings be primarily on water. On the basis of analytical studies and impact tests, the contractor had determined that "land impact problems are so severe that they require abandoning this mode as a primary landing mode." In these landings, North American had advised, it was highly probable that the spacecraft's impact limits would be surpassed. In fact, even in water landings "there may be impact damage which would result in leakage of the capsule." (See March 29–April 4.) ASPO Manager Joseph F. Shea, at this meeting, "stated that MSC concurs that land impact problems have not been solved, and that planning to utilize water impact is satisfactory." (See December 1962; February 1 and March 5, 1963.)

25

Three days later, Shea reported to the MSC Senior Staff that Apollo landings would be primarily on water. The only exceptions, he said, would be pad aborts and emergency landings. With this question of "wet" versus "dry" landing modes settled, Christopher C. Kraft, Jr., Assistant Director for Flight Operations, brought up the unpleasant problem of the CM's having two stable attitudes while afloat—and especially the apex-down one. This upside-down attitude, Kraft emphasized, submerged the vehicle's recovery antennas and posed a very real possibility of flooding in rough seas. Shea countered that these problems could be "put to bed" by using some type of inflatable device to upright the spacecraft. (See April 15 and August 16–September 15.)

"Minutes of NASA–NAA Technical Management Meeting, February 25, 1964," February 26, 1964, p. 3; MSC, "Minutes of Senior Staff Meeting, February 28, 1964," p. 4.

Apollo spacecraft simulator built by Honeywell for testing of its Apollo stabilization and control system (SCS) was described as the largest and most sensitive device of its kind. Cold-gas reaction jets maneuvered the huge circular platform at the company's Aeronautical Division, Minneapolis, as it simulated characteristics of the Apollo command module in flight to and from the moon. A single stainless steel bearing (center) resting on a paper-thin cushion of gas supported some eight metric tons (nine U.S. tons) of equipment, virtually isolating the platform from friction and vibration. Engineer Leonard Aske inspected one of the jets (upper right) while engineer Dick McKinley adjusted part of the bearing mechanism. Both SCS and simulator were developed for North American Aviation under NASA guidance.
—Minneapolis-Honeywell photo.

1964

February

25

Grumman and RCA signed a contract on the LEM communications subsystem. (See June 28, 1963.)

MSC, "ASPO Status Report for Period March 12–18, 1964"; MSC, "Project Apollo Quarterly Status Report No. 7 for Period Ending March 31, 1964," p. 3.

25

At a NASA–North American technical management meeting, the tower flap versus canard configuration for the launch escape vehicle was settled. ASPO Manager Joseph F. Shea decided that canards should be the approach for Block I vehicles, with continued study on eliminating this device on Block II vehicles. (See January 18 and November 12, 1963, and February 7, 1964.)

PART II: DEVELOPING HARDWARE DISTINCTIONS

"Minutes of NASA-NAA Technical Management Meeting, February 25, 1964"; "Apollo Monthly Progress Report," SID 62-300-23, p. 3.

1964

February

25

MSC conducted a Design Engineering Inspection of the LEM timing equipment at the Elgin National Watch Company.

MSC, "ASPO Status Report for Period February 20-26, 1964."

27

MSC Crew Systems Division (CSD) received an improved version of the Apollo space suit (the A-3H-024 Phase B). In the course of the following week, CSD engineers examined the suit for weight, leakage, donning, and mobility.

MSC, "ASPO Status Report for Period Ending February 27-March 4, 1964."

27

Boilerplate (BP) 19 was drop tested at El Centro, Calif., simulating flight conditions and recovery of BP-12. (See May 13.) A second BP-19 drop, on April 8, removed all constraints on the BP-12 configuration and earth landing system. Another aim, to obtain information on vehicle dynamics, was not accomplished because of the early firing of a backup drogue parachute.

"Apollo Quarterly Status Report No. 7," p. 5; "Apollo Monthly Progress Report," SID 62-300-23, p. 19; NAA, "Apollo Monthly Progress Report," SID 62-300-24, May 1, 1964, p. 28; MSC, "ASPO Management Report for Period April 9-16, 1964."

27-29

MSC and AC Spark Plug negotiated amendments to AC's contract for a research and development program for inertial reference integrating gyroscopes. The amendments covered cost overruns, an additional 30 pieces of hardware, and conversion of the contract to an incentive-fee type (target price, $3.465 million; ceiling price, $3.65 million).

MSC, "Consolidated Activity Report for the Office of the Associate Administrator, Manned Space Flight, February 16-March 21, 1964," p. 45; MSC, "ASPO Status Report for Period Ending February 27-March 4, 1964."

27-March 4

Representatives from MSC Crew Systems Division (CSD) visited Hamilton Standard to discuss space suit development. The prototype suit (024) was demonstrated and its features compared with the Gemini suit. Deficiencies in the Apollo helmet were noted and suggestions were made on how to improve the design. [At this time, CSD began looking into the possibility of using Gemini suits during Apollo earth orbital flights, and during the next several weeks began testing Gemini suits in Apollo environments. (See April 28-30.)]

MSC, "ASPO Status Report for Period Ending February 27-March 4, 1964;" MSC, "ASPO Management Report for Period April 2-9, 1964."

27-March 4

A joint Grumman, RCA, Ryan Aeronautical Company, ASPO, and Flight Crew Support Division (FCSD) meeting was held at Bethpage to review capability of the LEM landing radar to meet FCSD's requirements for ascent and for orbit circularization. A preliminary (unfunded) Ryan study (re-

141

To learn more about the meteoroid penetration hazard to spacecraft, scientists of Langley Research Center launched satellites to gather first-hand meteoroid-penetration data. In the photo a micrometeoroid satellite was prepared for an environmental test in Langley's 8- x 15-foot Thermal Vacuum Facility, in which the widely varying temperatures to be experienced in orbit were simulated. Installed around the fourth stage of a Scout launch vehicle, the satellite contained highly sensitive detectors to record penetrations by high-velocity space particles and to study the effects of the space environment on spacecraft systems and components, such as solar cells and thermal coatings.

1964

February

quested by ASPO earlier in the month) indicated some doubt that those accuracy requirements could be met. RCA advised that it would be possible to make these measurements with the rendezvous radar, if necessary. A large weight penalty, about 38 to 56 kilograms (84 to 124 pounds), would be incurred if the landing radar were moved from the descent to the ascent stage to become part of the abort guidance system. Adding this weight to the ascent stage would have to be justified either by improved abort performance or added crew safety. MSC authorized RCA and Ryan to study this problem at greater length. In the meantime, ASPO and FCSD would analyze weights, radar accuracies, and abort guidance performance capability. (See March 16 and May 22.)

MSC, "ASPO Status Report for Period Ending February 27–March 4, 1964"; "ASPO Status Report for Period March 19–26, 1964."

PART II: DEVELOPING HARDWARE DISTINCTIONS

The MSC Primary Propulsion Branch (PPB) completed a study on the current LEM ascent engine and performance that might be gained if the chamber pressure and characteristic exhaust velocity efficiency were increased. PPB also evaluated the use of hard versus soft chamber throats. A study by Bell Aerosystems Company had predicted a slightly lower performance than the MSC investigation (which estimated a drop of about six points below specification values if the current design were retained). PPB thought that specifications might be reached by increasing the chamber pressure to 82.7 newtons per square centimeter (120 psia) and the exhaust velocity efficiency to 97.3 percent, and by using a hard, rather than a soft, throat.

MSC, "ASPO Status Report for Period Ending February 27–March 4, 1964."

1964
February 27–March 4

At North American, a mockup of the crew transfer tunnel was reviewed informally. The mockup was configured to the North American-proposed Block II design (in which the tunnel was larger in diameter and shorter in length than on the existing spacecraft). MSC asked the contractor to place

March 2–9

An Apollo service module mockup showing the portion that contained the main rocket engine and propellant supply to be used for maneuvers to and from the moon. Produced by Aerojet-General Corporation under contract to NAA, the engine could provide more than 89 000 newtons (20 000 pounds) of thrust to keep Apollo on course and to perform other missions. Standing by the multiple-start engine's flaring skirt were NAA and Aerojet rocket engineers.

1964
March

an adapter in the tunnel to represent the physical constraints of the current design, which would permit the present design to be thoroughly investigated and to provide a comparison with the Block II proposal.

<small>MSC, "ASPO Status Report for Period Ending March 5–11, 1964."</small>

9

MSC received an additional $1.035 million in Fiscal Year 1964 funds to cover development of equipment and operational techniques for scientific exploration of the moon:

• Power supplies for long-life equipment to be installed on the lunar surface during Apollo missions

• Telemetry and Deep Space Instrumentation Facility requirements for this equipment

• Tools and materials needed for examining, packaging, and transporting lunar samples

• Cameras and film suitable for use on the moon by a space-suited astronaut

• Methods of obtaining and returning lunar samples without contaminating or changing them

• Techniques and instrumentation for geological mapping in the lunar environment

• Processes for obtaining water, hydrogen, and oxygen from indigenous material on the moon.

Additionally, MSC would evaluate current techniques in seismology used to determine subsurface structural conditions.

<small>Memorandum, Homer E. Newell, NASA, to Dir., MSC, through Assoc. Adm. for Manned Space Flight, "Funding for Development of Scientific Instruments for Apollo Lunar Missions," March 9, 1964.</small>

10

Grumman completed negotiations with Yardney Electric Corporation for an auxiliary battery for the LEM. A contract would be awarded when size requirements were determined by Grumman and MSC.

<small>MSC, "ASPO Status Report for Period Ending March 5–11, 1964."</small>

10

Grumman and North American began working out ways for common usage of ground support equipment (GSE). Through informal meetings and telephone discussions, the two prime contractors agreed to a formal procedure for the GSE's use, maintenance, and training procedures.

<small>"Monthly Progress Report No. 14," LPR-10-30, p. 32.</small>

12

Goddard Space Flight Center awarded a $1.963 million contract to the Commonwealth of Australia's Department of Supply to construct and install a data acquisition facility, including an antenna 26 meters (85 feet) in diameter, at Canberra, Australia. The station would become part of the NASA Space Tracking and Data Acquisition Network to track unmanned

PART II: DEVELOPING HARDWARE DISTINCTIONS

satellites and part of the Deep Space Network to track lunar and planetary probes. Unified S-band equipment was later installed to support the Manned Space Flight Network during Apollo lunar missions.

> *The New York Times,* March 12, 1964; NASA, *Twelfth Semiannual Report to Congress, July 1–December 31, 1964* (1965), pp. 129–130, 134; NASA, *Thirteenth Semiannual Report to Congress, January 1–June 30, 1965* (1966), p. 137; NASA, *Fourteenth Semiannual Report to Congress, July 1–December 31, 1965* (1966), p. 146.

1964
March

North American was directed by NASA to study feasibility of using the LEM propulsion system as backup to the SM propulsion system. The most important item in the contractor's analysis was strength of the docking structure and its ability to withstand LEM main-engine and reaction control system thrusting.

> Letter, H. P. Yschek, MSC, to NAA, Space and Information Systems Div., "Contract Change Authorization No. 161," March 12, 1964.

12

NASA completed formal negotiations with Aerojet-General Corporation for 12 Algol 1–D solid rocket motors, to be used in the Little Joe II vehicles. The contract was a fixed-price-plus-incentive-fee type with a target price of about $1.4 million. A maximum price of 20 percent more than the target cost was allowed.

> MSC, "Consolidated Activity Report for the Office of the Associate Administrator, Manned Space Flight, February 16–March 21, 1964," p. 46.

12

Grumman completed negotiations with Kearfott Products Division, General Precision, Inc., for the LEM rate gyro assembly, and a contract was awarded later in the month.

> MSC, "ASPO Status Report for Period March 12–18, 1964;" "Apollo Quarterly Status Report No. 7," p. 23.

12–18

Primarily as a weight-saving measure, the gas storage pressure in the LEM's descent stage helium tank was reduced from 3103 to 2413 newtons per square centimeter (4500 to 3500 psia). This allowed the thickness of the tank wall to be reduced.

> MSC, "ASPO Status Report for Period March 12–18, 1964;" MSC, "Consolidated Activity Report for the Office of the Associate Administrator, Manned Space Flight, February 16–March 21, 1964," p. 24.

12–18

ASPO notified Grumman that certain items were no longer to be considered in the weight saving program: guidance and navigation components, drinking water tankage, scientific equipment, pyrotechnic batteries, among others.

> Letter, W. F. Rector III, MSC, to GAEC, Attn: R. S. Mullaney, "Contract NAS 9-1100, weight reduction items," March 13, 1964.

13

Ryan Aeronautical Company signed a contract with RCA for the LEM lunar landing radar. Ryan was instructed to design for altitudes of 21 300

16

145

1964
March

meters (70 000 feet) and accuracies of 0.5 percent. (See February 27–March 4, and May 22.)

MSC, "ASPO Status Report for Period March 19–26, 1964."

16–April 15

AiResearch Manufacturing Company completed testing on development components of the CM environmental control system. Specifications for components had been submitted to North American.

The Garrett Corporation, AiResearch Manufacturing Division, "Monthly Progress Report, Environmental Control System, NAA/S&ID, Project Apollo, 16 March 1964–15 April 1964," SS-1013-R(23), April 30, 1964, p. 7.

16–April 15

North American held a design review of the CM heatshield substructure. Use of titanium in place of stainless steel was being evaluated as part of a weight reduction study for the Block II spacecraft. Added reliability and a weight saving of several hundred pounds might be achieved thereby. Three factors would be considered: the brittleness of stainless steel at extremely cold temperatures, the higher cost of titanium, and the verification of diffusion bonding of titanium honeycomb.

"Apollo Monthly Progress Report," SID 62-300-24, p. 14.

Intense heat like that experienced by spacecraft during atmosphere entry was generated in the laboratory by scientists of Langley Research Center. Tests in the Structures Laboratory 2500-kilowatt, subsonic electric-arc heater evaluated materials for heat shielding of reentry vehicles. In the 5800 K (10 000°F) air stream produced by the arc heater, an ablation material shaped like an Apollo spacecraft was placed for the experiment. The stream of gas impinging on the model issued from the nozzle at more than eight tenths kilometer per second (one half mile per second).

PART II: DEVELOPING HARDWARE DISTINCTIONS

The first prototype of the CM battery for use during reentry was delivered to North American by Eagle-Picher Industries, Inc.

1964

March

16–April 15

"Apollo Quarterly Status Report No. 7," p. 7; "Apollo Monthly Progress Report," SID 62-300-24, p. 14.

Texas Instruments, Inc., presented a progress report on their lunar surface experiments study to the MSC Lunar Surface Experiments Panel. (See September 30, 1963.) Thus far, the company had been surveying and rating measurements to be made on the lunar surface. Areas covered included soil mechanics, mapping, geophysics, magnetism, electricity, and radiation. Equipment for gathering information, such as hand tools, sample return containers, dosimeters, particle spectrometers, data recording systems, seismometers, gravity meters, cameras, pentrometers, and mass spectrometers had been considered. The next phase of the study involved integrating and defining the measurements and instruments according to implementation problems, mission needs, lunar environment limitations, and relative importance to a particular mission. Texas Instruments would recommend a sequence for performing the experiments.

17

Memorandum, H. R. Largent, MSC, to Instrumentation and Electronics Systems Div. Files, "Lunar surface experiments study (NAS 9-2115)," March 17, 1964.

NASA instructed North American to fix the CM crew couches along all axes during normal and emergency acceleration, except at impact. During nonacceleration mission phases, the couches would be adjustable for crew comfort.

19

Letter, H. P. Yschek, MSC, to NAA, Space and Information Systems Div., "Contract Change Authorization No. 167," March 19, 1964.

Grumman reported to MSC the current load status and projected load growth for the LEM's electrical power system, requesting a mission profile of 121 kilowatt-hours total energy. (See January 28 and August 15, 1963.) The company also presented its latest recommendation for the LEM power generation subsystem configuration: two 900-watt fuel cells, a descent stage peaking battery, an ascent stage survival battery, and four cryogenic storage tanks. To compensate for voltage drops in the power distribution subsystem, Grumman recommended that two cells be added to the current fuel cell stack; however, on March 23 ASPO directed the contractor to continue development of the 900-watt, three-fuel-cell assembly and a five-tank cryogenic storage system. MSC's position derived from the belief that the load growth would make the two-cell arrangement inadequate. Also the three-cell configuration, through greater redundancy, afforded greater safety and chances of mission success: the mission could continue in spite of a failure in one of the cells; should two cells fail, the mission could be aborted on

19–20

147

1964

March

the final power source. The cryogenic tanks should be sized for a usable total energy of 121 kilowatt-hours to permit immediate tank procurement.

> MSC, "ASPO Status Report for Period March 19–26, 1964"; letter, W. F. Rector III, MSC, to GAEC, Attn: R. S. Mullaney, "Contract NAS 9-1100, Electrical Power Generation Section (PGS) Configuration," March 23, 1964; "Apollo Quarterly Status Report No. 7," p. 26; interview, telephone, William E. Rice, MSC, March 2, 1970.

19–26

After the decision to use canards instead of tower flaps (see February 25), North American returned to the concept of a hard boost protective cover. The tower jettison motor would remove the cover along with the tower. (See July 24.)

> MSC, "ASPO Status Report for Period March 19–26, 1964."

19–26

MSC Crew Systems Division (CSD) evaluated a CM couch width of 58.4 centimeters (23 inches). CSD found that the couch hampered an astronaut's movement in an unpressurized suit and totally restricted him if his suit was pressurized.

> *Ibid.*

20

NASA's Office of Space Science and Applications began organizing several groups of scientists to assist the agency in defining more specifically the scientific objectives of Project Apollo. (See October 8 and December 15, 1963.) In a number of letters to prominent American scientists, Associate Administrator for Space Science and Applications Homer E. Newell asked them to propose suitable experiments in such fields as geology, geophysics, geochemistry, biology, and atmospheric science. This broadly based set of proposals, Newell explained, is "for the purpose of assuring that the final Apollo science program is well balanced, as complete as possible, and that all potential investigators have been given an opportunity to propose experiments." The proposals would then be reviewed by subcommittees of NASA's Space Sciences Steering Committee.

> Letter, Homer E. Newell, NASA, to Dr. S. P. Clark, Yale University, March 20, 1964. Twenty-eight nearly identical letters were sent to other members of the scientific and academic community.

20

Tests at North American demonstrated the possibility of using onboard tools to break the CM hatch windows for postlanding ventilation of the spacecraft.

> "Apollo Monthly Progress Report," SID 62-300-24, p. 8.

23

Members of the Gemini Flights Experiments Review Panel discussed procedures for incorporating Apollo-type experiments into the Gemini program, experiments that directly supported the three-man space program. These experiments encompassed crew observations, photography, and photometry.

> MSC, "ASPO Status Report for Period March 19–26, 1964."

PART II: DEVELOPING HARDWARE DISTINCTIONS

Mission Control Center (Building 30) at MSC was physically completed, if not yet operationally ready, March 21, 1964.

OMSF outlined launch vehicle development, spacecraft development, and crew performance demonstration missions, using the Saturn IB and Saturn V:

(1) Launch vehicle and unmanned CSM (at least two flights planned)
(2) CSM long-duration
(3) CSM and LEM (two flights planned)
(4) Launch vehicle and heatshield (at least two flights)
(5) Lunar mission simulation
(6) Lunar exploration.

Missions (1) through (3) would use the Saturn IB and (4) through (6) the Saturn V. Additional launch vehicles and spacecraft would be provided for contingency or repeated flights. If necessary, repeat flights could provide additional crew training.

NASA OMSF, "Apollo Flight Mission Assignments," Program Directive M-DE 8000.005B, March 23, 1964.

1964
March
23

To verify a narrower hatch configuration proposed for Block II spacecraft, North American evaluated the capability of an astronaut wearing a pressurized space suit and a portable life support system to pass through the main hatch of the CM for extravehicular activities. Subjects were able to enter and leave the mockup without undue difficulty despite the presence of gravity.

"Apollo Monthly Progress Report," SID 62-300-24, pp. 6-7.

24

The first formal inspection and review of the LEM test mockup TM-1 was held at Grumman. TM-1 allowed early assessment of crew mobility, ingress,

24-26

1964
March

and egress. It was a full-size representation of crew stations, support and restraint systems, cabin equipment arrangement, lighting, display panels and instrument locations, and hatches. The TM-1 evaluation became the basis for the final LEM mockup, TM-5, from which actual hardware fabrication would be made.

The TM-1 Review Board (comprising Chairman Owen E. Maynard, Maxime A. Faget, Donald K. Slayton, and William F. Rector III, all of

The LEM TM-1 mockup was displayed at the Grumman plant at Bethpage, N.Y., during its first inspection. Engineer Bill Peterson was photographed climbing the ladder to the entry platform.

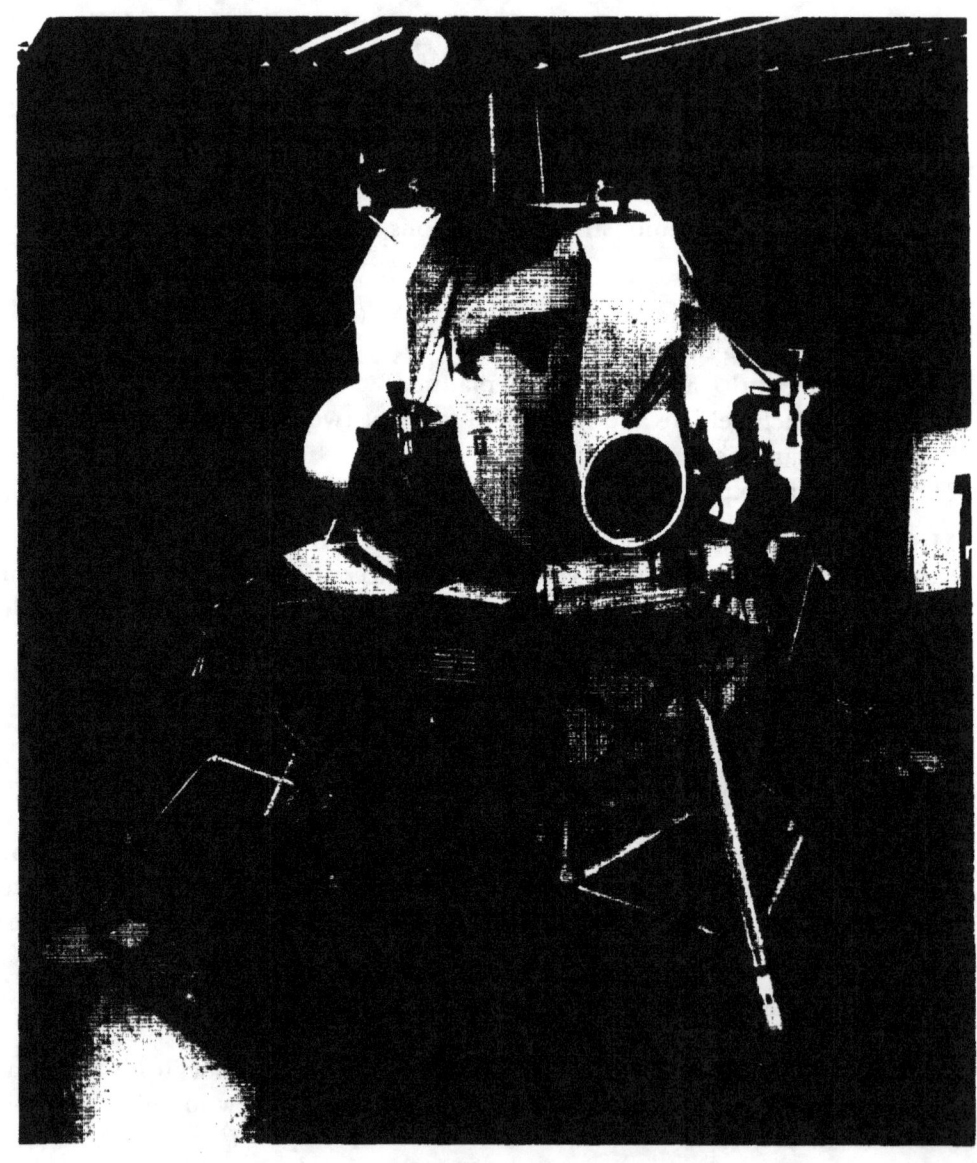

PART II: DEVELOPING HARDWARE DISTINCTIONS

MSC; and Tom J. Kelly and Robert M. Carbee of Grumman) approved 28 requests for change; 15 others were marked for further investigation.

1964
March

NASA, "Lunar Excursion Module, Project Apollo, Board Report for NASA Inspection and Review of TM-1 Mockup, March 19-26, 1964," pp. 1, 3, 4.

The Boeing Company received NASA's go-ahead to develop the Lunar Orbiter spacecraft. (See December 20, 1963.) Two significant changes were made in the original Statement of Work: (1) for the selenodetic part of the mission, the spacecraft lifetime was extended from 60 days to one year; and (2) to expand the area of photographic coverage, the film capacity was increased.

25

Lee R. Scherer, NASA, "Lunar Orbiter Program Status Report," March 26, 1964.

The General Electric (GE) Company submitted its cost quotations to NASA, starting the final phase of a program to provide Acceptance Checkout Equipment (ACE—formerly PACE [see February 1963]) ground stations for Apollo spacecraft. The overall "ACE" plan slated three ground stations for North American, two for Grumman, four for Cape Kennedy, and one for MSC. GE's contract called for spacecraft systems integration and checkout and for maintenance of the ACE stations. Much of the ACE equipment was government furnished and had been procured by NASA from several sources: Control Data Corporation—computer; Radiation, Inc.—"decommutators and pulse code modulation simulators." By May, GE had set up and commenced operating an experimental ACE station at Cape Kennedy. (See August 23–September 19.)

25

MSC, "ASPO Status Report for Period March 26–April 2, 1964;" "Apollo Quarterly Status Report No. 7," p. 61; "Apollo Quarterly Status Report No. 8," pp. 59–60; MSC, "Consolidated Activity Report for the Office of the Associate Administrator, Manned Space Flight, February 16–March 21, 1964," pp. 9, 78; "Consolidated Activity Report for the Office of the Associate Administrator, Manned Space Flight, April 19–May 16, 1964," p. 46; MSC, "Weekly Activity Report for the Office of the Associate Administrator, Manned Space Flight, May 17–23, 1964," p. 3; NASA News Release 63–286, "NASA to Extend Contract with Control Data Corporation," December 26, 1963; MSC News Release 64–108, June 8, 1964.

Because of the pure oxygen atmosphere specified for the spacecraft, North American reviewed its requirements for component testing. Recent evaluation of the CM circuit breakers had indicated a high probability that they would cause a fire. The company's reliability office recommended more flammability testing, not only on circuit breakers but on the control and display components as well. The reliability people recommended also that procurement specifications be amended to include such testing.

26–April 1

MSC, "ASPO Management Report for Period April 2–9, 1964."

Impact tests indicated that, because of oscillations and consequent high angles of attack, the CM might not withstand water impact and could sink.

29–April 4

Three generations of U.S. spacecraft and the ever-growing rockets that would propel astronauts into the unknown: Flying solo, orbital Mercury spacemen rode a craft 3 meters long and 1.8 meters in diameter (10 feet by 6 feet in diameter), hurled aloft by an Atlas launch vehicle. Gemini's two-man teams were to circle the earth in a capsule with 50 percent more cabin space than Mercury's. A 27-meter (90-foot) Titan II would boost them into an orbit 298 kilometers (185 miles) high. Apollo astronauts would fly to the moon in a command module twice the size of the Gemini capsule. Two of the three men aboard would descend to the moon's surface in the lunar excursion module (LEM), an 11-metric-ton (12½-U.S.-ton) craft. A third section of the Apollo vehicle, the service module, contained the vital flight equipment. Powerful, 110-meter (362-foot) Saturn V would launch the Apollo team into orbit. (Artist's concept) —National Geographic Society photo.

1964 **March**	North American planned a series of water impact tests using boilerplate 28 to study the problem. MSC, "Weekly Activity Report for the Office of the Associate Administrator, Manned Space Flight, March 29–April 4, 1964," p. 5; MSC, "ASPO Status Report for Period March 26–April 2, 1964."
30	MSFC awarded Rocketdyne a definitive contract (valued at $158.4 million) for the production of 76 F–1 engines for the first stage of the Saturn V launch vehicle and for delivery of ground support equipment.

PART II: DEVELOPING HARDWARE DISTINCTIONS

David S. Akens, Leo L. Jones, and A. Ruth Jarrell, *History of the George C. Marshall Space Flight Center from January 1 through June 30, 1964* (MHM-9, May 1965), Vol. I, p. 130.

1964

March

CSM boilerplate 12 (with launch escape system) was mated to its Little Joe II launch vehicle. (See May 13.)

30

MSC, "Postlaunch Report for Apollo Mission A-001 (BP-12)," MSC-R-A-64-1 (May 28, 1964), p. 5-2.

MSC negotiated a cost-plus-incentive-fee contract, valued at $1.65 million, with Hamilton Standard for 27 prototype Apollo space suits and 12 pairs of gloves.

April

1

MSC, "Consolidated Activity Report for the Office of the Associate Administrator, Manned Space Flight, March 22-April 18, 1964," p. 56.

Space Technology Laboratories (STL) began using its new San Juan Capistrano, Calif., test facility to static fire the firm's LEM descent engine. Hereafter, the bulk of STL's development firings were made at this site.

2-9

MSC, "ASPO Management Report for Period April 2-9, 1964"; MSC, "Weekly Activity Report for the Office of the Associate Administrator, Manned Space Flight, June 7-13, 1964," p. 2.

The MSC Operations Planning Division (OPD) reviewed recent revisions by OMSF to Apollo's communications requirements:

2-9

The first Apollo boilerplate to fly during the program was BP-13, shown here in Hangar AF at Cape Kennedy before being taken to the launch complex to be mated with the Saturn SA-6 launch vehicle. The Apollo escape rocket and tower are in the foreground.

1964
April

- Elimination of the requirement for continuous tracking of the spacecraft during translunar injection
- Sequential rather than simultaneous transmission of data from the ground to the two spacecraft (to be compatible with the Manned Space Flight Network)
- A five-kilometer (three-nautical-mile) communications range on the lunar surface (to be compatible with the design of the portable life support system)
- Elimination of the requirement for direct transmission to the CSM from an extravehicular astronaut; instead, such transmission would be relayed via the LEM.

Thus were resolved, OPD reported, a number of conflicting items (i.e., incompatibilities between OMSF's requirements and the capabilities of the two spacecraft). Two other items that OMSF made into firm requirements were already compatible with the design of the spacecraft:

(1) A radar in the CSM capable of tracking the LEM (provided the LEM had a compatible transponder)

(2) Three-way communications between an astronaut on the moon, his fellow crewman inside the LEM, and with mission control.

MSC, "ASPO Management Report for Period April 2–9, 1964."

6–13

Grumman issued a letter contract to AiResearch Manufacturing Company to start design of cryogenic tank assemblies for the LEM fuel cells. AiResearch received the formal contract on June 23.

MSC, "ASPO Management Report for Period April 9–16, 1964"; "ASPO Weekly Management Report, June 18–25, 1964"; "ASPO Weekly Management Report, July 23–30, 1964."

7

Bell Aerosystems Company completed the first of two lunar landing research vehicles, to be delivered to the NASA Flight Research Center for testing. (See January 18, 1963.)

MSC News Release 64–68, April 7, 1964.

7–8

At the April 7–8 NASA–North American Technical Management Meeting (the first of these meetings to be held at MSC's new home, "NASA Clear Lake Site 1"), ASPO Manager Joseph F. Shea summarized his office's recent activities concerning the Block II spacecraft. He spelled out those areas that ASPO was investigating—which included virtually the whole vehicle between escape tower and service engine bell. Shea outlined procedures for "customer and contractor" to work out the definitive Block II design, aiming at a target date of mid-May 1965. These procedures included NASA's giving North American descriptions of its Block II work, estimates of weight reduction, and a set of ground rules for the Block II design (see April 16). And to ensure that both sides cooperated as closely as possible in this work, Shea named Owen E. Maynard, Chief of MSC's Systems Engineering

PART II: DEVELOPING HARDWARE DISTINCTIONS

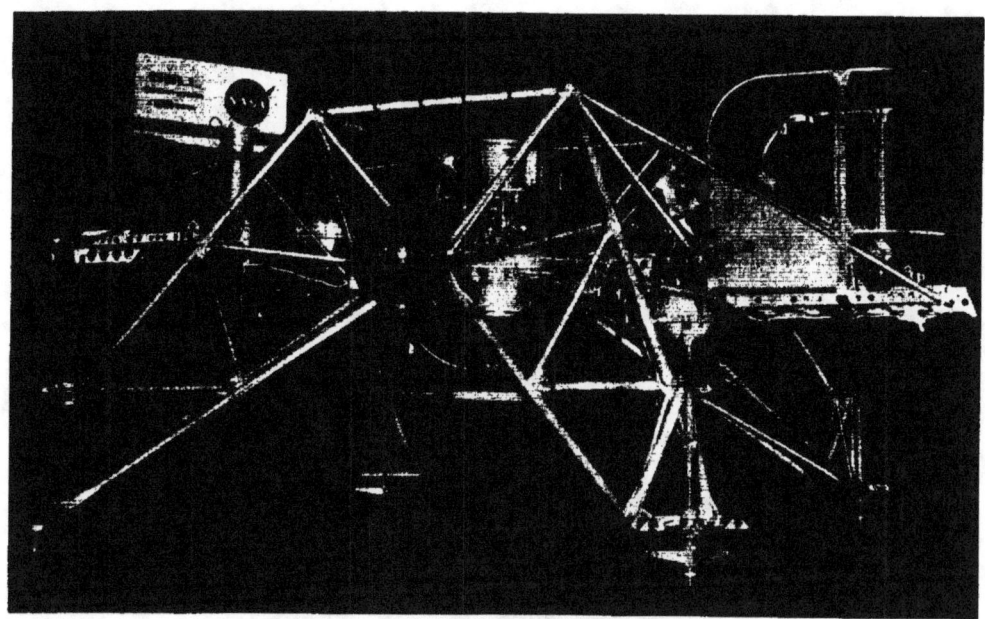

The lunar landing research vehicle.
—Bell Aerosystems photo.

Division, and his counterpart at Downey, Norman J. Ryker, Jr., to "honcho" the effort.

"Minutes of NASA–NAA Technical Management Meeting, April 7–8, 1964," pp. 3–5.

The first Gemini mission, *Gemini–Titan I*, was launched from Complex 19 at Cape Kennedy at 11:00 a.m., e.s.t. This was an unmanned flight, using the first production Gemini spacecraft and a modified Titan II Gemini launch vehicle (GLV). The mission's primary purpose was to verify the structural integrity of the GLV and spacecraft, as well as to demonstrate the GLV's ability to place the spacecraft into a prescribed earth orbit. Mission plans did not include separation of the spacecraft from the second stage of the vehicle, and both were inserted into orbit as a unit six minutes after launch. The planned mission encompassed only the first three orbits and ended about four hours and 50 minutes after liftoff. No recovery was planned. The flight qualified the GLV and the structure of the spacecraft.

James M. Grimwood and Barton C. Hacker, with Peter J. Vorzimmer, *Project Gemini Technology and Operations: A Chronology* (NASA SP–4002, 1969), p. 139.

ASPO gave Grumman specific instructions on insulating wiring in the LEM: Teflon-insulated wiring was mandatory in a pure oxygen atmosphere. If the standard-thickness Teflon insulation was too heavy, a thin-wall Teflon-insulated wiring with abrasion-resistant coating should be considered. Teflon-insulated wiring should also be used outside the pressurized cabin,

1964
April

8

13

The first *Gemini-Titan I* space vehicle lifted off Pad 19 at Cape Kennedy April 8, 1964.

1964

April

wherever that wiring was exposed. Any approved spacecraft insulation could be used within subsystem modules which were hermetically sealed in an inert gas atmosphere or potted within the case.

Letter, W. F. Rector III, MSC, to GAEC, Attn: R. S. Mullaney, "Contract NAS 9-1100, Spacecraft Electrical Wiring Insulation," April 13, 1964.

PART II: DEVELOPING HARDWARE DISTINCTIONS

The 56-metric-ton, 11.6-meter-diameter (62-U.S.-ton, 38-foot-diameter) stainless steel door for Chamber A of MSC's Space Environmental Simulation Laboratory was swung into place early in April 1964 to finish enclosing the large test facility. Chamber A was tubular, 35.7 meters high, 19.8 meters in diameter (117 feet high, 65 feet in diameter), and extended 7.3 meters (24 feet) into the ground. It was designed to simulate lunar trips with realism and detail close to those of actual flight.

Firings at the Arnold Engineering Development Center (AEDC) and at Aerojet-General Corporation's Sacramento test site completed Phase I development tests of the SM propulsion engine. The last simulated altitude test at AEDC was a sustained burn of 635 seconds, which demonstrated the engine's capability for long-duration firing. Preliminary data indicated that performance was about three percent below specification, but analysis was in progress to see if it could be improved.

1964
April
14

NAA, "Apollo Monthly Progress Report," SID 62-300-25, June 1, 1964, p. 11; MSC, "ASPO Management Report for Period April 23-30, 1964"; "ASPO Management Report for Period April 30-May 7, 1964."

An Atlas D launch vehicle lifted a Project Fire spacecraft (see November 27, 1962) from Cape Kennedy in the first test of the heat that would be encountered by a spacecraft reentering the atmosphere at lunar-return velocity. During the spacecraft's fall toward earth, a solid-fuel Antares II

14

THE APOLLO SPACECRAFT: A CHRONOLOGY

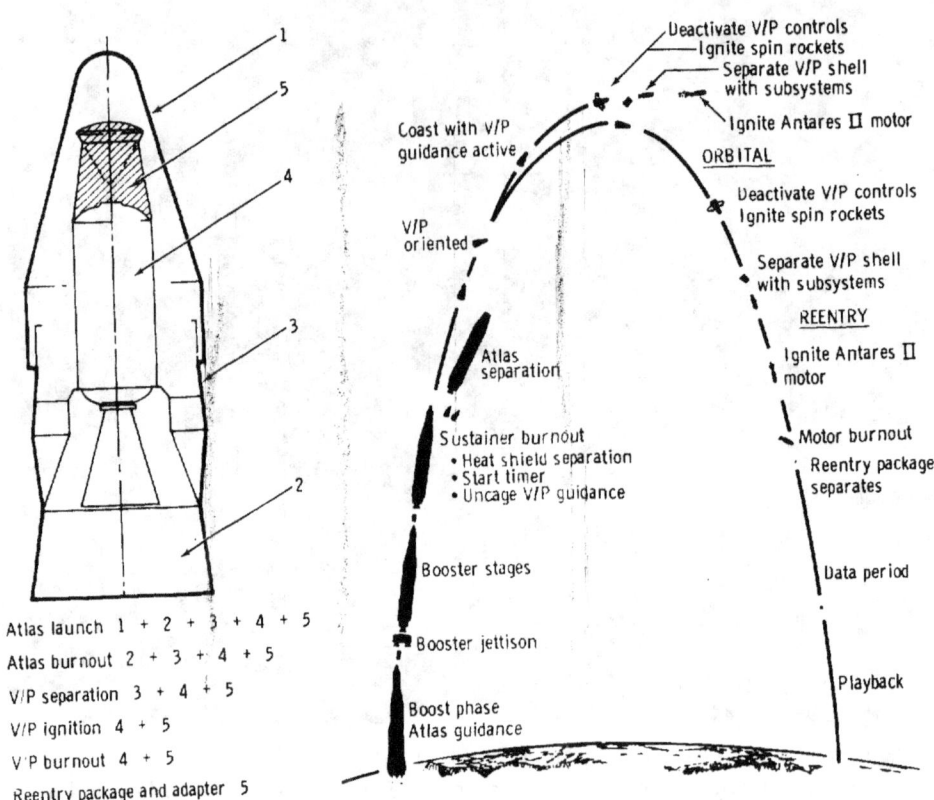

A typical Project Fire reentry or orbital mission. The weight sequence is at left.
—LTV report.

1964

April

rocket behind the payload fired for 30 seconds, increasing the descent speed to 40 501 kilometers (25 166 miles) per hour. Instruments in the spacecraft radioed temperature data to the ground. The spacecraft exterior reached an estimated temperature of 11 400 K (20 000 degrees F). About 32 minutes after launch, the spacecraft impacted into the Atlantic Ocean. The mission, sponsored by Langley Research Center, provided reentry heating measurements needed to evaluate heatshield materials and information on the communications blackout during reentry.

NASA News Release 64-69, "NASA Schedules Project Fire Launch," April 1, 1964; *Astronautics and Aeronautics, 1964*, p. 135.

15

Dale D. Myers, North American's Space and Information Systems Division vice president, succeeded John W. Paup as the contractor's program manager for the CM.

Oakley, *Historical Summary, S&ID Apollo Program*, p. 10.

PART II: DEVELOPING HARDWARE DISTINCTIONS

ASPO gave Grumman a go-ahead on procurement of the flight attitude indicator ("8-ball") and associated equipment for the LEM.

1964
April
15

Letter, W. F. Rector III, MSC, to GAEC, Attn: R. S. Mullaney, "Contract NAS 9-1100, Lunar Excursion Module, Attitude Indicator," April 15, 1964.

ASPO asked North American to investigate the possibility of designing apex-upright, stable flotation attitude into Block I and Block II CM's.

15

MSC, "ASPO Management Report for Period April 9-16, 1964."

Grumman completed an environmental control system water management configuration study and concluded that a revised design would significantly improve the probability of mission success and crew safety. This design would combine water tanks for the water management functions into one easily accessible package.

15

MSC, "ASPO Weekly Management Report, May 21-28, 1964."

MSC Crew Systems Division representatives attended a demonstration at Grumman of Apollo Phase B and Gemini space suits using the LEM TM-1 mockup and a mockup portable life support system. Tests demonstrated ingress/egress capability through the forward and top hatches, operation of controls and displays, and methods of getting out on the lunar surface and returning to the spacecraft. Generally, the Apollo suit proved sufficiently mobile for all these tasks, though there was no great difference between its performance and that of the Gemini suit during these trials.

15-16

MSC, "ASPO Management Report for the Period April 16-23, 1964"; GAEC, "Monthly Progress Report No. 15," LPR-10-31, May 10, 1964, p. 9.

NASA's Office of Space Science and Applications (OSSA) and the National Academy of Sciences (NAS) were planning a scientist-astronaut program. NAS people had met in Houston with MSC officials in February to help draft a formal plan to develop a "scientist astronaut program for NASA." This plan also placed the responsibility on NAS to define what scientific qualifications a person would need; MSC agreed to define "other qualifications."

16

OSSA Associate Administrator Homer E. Newell asked Harry H. Hess, Chairman of the Space Science Board, NAS, and his group to pursue this plan and be ready with a qualification list (both NAS and NASA requirements) by August for advertisement. Newell said the screening-for-selection process could be scheduled for February 1965. (See August 19.)

Letter, Newell, NASA, to Harry H. Hess, Chairman, Space Science Board, National Academy of Science, April 16, 1964.

Joseph F. Shea, ASPO Manager, in a letter to North American's Apollo Program Manager, summarized MSC's review of the weight status of the

16

159

1964

April

Block I and the design changes projected for Block II CSM's. (See April 7-8.)

The Block II design arose from the need to add docking and crew transfer capability to the CM. Reduction of the CM control weight (from 9500 to 9100 kilograms [21 000 to 20 000 pounds]) and deficiencies in several major subsystems added to the scope of the redesign.

Redesign of the CM would cause a number of changes above the deck, although ASPO believed that the 73.7-centimeter (29-inch)-diameter tunnel could be retained and tunnel access might be improved if the restrictions for seating the hatches were removed. Other changes not related to the docking and transfer requirement would be considered as long as they did not affect the structure below the deck.

Changes below the deck would be kept to a minimum on both the inner and the outer structure. Anything which might invalidate the applicability of the Block I lunar reentry tests to the Block II design would not be changed.

ASPO wanted to evaluate a preliminary design of the CM in which the only access to the LEM would be by extravehicular transfer. Although this approach was not currently considered operationally acceptable, any gains from such a design should be studied.

ASPO agreed that the CM thermal protection would be enhanced by addition of a boost protective cover for both Block I and Block II. A "soft" cover should be simple to design and operate, and a boost cover would permit coating the CM with a thermally efficient surface. This, with the help of attitude programming, should permit North American to reduce the initial ablator bond line temperature from 394 K (250 degrees F) to below 338 K (150 degrees F). ASPO also asked the contractor to consider raising the bond line temperature on the blunt face from 590 K (600 degrees F) to 700 K (800 degrees F). These changes would reduce ablator weight significantly.

To eliminate the humidity problem in the Block I subsystems, ASPO believed that electronic repackaging would be required. Such a redesign should take advantage of ASPO's decision to eliminate onboard maintenance as an acceptable means of achieving mission reliability. A more efficient mounting arrangement should be considered in conjunction with electronic system repackaging. Elimination of onboard maintenance would change requirements on the inflight test system; perhaps that system could be eliminated from the spacecraft.

The biggest uncertainty in weight requirements was meteoroid protection. The design approach to this problem should be incorporated with a redesign of the SM to reduce both the tank size and structure (but see August 6 statement of Robert O. Piland) consistent with a 16 800-kilogram (39 000-

pound) consumable fuel load, rather than the current 20 400-kilogram (45 000-pound) capacity. The SM design concept should remain the same, but North American should use this opportunity to clean up several structural details.

The SM thermal control system should be passive. Spacecraft orientation, either on a semicontinuous or discrete attitude program, would be permissable to maintain necessary temperature limits. To reach acceptable thermal time constants, the reaction control system (RCS) might have to be modified. It might also be desirable to change the RCS fuel to monomethylhydrazine.

Because of the large amount of spacecraft wiring, North American was asked to study using smaller sizes and reduced insulation thicknesses.

Another consideration was reducing the lunar mission time from 14 days to the reference mission length of about 10 days. But the current tank sizes should be maintained and the spacecraft should be capable of 14-day earth orbital missions with three men. The velocity reserve in the RCS might be decreased if the attitude requirements for guidance and navigation were eased. Here, also, the current tank sizes should not be changed.

Other major changes (such as redesign of the fuel cell, incorporation of new heatshield material, cryogenic helium pressures, and adapter staging) could be considered in the redesign; they would, however, be approved only if the foregoing changes did not provide sufficient weight margin.

ASPO would require a complete preliminary design and impact assessment of the Block II spacecraft before its incorporation into the program would be authorized.

Letter, Joseph F. Shea, MSC, to John W. Paup, NAA, April 16, 1964.

1964

April

North American conducted a preliminary study on removal of one of three fuel cells from the Block II CSM. The contractor predicted a total weight saving of about 168 kilograms (370 pounds), with potential indirect reductions in the cryogenic systems, but this change would require a significant increase in reliability.

MSC, "ASPO Management Report for Period April 23–30, 1964."

16–22

MSC, North American, and Grumman reviewed development problems in the LEM and SM reaction control thrust chambers. They agreed that a reassessment of the chambers' operational and thermal parameters was necessary.

MSC, "ASPO Management Report for Period April 16–23, 1964."

16–23

North American completed the first of a series of simulations to evaluate the astronauts' ability to perform attitude change maneuvers under varying rates and angles. Subjects were tested in a shirtsleeve environment and in vented and pressurized International Latex Corporation state-of-the-art

16–May 15

1964

April

pressure suits. The subjects had considerable difficulty making large, multi-axis attitude corrections because the pressurized suit restricted manipulation of the rotational hand controller.

"Apollo Monthly Progress Report," SID 62-300-25, p. 5.

17

Grumman conducted manned drop tests to determine the LEM crew's ability to land the spacecraft from a standing position. (See September 16-18, 1963.) All tests were run with the subject in an unpressurized suit in a "hands off" standing position with no restraint system or arm rests.

"Monthly Progress Report No. 15," LPR-10-31, p. 10.

20

NASA selected IBM, Federal Systems Division, to develop and build the instrument units (IU) for the Saturn IB and Saturn V launch vehicles. [IBM had been chosen by NASA in October 1963 to design and build the IU data adapters and digital guidance computers and to integrate and check out the IUs.] Under this new contract, expected to be worth over $175 million, IBM would supply the structure and the environmental control system. NASA would furnish the telemetry system and the stabilized platform (ST-124M) of the guidance system. MSFC would manage the contract.

NASA News Release 64-89, "NASA Selects IBM as Lead Contractor for Saturn IB, V Instrument Unit," April 20, 1964.

21

ASPO directed Hamilton Standard to provide urine storage in the Apollo space suit for prelaunch and launch. The contractor was to investigate the suitability of a Mercury-Gemini type urinal for storage and subsequent disposal.

TWX, W. F. Rector III, MSC, to GAEC, Attn: Waste Management Program Manager, April 21, 1964.

21

Officials from ASPO, Flight Crew Operations Directorate, Crew Systems Division, and Hamilton Standard established the basic ground rules for Apollo space suit operation:

(1) At least one crewman would wear his space suit at all times

(2) All three crewmen would wears their suits continuously during launch through translunar injection, lunar operations, and reentry

(3) The three crewmen could remain suited at all times, although they could remove the suits during translunar and transearth phases

(4) The crew would be able to return from any point in the mission in pressurized suits

(5) Two men in the CM would be able to don their suits within five minutes.

Operations Planning Division reported that these rules required no modifications to the suit and only minor changes to the environmental control system.

PART II: DEVELOPING HARDWARE DISTINCTIONS

MSC, "ASPO Management Report for Period April 16–23, 1964"; "ASPO Management Report for Period April 23–30, 1964."

1964

April

23

After completing estimates of the heating conditions for a series of MIT guided reentry trajectories, the MSC Engineering and Development Directorate recommended that the heatshield design philosophy be modified from the current "worst possible entry" to the "worst possible entry using either the primary or backup guidance mode." North American had drawn up the requirements early in 1962, with the intent of providing a heatshield that would not be a constraint on reentry. However, it was now deemed extremely unlikely that an entry, employing either the primary or backup guidance mode, would ever experience the heat loads that the contractor had designed for earlier. The ablator weight savings, using the MIT trajectories, could amount to several hundred pounds.

Memorandum, C. H. Perrine, MSC, to Mgr., ASPO, "Modification of the heat shield design philosophy," April 23, 1964.

Grumman redesigned the LEM environmental control system to incorporate a replaceable lithium hydroxide cartridge with a portable life support system cartridge in parallel for emergency backup. The LEM cartridge would be replaced once during a two-day mission.

23–30

Also MSC advised Grumman that estimates of the metabolic rates for astronauts on the lunar surface had been increased. The major effect of this change was an increase in the requirements for oxygen and water for the portable life support system.

MSC, "ASPO Management Report for Period April 23–30, 1964."

Rocketdyne conducted the first firing of the prototype thrust chamber assembly for its LEM descent engine.

23–30

Ibid.

Representatives from a number of elements within MSC (including systems and structural engineers, advanced systems and rendezvous experts, and two astronauts, Edward H. White II and Elliot M. See, Jr.) discussed the idea of deleting the LEM's front docking capability (an idea spawned by the recent TM-1 mockup review [see March 24–26]). Rather than nose-to-nose docking, the LEM crew might be able to perform the rendezvous and docking maneuver, docking at the spacecraft's upper (transfer) hatch, by using a window above the LEM commander's head to enable him to see his target. A good many factors pointed to the merit of this approach:

24

• A rectangular window 18 by 38 centimeters (seven by 15 inches) above the commander's head could readily be incorporated into the LEM's structure, with only minimal design changes. The weight penalty would be between 4.5 and 6.8 kilograms (10 and 15 pounds) (excluding possible effects

163

1964
April

on the vehicle's environmental control system). On the other hand, eliminating the front docking mechanism would save about 11 or 14 kilograms (25 or 30 pounds). A docking aid on the CM was essential, but the device "would pay for itself in increased reliability and decreased design load requirements and fuel requirements." Additionally, instead of two docking aids on the LEM (as currently envisioned), only the upper one would be needed.

• The top-only docking arrangement would simplify the docking operation per se. The crew would no longer have to transfer the drogue from the top to the front hatch prior to rejoining the CM. [The need for depressurizing the spacecraft to perform this task thus was obviated.] As an additional "fringe benefit," the front hatch could possibly be reconfigured to make it easier for the crewmen to get out of and back into their craft while on the moon.

• The overhead window would enable the LEM commander to see the moon during powered descent and ascent portions of the flight, and thus would afford the crew a visual attitude and attitude reference.

There existed, naturally, some offsetting factors: the pilot's limited view of his target (thought to be of "no major consequence"); and his being unable quickly to scan his instrument panel (which was not essential). Also, the maneuver called for the pilot to fly his vehicle, for a considerable period, in a rather strained physical position (i.e., with his head tossed backward). But because of the many inherent advantages, the group concluded, LEM-active docking at the upper hatch was acceptable as a backup method for docking. (CM-active docking still would be the normal procedure, because that vehicle could "perform the docking maneuver more easily and more reliably than can the LEM . . . Deletion of the front docking capability on [the] LEM will not alter this relationship, therefore the LEM should be required to dock only when the CSM or the crew member inside is incapacitated. If the CSM is incapacitated returning to it is of questionable importance.") They recommended that Grumman be directed to proceed with this concept for the LEM. (See May 7-14 and May 22.)

Letter, Joseph P. Loftus, Jr., to Assistant Chief, Systems Engineering Division, "Disposition of TM-1 mockup review chit no. A9-4," April 28, 1964, with enclosure, attendance list.

24

To train astronauts in various mission procedures, LTV had completed simulations of manual abort and, within a week, would be able to conduct simulated final maneuver phases of a rendezvous. (See May 6, September 17, and October 10, 1963; also see June 1963.)

"Monthly Progress Report No. 15," LPR-10-31, p. 1.

24

The NASA Manned Space Science Division was planning a scientific experiments program for manned and unmanned earth orbital flights. The

PART II: DEVELOPING HARDWARE DISTINCTIONS

manned program would be a direct outgrowth of the Gemini experiments program. (See March 23.)

1964
April

<small>Memorandum, Willis B. Foster, NASA, to Assoc. Adm. for Manned Space Flight, "Science program for SIB's and SV's," April 24, 1964.</small>

NASA definitized the letter contract with the Philco Corporation Techrep Division for spacecraft flight control support. The definitive contract covered the period from September 16, 1963, through March 31, 1965, and the total cost-plus-fixed-fee was $720 624.

24

<small>MSC, "Consolidated Activity Report for the Office of the Associate Administrator, Manned Space Flight, April 19–May 16, 1964," p. 46.</small>

At Downey, Calif., MSC and North American officials conducted a mockup review on the Block I CSM. Major items reviewed were:
- Cabin interior (complete except for hatches, display panel lighting, survival equipment, umbilical connections, and zero-g restraints)
  - CM exterior (complete except for hatches and boost protective cover)
  - Earth landing system
  - Launch escape system
  - SM.

28–30

One hundred and eleven request for change forms were submitted to the mockup review board, composed of Robert O. Piland (Chairman), Christopher C. Kraft, Jr., Donald K. Slayton, Caldwell C. Johnson, Owen E. Maynard, and Clinton L. Taylor of MSC; and H. G. Osbon and Charles H. Feltz of North American.

For the first time, three representative Apollo space suits were used in the CM couches. Pressurized suit demonstrations, with three suited astronauts lying side by side in the couches, showed that the prototype suit shoulders and elbows overlapped and prevented effective operation of the CM displays and controls. Previous tests, using only one suited subject, had indicated that suit mobility was adequate. Gemini suits, tested under the same conditions, proved much more usable. (See February 27–March 4.) Moreover, using Gemini suits for Apollo earth orbital missions promised a substantial financial saving. As a result of further tests conducted in May, the decision was made to use the Gemini suits for these missions. The existing Apollo space suit contract effort was redirected to concentrate on later Apollo flights. A redesign of the Apollo suit shoulders and elbows also was begun.

<small>MSC, "Command and Service Modules, Project Apollo Board Report for NASA Inspection and Review of Block I Mock-Up, April 23–30, 1964," pp. 1–2; MSC, "ASPO Management Report for Period April 30–May 7, 1964"; MSC, "Weekly Activity Report for the Office of the Associate Administrator, Manned Space Flight, May 3–9, 1964," p. 5; "Apollo Quarterly Status Report No. 8," pp. 47–48; interview, telephone, Matthew I. Radnofsky, Houston, March 24, 1970.</small>

PART III

Developing Software Ground Rules

April 29, 1964, through September 30, 1964

PART III

The Key Events

1964

May 4: Apollo Mission Planning Task Force specified the program's mission objectives and ground rules.

May 13: First flight test of Little Joe II using a command module (CM) boilerplate (BP-12) at White Sands Missile Range, N. Mex.

May 28: Apollo Saturn Mission A-101, using CM BP-13 atop SA-6 Saturn I launch vehicle, launched at Cape Kennedy, Fla., to prove spacecraft/launch vehicle compatibility.

June 11: NASA directed North American Aviation, Inc. (NAA), to make certain mandatory changes to both Block I and Block II spacecraft systems.

July 28: Ranger VII mission finally succeeded in televising pictures of lunar surface up to impact.

August 18: Scout launch tested Apollo-type ablator materials at lunar reentry heating levels.

September 14: Ground rules for lunar excursion module guidance and control system firmly defined.

September 18: Apollo Mission A-102, using BP-15 for the command and service modules (CSM) and SA-7 for the launch vehicle, confirmed Saturn Block II and CSM compatibility as well as the launch escape vehicle system.

September 30: NAA conducted formal inspection and review of Block II CSM mockup.

PART III

Developing Software Ground Rules

April 29, 1964, through September 30, 1964

ASPO defined weight and volume allocations for scientific equipment. Exact location of this equipment could not be specified, but each module had to have the following capacities:

• CM and LEM ascent stage: 36 kilograms (80 pounds); 0.06 cubic meter (2 cubic feet)

• LEM descent stage: 95 kilograms (210 pounds); 0.27 cubic meter (9 cubic feet), minimum; 0.45 cubic meter (15 cubic feet), design objective.

Any additional space gained by jettisoning expendable equipment could also be used for storage. (See June 8.)

Requirements for thermal protection for the scientific equipment were not yet defined, nor was the packaging concept. Electrical outlets on the LEM, furnishing power to the equipment, would of course have to be within the reach of an astronaut while he was standing on the moon's surface outside the spacecraft.

> Letter, W. F. Rector III, MSC, to GAEC, Attn: R. S. Mullaney, "Contract NAS 9-1100, Scientific Equipment," April 29, 1964.

MSC established new LEM abort guidance ground rules, which defined the operation and reliability requirements of the stabilization and control system's abort guidance section. Grumman was to continue studies on the abort pitch programmer and on the capability of the LEM to perform rendezvous.

> Letter, W. F. Rector III, MSC, to GAEC, Attn: R. S. Mullaney, "Contract NAS 9-1100, Abort Guidance Section of the Stabilization and Control Subsystem," April 29, 1964.

MSC authorized major revisions in the CM communications system to provide better voice and data relay between the CM, the LEM, and ground stations.

> Letter, H. P. Yschek, MSC, to NAA, Space and Information Systems Div., "Contract Change Authorization No. 201," April 30, 1964.

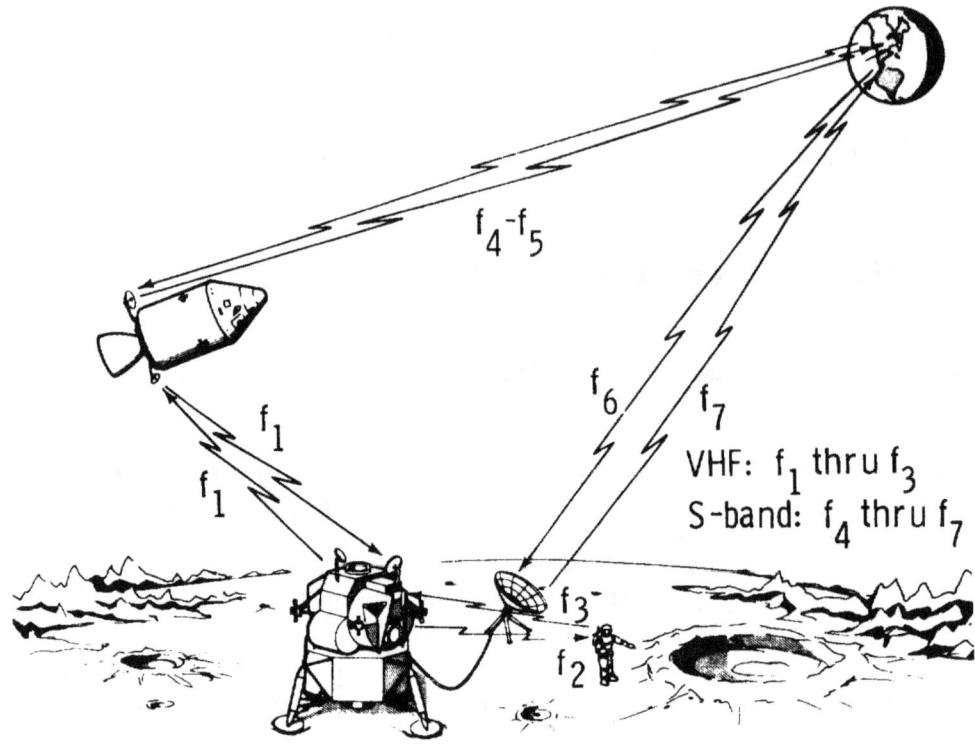

Communications links between CM, LEM, and earth stations.

1964

April 30

Following a series of 15 acceptance firings at Rocketdyne's Santa Susana test facility (conducted during March and April), the first hot-firing production J–2 engine was delivered to Douglas Aircraft Company (DAC). The engine then began "battleship" testing (i.e., fitted to a heavyweight stage of the vehicle built especially for static testing) at DAC's Sacramento test site.

Akens, et al., *History of Marshall . . . January 1 through June 30, 1964*, Vol. I, pp. 148, 224.

During the Month

Grumman awarded Bell Aerosystems Company the contract for the LEM ascent stage reaction control system propellant tanks. The contract was worth about $3.5 million.

Missiles and Rockets, 14 (April 27, 1964), p. 23.

During the Month

Grumman recommended using a self-stabilized trim gimbal system in the descent stage of the LEM, which would save about 34 kilograms (75 pounds) of reaction control system propellant.

"Monthly Progress Report No. 15," LPR-10-31, p. 24.

May 1

MSC Structures and Mechanics Division began vibration tests on SM boilerplate (BP) 22 to determine resonant frequencies, mode shapes, and

PART III: DEVELOPING SOFTWARE GROUND RULES

structural damping characteristics. The results would be used in evaluation of data from the BP-22 flight test of the launch escape system at WSMR, scheduled for 1965.

1964
May

> MSC, "Consolidated Activity Report for the Office of the Associate Administrator, Manned Space Flight, April 19–May 16, 1964," p. 56; MSC News Release 64–86, May 1, 1964.

ASPO Manager Joseph F. Shea reported to the Senior Staff that NASA was not imposing any requirement for the crew to get out of the CM quickly should some problem arise with the launch vehicle while on the pad. Given such an occurrence with the crewmen perched almost 122 meters (400 feet) high—and atop a fueled Saturn V—it was believed more rational to make a standard abort (using the launch escape system) or to hold the countdown until the vehicle could be made safe.

1

> MSC, "Minutes of Senior Staff Meeting, May 1, 1964," p. 3.

MSC Instrumentation and Electronic Systems Division personnel visited Jet Propulsion Laboratory to review the Surveyor landing radar test program and to investigate the use of either a reflector or a transponder on the Surveyor to help in the selection of landing sites for the LEM. At that time, the possibility did not appear promising because reflector usage seemed impractical and because power requirements were far above what was available. Additional study on the matter was planned.

1

> MSC, "ASPO Management Report for Period April 23–30, 1964"; "ASPO Management Report for Period April 30–May 7, 1964."

Grumman completed negotiations with RCA for the attitude and translation control assembly (ATCA) for the LEM. The ATCA imposed thrust demands on the vehicle's stabilization and control system based upon information from the guidance equipment.

1

> MSC, "Consolidated Activity Report for the Office of the Associate Administrator, Manned Space Flight, April 19–May 16, 1964," p. 45.

The Apollo Mission Planning Task Force presented its Phase I progress report to ASPO. (See November 29, 1963, and January 16, 1964.) ASPO, in assigning this task, had defined its principal objectives: the determination of mission-related, functional requirements for spacecraft subsystems; the examination of current subsystem capabilities to meet these requirements; the evaluation of the capability of the spacecraft to fly missions which met the program objectives; the determination of flexibilities available within established control weights; and the provision of mission plans which would be the basis for other analyses and reporting.

4

The task force further refined program objectives: (1) to land two astronauts and scientific equipment on the near-earth-side of the moon and return them safely to earth; and (2) to perform experiments within the restrictions

1964
May

of 113 kilograms (250 pounds) and 0.3 cubic meter (10 cubic feet) of scientific payload, which would be landed on the lunar surface, and 36 kilograms (80 pounds) and 0.06 cubic meter (two cubic feet), which would be returned to earth.

Mission related spacecraft design rules were studied. Seventeen rules for spacecraft operations and seven for contingencies were selected. Although trajectory ground rules were considered more operational than design in nature, the group included 16 as necessary to define the performance capabilities of the spacecraft design. A reference trajectory, provided by MSC, assumed a launch date of May 8, 1968, and a 41 000-kilogram (90 000-pound spacecraft injected into a 66.4-hour translunar-coast/free-return trajectory.

GAEC, "Apollo Mission Planning Task Force, Phase I Progress Report," LED-540-7, Vols. I, II, III, May 4, 1964.

4–11

MSC ordered Grumman to halt all work on a radiatively cooled nozzle for the LEM's ascent engine. (See January; also see September 19–October 16, 1963.) The Center took this action largely to avoid schedule slippage (because the work was drawing valuable people away from the "mainstream" effort, an ablative nozzle). Also involved in the cancellation were such factors as high risk and cost; the lack of previous experience with this type; and the minor saving in weight at best.

MSC, "ASPO Weekly Management Report, May 7–14, 1964."

5

MSC Operations Planning Division (OPD) reviewed power usage aboard the LEM if the fuel cell assembly (FCA) failed. OPD concluded that Grumman's requirements were too stringent (i.e., turning off all equipment not needed for lunar landing should one FCA fail and turning off everything not needed for crew safety following an abort should two FCA's fail). OPD planned to review all subsystems to determine their duty cycles after an FCA-dictated abort.

MSC, "ASPO Management Report for Period April 30–May 7, 1964."

6

NASA selected RCA for negotiation of a contract for C-band radar equipment to be used on tracking ships by NASA and the Department of Defense, under the U.S. Navy Instrumentation Ships Project Office, during lunar missions.

NASA News Release 64-107, "NASA Selects RCA Radar for Tracking Ships," May 6, 1964.

7

ASPO notified Grumman that a number of components must remain as common-use items, because they were used in conjunction with government furnished equipment that was interchangeable between the two spacecraft: oxygen and water disconnects on the portable life support system and quick-disconnects for the suit umbilicals. ASPO added suit umbilicals and carbon dioxide sensors to the common-use list.

PART III: DEVELOPING SOFTWARE GROUND RULES

ASPO decided that the Gemini pressure suit would be used in Apollo Block I earth orbital flights and, on May 19, notified North American accordingly. This decision grew out of continuing mobility problems with Apollo prototype suit, especially restrictive inside the spacecraft. (See April 28–30.)

1964
May

MSC, "Minutes of Senior Staff Meeting, May 8, 1964," p. 4; MSC, "ASPO Weekly Management Report, May 14–21, 1964."

At MSC's request, Grumman studied the use of the LEM stabilization and control system in aligning that vehicle's inertial measurement unit before spacecraft separation. The company found that the maneuver would consume 5.33 kilograms (11.74 pounds) of fuel from the vehicle's stabilization and control system (SCS), compared with 2.83 kilograms (6.24 pounds) for the same alignment with a free LEM. Grumman advised that the best procedure would be to use the CSM to position the LEM telescope field of view. The LEM could then begin the necessary drift for sighting, using less than 0.23 kilogram (0.5 pound) of SCS fuel.

7–14

Also, Grumman studied the feasibility of an overhead window at the command pilot's station in the LEM. The contractor was pursuing the question of the optimum window size and location and the type of reticle required. (See April 24 and May 22.)

MSC, "ASPO Weekly Management Report, May 7–14, 1964."

North American completed the environmental requirements for the CM television camera. The camera must be able to function under conditions of 100 percent humidity, including unhooking and reconnecting the cable. Also, because of the humidity requirement and the "outgassing" properties of commercial lenses (that is, the gases which they could possibly give off inside the spacecraft's cabin), North American decided that a special zoom lens would have to be developed, which would cost around $110 000.

7–14

Ibid.

NASA and The Boeing Company signed a contract for five Lunar Orbiter spacecraft. Under the incentive provisions, Boeing could receive up to $5.3 million more than the basic $80 million cost if all Lunar Orbiter missions were successful. (See December 20, 1963.)

8

NASA News Release 64–109, "NASA Signs Contract with Boeing for Lunar Orbiter," May 8, 1964.

ASPO Manager Joseph F. Shea told the Center's Senior Staff that it was imperative to decide whether to use the gas-cooled space suit or the liquid-cooled undergarment. (See February 1.) Studies had shown that the current gas-cooled suit would not meet the heat load requirements and improvement would be difficult. Shea felt that parallel developments should not be carried out. A more conservative approach might be to adopt the liquid-cooled

8

1964
May

garment, which could readily handle the heat load, although it entailed some increase in weight and cost, if it could be developed and qualified within the next four years. On May 22, Robert O. Piland, Shea's Deputy, reported to the Staff that liquid-cooled undergarments had been selected for the Block II spacecraft. (See July.)

In line with selection of the liquid-cooled undergarment, Hamilton Standard was directed to stop work on the gas-cooled and begin work on a water-cooled portable life support system (PLSS). On June 3, Grumman was officially notified that the PLSS was being redesigned to include a liquid transport loop for removal of heat from inside the space suit. This would be done by the liquid-cooled garment and incorporation of flexible tubing through which a coolant would be circulated. Current PLSS interfaces would be used to the greatest practical extent. It was expected that the new undergarments would first be used in manned flight about mid-1967.

> MSC, "Minutes of Senior Staff Meeting, May 8, 1964," p. 4; "Minutes of Senior Staff Meeting, May 22, 1964," p. 4; MSC, "ASPO Weekly Management Report, May 14–21, 1964"; letter, W. F. Rector III, MSC, to GAEC, Attn: R. S. Mullaney, "Contract NAS 9-1100, Portable life support system changes," June 3, 1964.

11–18

After a 444-second firing, Rocketdyne's first LEM descent engine prototype thrust chamber developed a hot gas leak at the injector flange. Studies were under way by the contractor to determine the cause of the leak.

> MSC, "Consolidated Activity Report for the Office of the Associate Administrator, Manned Space Flight, May 17–June 20, 1964," p. 24; MSC, "ASPO Weekly Management Report, May 14–21, 1964."

12

Verne C. Fryklund, Jr., Chief of the Lunar and Planetary Branch in NASA's Office of Space Science and Applications, reported that the Lunar Orbiter program was being coordinated with Apollo's requirements for moon maps. This agreement was reached through a series of meetings of Fryklund with William B. Taylor, of OMSF's Advanced Manned Missions Program Directorate; and Lee R. Scherer, Lunar Orbiter Program Manager. Fryklund set forth general requirements for maps for the Apollo program. Because most Lunar Orbiter data were intended for Apollo's use, Fryklund said, these requirements must be borne in mind when Lunar Orbiter's information was analyzed and distributed. MSC was interested primarily in the equatorial area of the moon (10 degrees above and below the equator), and established rather stringent demands for accuracy around selected landmarks. These requirements were dictated by Apollo's need for selenodetic and topographic information, essential for lunar navigation and landing site selection and for scientific activities by the astronauts on the lunar surface. Although each mission might ultimately require special maps, Fryklund advised, major requirements could be met by a common series of charts and photomosaics.

> Memorandum, Fryklund, NASA, to Distr., "The Lunar Orbiter Program and the lunar mapping requirements of Project Apollo," May 12, 1964.

PART III: DEVELOPING SOFTWARE GROUND RULES

Apollo's first flight test using the Little Joe II launch vehicle, Mission A–001, using CSM boilerplate (BP) 12, was launched from WSMR. The test was conducted to determine aerodynamic characteristics of the launch escape system (LES) and its capability to pull the spacecraft away from the launch vehicle during an abort at transonic speeds and high dynamic pressure. Thrust termination subjected the spacecraft to an environment more severe than expected, above the qualification test level of many of the CM's components.

Except for a parachute failure, spacecraft and LES functioned flawlessly. All but one test objective was met: because of excessive spacecraft oscillation at the time the main parachutes were deployed, one riser was dragged across the spacecraft structure and severed. The shroud lines of the now-freed parachute burned a gore in one of the two remaining parachutes. Although the damaged gore failed, these two main parachutes deployed normally. BP–12 landed 828 meters (22 400) feet downrange about five minutes and 50 seconds after liftoff. At impact, its rate of descent was 7.9 meters (26 feet) per second, 0.06 meters (two feet) per second faster than planned but still within human tolerances.

> "Postlaunch Report for Apollo Mission A–001 (BP–12)," pp. 1-1, 2-1, 3-1, 6-1.

MSC decided to provide equipment in the LEM for recording the astronauts' voices, and was studying ways to achieve a capability for time correlation with a minimum increase in power and weight.

> MSC, "ASPO Weekly Management Report, May 14–21, 1964."

The first test of a fully ablative thrust chamber for the LEM descent engine was held at Space Technology Laboratories. The chamber, with a wall thickness of 22.4 millimeters (0.88 inch), was fired for 488 seconds. Although some charring occurred, there was no streaking or gouging. Data showed good performance at low thrust.

> MSC, "ASPO Weekly Management Report, May 21-28, 1964"; MSC, "Weekly Activity Report for the Office of the Associate Administrator, Manned Space Flight, May 24–30, 1964."

General Electric (GE) issued a report on postlanding tilt angles for the LEM (the result of a study ordered by ASPO). The Apollo Systems Specification, put out by OMSF, called for the LEM's ability to lift off from the moon from an angle of 30 degrees; MSC's LEM Technical Approach stated that "the Lunar Touchdown System [i.e., the landing gear] will be required to land the LEM in a near vertical position satisfactory for lunar launch and normal egress." GE's study was an attempt to reconcile this difference. There was some concern that, for a variety of reasons, a 30-degree tilt might be undesirable: the spacecraft could tip over; once stage separation occurred, the vehicle's ascent portion could shift slightly; and the crew's visibility and mobility—including their ability to get in and out of the craft—might

1964
May
13

14–21

18–25

21

THE APOLLO SPACECRAFT: A CHRONOLOGY

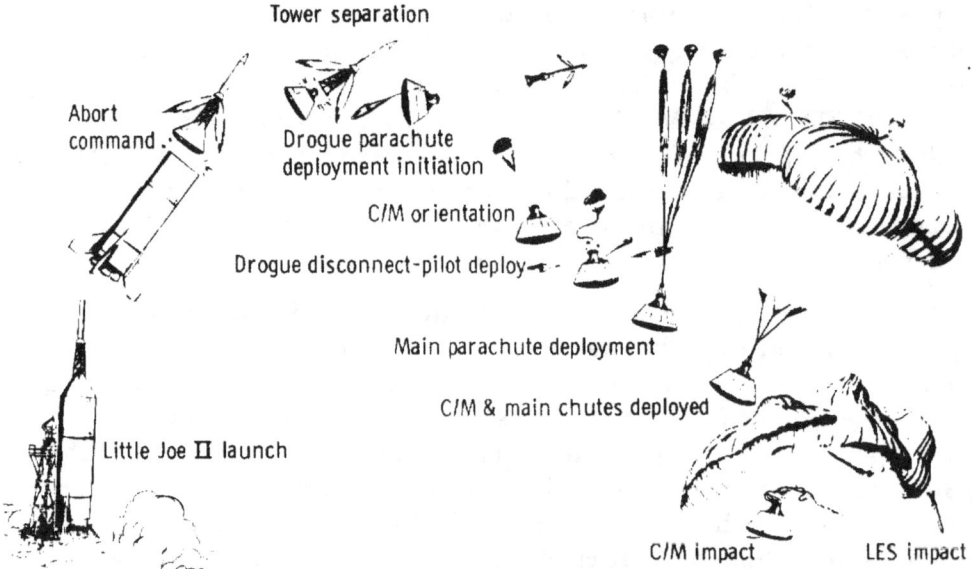

The planned sequence of events for the BP-12 suborbital flight is shown above. At right is the liftoff.

PART III: DEVELOPING SOFTWARE GROUND RULES

be impaired. Added to this were possible constraints imposed by the performance of many of the LEM's operational systems (e.g., communications, ascent propulsion, stabilization and control). In sum, GE reported that it had found no constraints that negated the 30-degree figure, and recommended that MSC's Technical Approach be revised to correspond with OMSF's specification.

1964

May

> General Electric Company, Apollo Support Department, "Study of the Postlanding Tilt Angle of the LEM," TIR 545-S64-03-006, May 21, 1964, *passim*, but especially pp. 1-4, 32-34; MSC, "ASPO Weekly Management Report, May 21-28, 1964"; interview, telephone, Richard H. Kohrs, Houston, March 9, 1970.

NASA completed negotiations with General Dynamics/Convair (GD/C) for two additional Little Joe II test vehicles and associated ground equipment. (See February 18, 1963.) The amendment (worth $1 352 050) increased the contract's total estimated cost and fee to $12 478 205, and brought to eight the total number of Little Joes (excluding the qualification vehicle) that NASA bought from GD/C.

21

> MSC, "Consolidated Activity Report for the Office of the Associate Administrator, Manned Space Flight, May 17-June 20, 1964," p. 42; *Little Joe II Test Launch Vehicle, NASA Project Apollo: Final Report*, Vol. I, pp. 1-7, 4-4.

North American completed zero-g egress tests, using the proposed small configuration CM side entry hatch with a crewman wearing a pressurized Gemini space suit and an operational portable life support system. Weightless tests were also conducted on the crew couch zero-g restraint harness. The subjects had considerable difficulty attaching the harness; additional development and testing were necessary.

21-28

> NAA, "Apollo Monthly Progress Report," SID 62-300-26, July 1, 1964, p. 7; MSC, "ASPO Weekly Management Report, May 28-June 4, 1964."

ASPO directed Grumman to provide an overhead window in the LEM to permit the pilot to dock at the upper docking hatch. The forward access hatch was retained for lunar surface ingress and egress and on-the-pad access capabilities. The contractor would remove the forward docking interface and tunnel.

22

> MSC, "ASPO Weekly Management Report, May 21-28, 1964"; MSC, "Minutes of Senior Staff Meeting, May 22, 1964," p. 4.

MSC received results of RCA and Ryan Aeronautical Company studies on modifying either the LEM landing or rendezvous radar to achieve the high accuracies needed to circularize the LEM's lunar orbit. The contractors concluded that, as currently designed, radar performance would be marginal. Attempts to obtain this degree of accuracy could cause schedules to slip, because of the lack of knowledge of lunar reflectivity. As a means of reducing

22

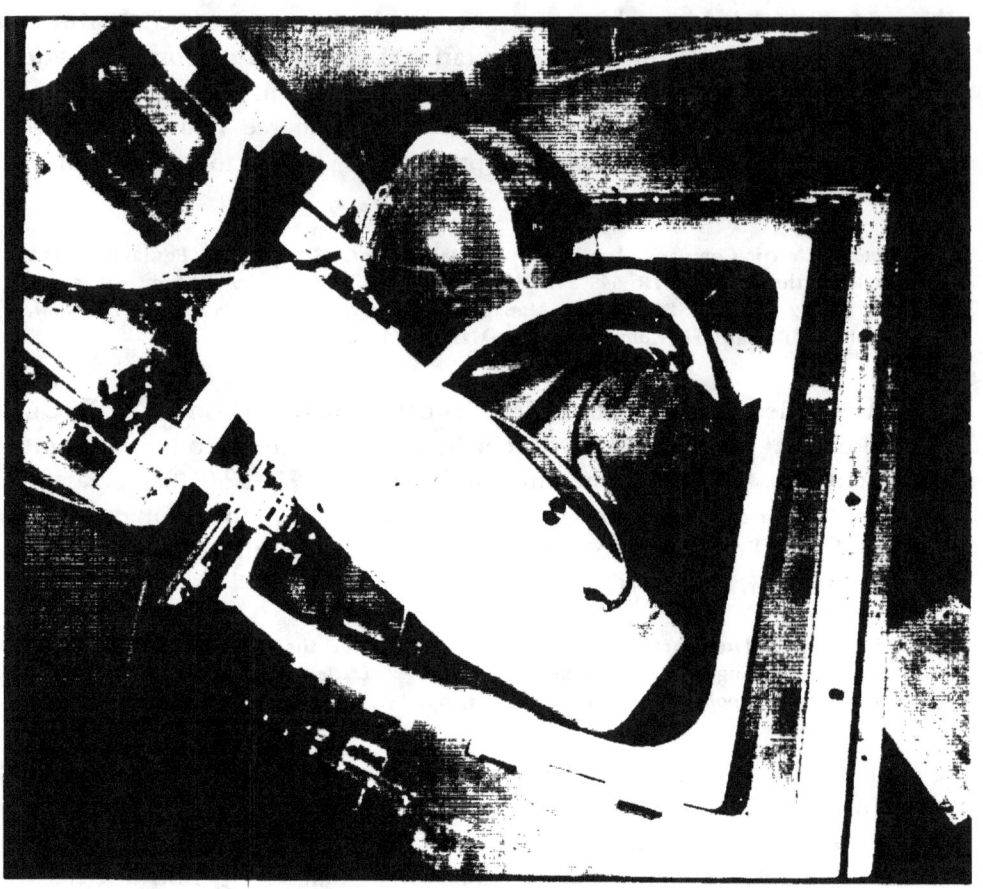

Command module zero-g mockup.

1964

May

the effects of surface variations, RCA and Ryan recommended lessening the spectrum of the radar. (See February 27–March 4 and March 16.)

> MSC, "ASPO Weekly Management Report, May 21–28, 1964"; MSC, "Consolidated Activity Report for the Office of the Associate Administrator, Manned Space Flight, May 17–June 20, 1964," p. 58.

22

MSC informed Grumman of two major revisions to the ground rules for crew transfer between the two spacecraft:

(1) Definite tasks were replaced with a general requirement that a "pressurized crew" should be able to prepare the docked spacecraft for translunar operations.

(2) The requirement for a crewman to pressurize his space suit and, with the aid of a second crewman, move through the transfer tunnel without damage to the suit was changed: the crew must be able to transfer through the tunnel in a pressurized suit as a degraded mode of operation.

PART III: DEVELOPING SOFTWARE GROUND RULES

Apollo system check was performed at Honeywell's Minneapolis Aeronautical Division on manual controls and panel displays of the spacecraft's stabilization and control system. Engineer Bruce Lockhart held one set of manual control sticks for translation maneuvers (left hand) and rotation maneuvers (right hand). The instruments were (clockwise from upper left) the flight director attitude indicator, attitude set/gimbal position indicator, SCS control panel, and velocity change indicator. —Minneapolis-Honeywell photo.

1964

May

Transfer in an unpressurized suit continued to be the primary and extravehicular transfer the emergency mode. Crew transfer tests at North American indicated that no significant hardware changes were necessary to implement these revisions.

> Letter, W. F. Rector III, MSC, to GAEC, Attn: R. S. Mullaney, "Revision of the Apollo Docking Interface and Ground Rules," May 22, 1964.

26

At Hamilton Standard, MSC representatives reviewed status of the Apollo space suit (A3H–024). Tests showed that a suited astronaut could not put on the thermal coverall while wearing a portable life support system.

> MSC, "ASPO Weekly Management Report, May 28–June 4, 1964."

26

ASPO notified Grumman that the carbon dioxide sensor was a crew safety item. Since failure of this component could cause loss of the crew, it must be designed to meet crew safety reliability. NASA's contract with The Perkin-Elmer Corporation, manufacturer of the sensor, had been amended to include testing required for crew safety items.

> Letter, W. F. Rector III, MSC, to GAEC, Attn: R. S. Mullaney, "Contract NAS 9-1100, carbon dioxide (CO_2) sensor requirement," May 26, 1964.

26

ASPO directed North American to provide a station in the CM where the astronauts could put on and remove the portable life support systems.

> MSC, "ASPO Weekly Management Report, May 21–28, 1964."

27

Meetings at Grumman (on May 21–22) had disclosed that the contractor had changed from an all-welded LEM cabin to one that was partially riveted. Although this change had not been coordinated with MSC, the Center nonetheless agreed to it, provided the structural integrity of a cabin thus fabricated could be demonstrated under all load, temperature, and vacuum conditions. MSC recommended that representatives from Grumman visit MSFC to review welding and sealant techniques developed for Saturn launch vehicles.

> MSC, "ASPO Weekly Management Report, May 28–June 4, 1964."

28

Apollo Mission A–101, the first flight of an Apollo spacecraft with a Saturn launch vehicle, was launched from Cape Kennedy. The purpose of the flight was to demonstrate the compatibility of the spacecraft with the launch vehicle for earth orbital flights. A–101 also was the first Apollo flight test conducted at Cape Kennedy, and consisted of CSM boilerplate (BP) 13 and the Saturn SA–6 vehicle.

Launch azimuth was 105 degrees. S–I's first stage number eight engine shut down prematurely at T + 116.9 seconds, delaying S–I cutoff and separation, which occurred at T + 148.8 seconds (2.7 seconds late). The S–IV second stage ignited at T + 150.9 seconds, and the LES was jettisoned 10.3 seconds later and was propelled safely from the flight path. S–IV cutoff took place

PART III: DEVELOPING SOFTWARE GROUND RULES

at T + 624.5 seconds (1.26 seconds earlier than predicted). Orbit insertion was completed at T + 629.5 seconds, with a 31.78 degree equatorial plane. The payload weight at orbit insertion was 7622 kilograms (17 023 pounds). Deviations from planned flight path angle and velocity were minus 0.05 degrees and plus 3.6 meters (11 feet) per second, respectively. Orbital parameters were 182 and 227 kilometers (98.4 and 122.5 nautical miles); the orbital period was 88.62 minutes.

Although there were a few cases of excessive delay in transmission, data coverage and availability were, in general, quite good. Electromagnetic

Engineering test pilot Charles Smythe wore a Gemini pressure suit as he stood on the ladder of the all metal mockup of the LEM. This mockup was the final design version (including rivets) established as a basis for tooling and fabrication.

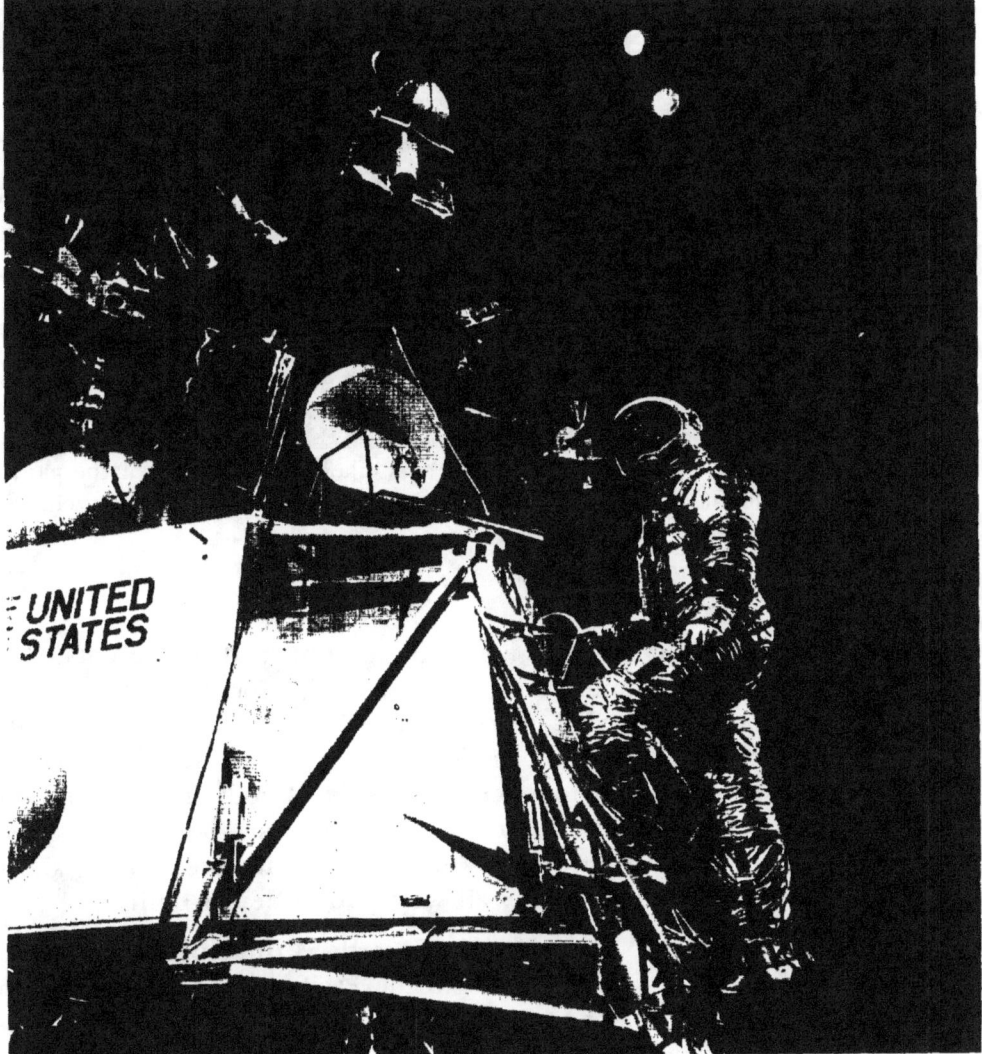

An unusual view of the SA-6 space vehicle was taken from the top of the gantry at Pad 34 during a radio frequency interference test before the launch.

1964

May

interference was minor and did not degrade or invalidate the data. The instrumentation and communications systems performed satisfactorily; battery performances exceeded expectations. LES separation caused no detectable disturbance of the flight vehicle. The sequencer system, explosive bolts, and tower jettison all functioned properly. Aerodynamic, thermodynamic, acoustic, and vibration data contained no surprises. As expected, stresses on the LES were considerably less than those imposed during abort; loads on other spacecraft structures all were within design limits.

BP-13 and the spent S-IV stage circled the earth 54 times before reentering the atmosphere east of Canton Island in the Pacific Ocean on June 1. No spacecraft recovery was planned.

PART III: DEVELOPING SOFTWARE GROUND RULES

NAA, "Project Apollo Flight-Test Report, Boilerplate 13," SID 63-1416-3, August 1964, pp. 2-1, 2-2; "Postlaunch Report for Apollo Mission A-101 (BP-13)," pp. 2-1, 3-2 through 3-5, 4-1 through 4-3, 7-1.

1964
May

MSC issued a cost-plus-fixed-fee contract to Bissett-Berman Corporation of Santa Monica, Calif., for studies of Apollo mission planning, guidance and navigation system analysis, and related tasks. The contract was valued at $915 357.

MSC, "Consolidated Activity Report for the Office of the Associate Administrator, Manned Space Flight, May 17–June 20, 1964," p. 42.

28

MSC instructed North American to continue the Apollo food studies (being done under subcontract by the Stanford Research Center) on diet selection, nutritional value, packaging design and materials, and rehydration. North American was asked to furnish a final report documenting the project and to provide MSC Crew Systems Division with one set (i.e., food supply for three crewmen for a two-week Apollo mission) for evaluation of both the food itself and of packaging concepts. The contractor also was asked to report its findings on studies of snacks for the crewmen.

Letter, H. P. Yschek, MSC, to NAA, Space and Information Systems Div., "Contract Change Authorization No. 174, Revision 1," May 28, 1964.

28

MSC reported that Grumman was studying how much restraint the LEM crew needed during lunar landing, and was conducting manned drop tests to help define requirements. The program was divided into two phases, one on vertical and the other on off-axis landing. In the first part, already completed, the subject had needed no restraints. The second phase, however, was much more severe, and it was believed that restraint would probably be essential.

MSC, "ASPO Weekly Management Report, May 28–June 4, 1964"; "Quarterly Status Report No. 8," p. 35.

28–June 4

At the CSM mockup review at North American on April 28–30, MSC officials were concerned about the complexity of the couch/restraint system. Because of the decision that primary landing would be on water (see February 28), the system was reviewed. Based upon load analyses, supplemented by manned tests at Holloman Air Force Base, a simpler system (principally a combination lap belt and shoulder harness) was found acceptable.

MSC, "ASPO Weekly Management Report, May 14–21, 1964"; "Quarterly Status Report No. 8," pp. 12–13.

During the Month

MSC notified Grumman that primary LEM ingress and egress was through the forward hatch. To aid the LEM crew in getting down to the lunar surface and in climbing back into their vehicle, the Center said, a narrow platform must be provided from the hatch to the landing gear knuckle

June 1–5

183

1964

June

(which became the "front porch"), and a handrail and ladder down the strut to the foot pad.

MSC, "ASPO Weekly Management Report, June 4–11, 1964."

1–5

Technicians of MSC's Landing and Recovery Division began initial testing with a prototype flotation collar (similar to those used with both Mercury and Gemini spacecraft). Boilerplate 25 served as the test vehicle.

MSC, *Space News Roundup*, June 24, 1964, p. 3.

2

NASA signed a production contract worth $1.82 million with Sperry Gyroscope for accelerometers for the CSM's navigation and guidance system. (See Volume I, May 8, 1962.) [Sperry Gyroscope had been chosen during the first half of 1962 to develop these devices, and a developmental contract had been signed on June 1 of that year.]

NASA Contract NAS 9–2847, June 2, 1964.

Technical Services Division supported the tests of the flotation collar at Ellington AFB with scuba divers and other personnel. Also present for the tests were representatives of the Overhaul and Repair Department of the Naval Air Station, Pensacola, Fla., who fabricated the test collar and were aiding in its development.

PART III: DEVELOPING SOFTWARE GROUND RULES

ASPO confirmed for Grumman that no conclusive requirement for a LEM emergency detection system (EDS) had been established. The LEM should be designed to preclude any potential failure which could cause a time-critical emergency. Malfunctions which were not time-critical would be monitored by the caution and warning system while the LEM was manned. Equipment which operated during unmanned periods should be designed to present minimum hazard and to shut down or discharge in a safe condition in cases of malfunction.

ASPO therefore directed Grumman to take no further action on an EDS for the LEM; to analyze possible failures continuously to ensure that safety requirements were met; and to advise ASPO if, at any time, those analyses indicated increased criticality which might warrant reconsideration of an EDS.

<small>Letter, W. F. Rector III, MSC, to GAEC, Attn: R. S. Mullaney, "Contract NAS 9–1100, Lunar Excursion Module Recommendation Concerning LEM Emergency Detection," June 3, 1964.</small>

After studying several configurations for the probe and drogue docking concept, North American recommended one particular design: three radial attenuators attached to three pitch arms, a probe head, a sliding center probe, a stored gas retracting mechanism, and three probe-to-tunnel mounting

Docking concept for the probe and drogue.

1964

June

3

4

1964
June

arms. This configuration would be about 15 percent lighter than the single, center probe, attenuator configuration.

>MSC, "ASPO Weekly Management Report, June 4–11, 1964."

4–11

North American assessed the ultraviolet energy emitted from the shock layer surrounding a spacecraft during reentry. The contractor sought to determine how much that energy added to the radiative heat load imposed on the vehicle, and what effect it would have on the amount of ablative material on the CM. North American's first estimates placed the figure at about 20 percent for lunar return velocities (a figure that thermodynamics experts at MSC called "very conservative"), which would cause about a 4.5-kilogram (10-pound) increase in ablator weight. Because ultraviolet emissions were insignificant at orbital speeds, MSC's Structures and Mechanics Division recommended that their effect be considered only for the design of the Block II CM's heatshield.

>Ibid.

8

ASPO redefined the allowances for scientific equipment in the LEM ascent stage. Major changes were the increase of storage space from 0.06 to 0.09 cubic meter (two to three cubic feet) and of weight from 36 to 45 kilograms (80 to 100 pounds). (See April 29.)

>Letter, W. F. Rector III, MSC, to GAEC, Attn: R. S. Mullaney, "Contract NAS 9-1100, Scientific Equipment," June 8, 1964.

8

A test of the landing impact and stability test program was conducted at North American's drop facility. CM boilerplate 2 was tested with the centerline perpendicular to the water at a vertical speed of 10.4 meters (34 feet) per second. For the first time, a self-contained instrumentation package was installed in the dummy in the center couch. The other two dummies were not instrumented. Onboard cameras documented the general motions and responses during impact. No motion of the dummies in couches or restraint harnesses was observed, indicating that support and restraint were excellent. The simulated heatshield ruptured, as expected.

>NAA, "Apollo Monthly Progress Report," SID 62-300-27, August 1, 1964, pp. 5-7, 17; MSC, "ASPO Weekly Management Report, June 4–11, 1964"; interview, telephone, Glenn W. Briggs, RASPO/NAA, January 12, 1970.

9

In response to a Grumman request, ASPO provided information on LEM crew provision requirements. Caloric requirements, management, packaging, and reconstitution of food supplies were spelled out in detail.

>Letter, W. F. Rector III, MSC, to GAEC, Attn: R. S. Mullaney, "Contract NAS 9-1100, LEM crew provisions," June 9, 1964.

9

MSC announced the letting of a $67 261 contract to Geonautics, Inc., for a study of LEM navigation using lunar landmarks for reference. Geonautics

PART III: DEVELOPING SOFTWARE GROUND RULES

would evaluate crew techniques and procedures for choosing safe landing sites, navigational devices and displays in the LEM, navigational data on the spacecraft's position and trajectory, errors to be expected using various methods of navigation, and the value of available lunar maps.

MSC News Release 64–109, June 9, 1964.

Micro Systems, a subsidiary of Electro-Optical Systems, received two North American contracts valued at $1.85 million to provide temperature and pressure transducer instrumentation for the CM.

Space Business Daily, June 9, 1964, p. 212.

Intending to rely on redundant and backup systems to ensure the spacecraft's reliability, MSC ordered North American to discontinue all effort on the inflight test and maintenance concept for the CM, including spare parts.

Letter, H. P. Yschek, MSC, to NAA, Space and Information Systems Div., "Contract Change Authorization No. 213," June 9, 1964.

MSC clarified design criteria for the launch escape vehicle (LEV). During initial portions of the first-stage flight, when range safety considerations precluded thrust termination (estimated to be 40 seconds), the LEV must

1964
June

9

9

9

Spacesuits and computers were used in combination with a simplified mockup of NASA's Apollo moonship (background) at the Aeronautical Division of Honeywell in Minneapolis, where the stabilization and control system for the three-man spacecraft was developed. In the photo engineer Bill Summers (left) made final adjustments on one of a number of computers which would feed simulated flight information to engineer-test pilot Jim O'Neil (right) when he was inside the command module mockup.
—Minneapolis-Honeywell photo.

1964

June

be capable of aborting safely. Also, the LEV structure must be designed to withstand loads arising from tumbling or oscillating.

MSC, "ASPO Weekly Management Report, June 4-11, 1964."

Early June

MSC geologist Ted H. Foss described a simulated lunar surface (modeled after the Kepler crater in the Oceanus Procellarum) to be constructed at MSC. It would be used for geological training of astronauts and for studying their mobility in space suits. The 100-meter (328-foot)-diameter area would be covered mainly with slag. Plans for several craters about 15 meters (50 feet) in diameter and 4.6 meters (15 feet) deep were later altered to include a large crater 19.5 meters (64 feet) in diameter and 4.9 meters (16 feet) deep and a smaller crater 12.2 meters (40 feet) in diameter and 3 meters (10 feet) deep. There would be a major ridge, 102.4 meters (336 feet) long and 3.7 meters (12 feet) high, and about 75 small craters less than 1.2 meters (4 feet) in diameter. [The mock lunar surface was completed in December.]

MSC, *Space News Roundup*, June 10, 1964, p. 7; MSC News Release 64-194, December 21, 1964.

Early June

NASA notified Grumman, MIT, and North American that RCA would furnish the CSM rendezvous radar to be used with the radar equipment on the LEM. A purchase order for the additional units was issued.

"Apollo Quarterly Status Report No. 8," p. 46.

11

MSC directed North American to make a number of changes to the Block II CSM configuration, some of which were mandatory for Block I vehicles as well. This action followed reviews of the contractor's CSM Block II Technical Report at Houston and at NASA Headquarters (by Apollo Program Director Samuel C. Phillips and OMSF chief George E. Mueller) during May. (See April 16.)

Basically, these changes (including a number to the spacecraft's subsystems) were imposed by the requirements of a lunar mission. Most pertained to the CM per se: provisions for docking (including visual aids) and redesign of the transfer tunnel; capability for extravehicular transfer; and adding portable life support systems and scientific equipment. Micrometeoroid protection had to be added to the SM. (See September 30.)

Memorandum, Owen E. Maynard, MSC, to Addressees, "CSM Block II changes transmitted to NAA for implementation," June 19, 1964, with enclosure: letter, H. P. Yschek, MSC, to NAA, Space and Information Systems Div., Attn: E. E. Sack, "Block II changes," June 11, 1964, with enclosures.

11-18

North American canceled its contract with Avien, Inc., for the CSM S-band high-gain antenna system. (See June 21-27, 1963.) Between July 16 and August 15, North American awarded 90-day study contracts to Hughes Air-

PART III: DEVELOPING SOFTWARE GROUND RULES

craft Company and GE to determine the best approach for developing these antennas for Block II spacecraft. The studies were scheduled for completion in October.

1964
June

> MSC, "Apollo/E and D Technical Management Meeting No. 5," June 3, 1964, p. 1; MSC, "ASPO Weekly Management Report, June 11–18, 1964"; NAA, "Apollo Monthly Progress Report," SID 62-300-28, September 1, 1964, p. 8.

MSC and Space Technology Laboratories (STL) completed negotiations (begun May 12) on a $4.6 million cost-plus-fixed-fee contract for a Mission Trajectory Control Program, a continuing project begun in September 1963 to analyze Gemini missions. STL would develop computer programs for flight control trajectories, orbital maneuvers, and analyses of guidance systems, range safety, and mission error. NASA Headquarters approved the contract on August 18 and announced the contract award on August 20.

12

> MSC, "Consolidated Activity Report for the Office of the Associate Administrator, Manned Space Flight, July 19–August 22, 1964," p. 42; "Consolidated Activity Report for the Office of the Associate Administrator, Manned Space Flight, May 17–June 20, 1964," p. 43; NASA News Release 64-206, "STL to Compute Gemini, Apollo Missions Simulations," August 20, 1964.

MSC approved Grumman's subcontract (valued at $9 411 144) with Pratt and Whitney Aircraft for the LEM fuel cell assembly.

12

On this same day, the Center awarded a letter contract with a total estimated cost and fee of $3.315 million to AC Spark Plug for the LEM guidance and navigation and coupling display unit. (See October 18, 1963.)

> MSC, "Consolidated Activity Report for the Office of the Associate Administrator, Manned Space Flight, June 21–July 18, 1964," p. 37.

Space Business Daily reported that MSC was developing a packaging system for bringing back uncontaminated lunar specimens for study. First, the Center would explore methods for collecting, storing, and shipping geological, chemical, and biological specimens in their original conditions to earth laboratories. MSC then would award a contract for production of the system.

15

> *Space Business Daily*, June 15, 1964, p. 239.

ASPO notified Grumman that the use of reclaimed high explosives was undesirable, since this might reduce the reliability and quality of pyrotechnic systems. To trace any lot of reclaimed material to its point of origin was virtually impossible, nor could adulterants such as TNT, which might have been added for original military use, be easily removed. MSC therefore directed North American to use only virgin, newly manufactured high explosives in Apollo pyrotechnic devices and systems.

16

> Letter, W. F. Rector III, MSC, to GAEC, Attn: R. S. Mullaney, "Contract NAS 9-1100, High explosives in the Apollo Spacecraft," June 16, 1964.

1964

June 16

A realignment of CSM guidance and navigation subsystems functions was mandatory for Block II spacecraft. MSC therefore directed North American and MIT to conduct a program definition study of these systems. MSC outlined Block II responsibilities, systems changes (both required and desired), and implementation requirements and assigned responsibilities in these areas to the appropriate contractors.

> Letter, H. P. Yschek, MSC, to NAA, Space and Information Systems Div., "Contract Change Authorization No. 216," June 16, 1964, with enclosure: "Notes for CSM Block II, Definition Discussions," June 4, 1964.

17

NASA selected Collins Radio Company for an estimated $20 million fixed-price-plus-incentive-fee contract to fabricate, install, integrate, and test unified S-band tracking, data acquisition, and communications equipment for Manned Space Flight Network stations. Chosen from 14 competing firms, Collins would provide NASA with nine systems, each with a 9-meter (30-foot)-diameter parabolic antenna. Six of these would be integrated into facilities being prepared for Gemini flights and three would be installed at new Apollo stations. About 30 partial systems would also be integrated into existing ground stations for tracking Apollo flights.

> NASA News Release 64-116, "NASA Negotiating Apollo Communications Systems Contracts," May 14, 1964; NASA News Release 64-146, "NASA Selects Collins Radio to Provide Apollo Tracking Systems," June 17, 1964.

18-25

At MSC, tests were completed on the modified space suit with the new prototype helmet. Tests in the CM mockup indicated that the new helmet gave better visibility than previous helmets. The range of nodding provided by the neck joint, however, was not considered adequate. Both the suit and helmet were shipped back to Hamilton Standard for additional work.

> MSC, "ASPO Weekly Management Report, June 11-18, 1964"; "ASPO Weekly Management Report, June 18-25, 1964."

18-25

Beech Aircraft Corporation completed qualification testing of the hydrogen pressure vessel for the CSM electrical power system cryogenic storage. All four vessels exceeded burst pressure specification requirements. Two Inconel oxygen tanks also were burst tested, with satisfactory results.

> MSC, "Consolidated Activity Report for the Office of the Associate Administrator, Manned Space Flight, June 21-July 18, 1964," p. 19; MSC, "ASPO Weekly Management Report, June 18-25, 1964."

18-25

MSC and Honeywell studied feasibility of the astronauts' exercising manual control of the spacecraft during SM propulsion engine firing to eject from earth orbit. Investigators found that, although the task became increasingly difficult as the maneuver progressed from attitude to position changes, manual control nonetheless was entirely feasible. North American had studied six possible methods of providing electronic redundancy in the stabilization and control system (SCS) to perform just this function, but in

PART III: DEVELOPING SOFTWARE GROUND RULES

Beech Aircraft's cryogenic gas storage system—developed under contract let in 1964—supplied oxygen to the *Apollo 13* command module's environmental system in 1970 and hydrogen and oxygen to fuel cells for electrical power and drinking water. The system was in the *Apollo 13* service module. At upper left was one of two oxygen tanks. At lower right was a cylindrical housing jacket, rounded on each end, that enclosed two cryogenic hydrogen tanks.

—Beech Aircraft photo.

the end recommended manual rate command. Based upon this recommendation and the earlier study, on August 19 MSC decided to incorporate this manual rate control capability in Block I SCS systems.

1964
June

> MSC, "ASPO Weekly Management Report, June 18–25, 1964"; MSC, "Consolidated Activity Report for the Office of the Associate Administrator, Manned Space Flight, July 19–August 22, 1964," pp. 20, 47; NAA, "Apollo Monthly Progress Report," SID 62-300-29, October 1, 1964, p. 11; interview, telephone, Kenneth J. Cox, Houston, March 10, 1970.

Qualification testing on the launch escape motor began with a successful static firing by the Lockheed Propulsion Company. Twenty motors were tested during July and August; all performed satisfactorily. (See August 30.)

19

> Lockheed Propulsion Company, "Apollo Launch Escape and Pitch Control Motors, Monthly Progress Report No. 28," LPC No. 588-P-28, September 30, 1964, p. 5; "Apollo Monthly Progress Report," SID 62-300-27, p. 15.

NASA announced a realignment of CSM guidance and navigation system contractors, effective July 25. (See February 16–March 21.) Two of the prime contractors, Kollsman Instrument Corporation (supplier of the scanning telescope, sextant, and map and data viewer) and Raytheon Company (manufacturer of the onboard computer), became subcontractors to AC Spark Plug, prime contractor for the inertial measuring unit and for

20

1964

June

assembly and test of the complete system. Under separate contracts, MIT continued to direct overall design, development, and integration of the system, while Sperry Gyroscope provided accelerometers. All contracts for the guidance and navigation system were managed by MSC.

> NASA News Release 64-148, "AC Spark Plug Becomes Prime Contractor for Production of Apollo Guidance and Navigation System," June 20, 1964; MSC, "Weekly Activity Report for the Office of the Associate Administrator, Manned Space Flight, July 19-25, 1964," p. 3.

21–July 18

Two amendments to the LEM contract were forwarded to Grumman for signature. One, for $1.257 million, was for additional flight engineering support at MSC; the other, for $4.252 million, was for a data acquisition system to be installed in the Apollo Propulsion System Development Facility at WSMR.

> MSC, "Consolidated Activity Report for the Office of the Associate Administrator, Manned Space Flight, June 21–July 18, 1964," p. 37; MSC News Release 64-151, September 11, 1964.

24

NASA Headquarters approved the definitive contract with Rocketdyne for the production of 55 J–2 engines (used in the S–IVB stage of the Saturn IB and Saturn V launch vehicles). Negotiations had taken place from April 13 to May 15. Initial value of the contract was $89.5 million.

> Akens et al., *History of Marshall . . . January 1 through June 30, 1964*, Vol. I, pp. 145, 226; David S. Akens, Leo L. Jones, and A. Ruth Jarrell, *History of the George C. Marshall Space Flight Center from July 1 through December 31, 1964* (MHM-10, undated), Vol. I, p. 132.

24

The Army Map Service reported the completion for NASA of the first complete topographic map of the visible face of the moon.

> *The San Diego Union*, June 25, 1964.

24

North American conducted the first hot fire tests of the SM reaction control system, with steady and pulsed firings. Only one engine was fired. The only problem encountered was with the oxidizer shutoff valve, which would have to be completely redesigned.

> MSC, "ASPO Weekly Management Report, June 25–July 2, 1964."

25

Grumman engineers, meeting with ASPO officials in Houston, outlined the contractor's philosophy about onboard checkout of the LEM and equipment required to do the job. Scheduled at times when the astronauts were not heavily pressed with other activities, company engineers said there should be three major checkouts of the LEM to come: (1) after lunar orbit injection, (2) immediately after lunar landing, and (3) just before lunar launch. Of course, the astronauts would monitor the various systems during activity with the LEM to manage and operate its subsystems. The contractor did not

PART III: DEVELOPING SOFTWARE GROUND RULES

The hydrogen-fueled J-2 rocket engines for the upper stages of the Saturn IB and Saturn V launch vehicles were completed on the assembly line at the Canoga Park, Calif., plant of Rocketdyne Division of NAA. The J-2 developed a thrust of 1000 kilonewtons (225 000 pounds) at altitude. It operated in a cluster of five engines in the S-II stage and singly in the S-IVB stage of the Apollo launch vehicle. —Rocketdyne photo.

visualize any need for "centralized onboard checkout equipment"—caution and warning lights, controls and displays, help from the ground network, among others, should satisfy the needs. Grumman asked MSC for authority

1964
June

1964

June

to delete the requirement for centralized checkout equipment, and ASPO concurred with their recommendations on July 27.

> Letter, W. F. Rector III, MSC, to GAEC, Attn: R. S. Mullaney, "Contract NAS 9-1100, LEM on-board checkout equipment," July 14, 1964, with enclosure: "Minutes of Meeting At MSC Discussing LEM On-Board Checkout Equipment, June 25, 1964"; letter, Rector to Mullaney, "Contract NAS 9-1100, LEM On-Board Checkout Equipment," July 27, 1964.

25

LTV was awarded a $1 125 040 contract for a dynamic crew procedures simulator to study task assignments in simulated space flight. The trainer was capable of yaw, pitch, and roll movements and duplicated vibrations and noise incurred during liftoff, powered flight, and reentry. Visual displays simulated views of starfields, earth or moon horizons, rendezvous target vehicles, and landscapes.

> MSC News Release 64-122, July 1, 1964; MSC, "Consolidated Activity Report for the Office of the Associate Administrator, Manned Space Flight, June 21-July 18, 1964," p. 38.

25-July 1

Zero g tests of the CM/LEM crew transfer tunnel were performed in KC-135 aircraft at Wright-Patterson Air Force Base, verifying data obtained during crew-transfer zero-g simulations conducted at North American in February and March. The task of controlling equipment proved difficult. For example, the docking probe was temporarily lost during removal.

> MSC, "ASPO Weekly Management Report, July 2-9, 1964."

26

MSC awarded a letter contract (with a total cost and fee estimated at $1.234 million) to Kollsman Instrument Corporation for optical components for the LEM guidance and navigation system. (See October 18, 1963.) Negotiations for a definitive contract began July 10.

> MSC, "Consolidated Activity Report for the Office of the Associate Administrator, Manned Space Flight, June 21-July 18, 1964," p. 37.

26

ASPO, Bellcomm, Inc., and MSC's Mission Planning and Analysis Division completed a study on reentry range requirements. Because of the deceleration limit of 10 g's, the minimum reentry range was 2200 kilometers (1200 nautical miles [n.m.]). A range flexibility of about 1600 kilometers (1000 n.m.) was essential to allow for weather conditions. An additional 1600 kilometers (1000 n.m.) was required by the emergency reentry monitoring system. Therefore, the heatshield must be designed to withstand reentry heating over a 5920-kilometer (3200-n.m.) range.

During mid-July, ASPO learned from the Landing and Recovery Division that the minimum acceptable CM maneuverability during reentry was 1600 kilometers (1000 n.m.) for water landings. "This requirement was based on storm size, weather predictability, and reliability of storm location and direction of movement." Landing errors associated with reentry on backup

PART III: DEVELOPING SOFTWARE GROUND RULES

guidance demanded that the spacecraft be capable of a 6500-kilometer (3500-n.m.) reentry.

1964
June

> Memorandum, Aaron Cohen, MSC, to Owen E. Maynard, "Reentry Range Requirement," June 26, 1964; MSC, "ASPO Weekly Management Report, July 16–23, 1964"; memorandum, Claude A. Graves, MSC, to Chief, Mission Planning and Analysis Div., "Operational entry range requirement," June 18, 1964; memorandum, Carl R. Huss, MSC, to BE4/Historical Office, "Comments on Volume II of *The Apollo Spacecraft: A Chronology*," March 30, 1970.

MSC authorized Grumman to procure a "voice only" tape recorder with time correlation for use in the LEM data storage electronic assembly. The unit would be voice operated and have a capacity of 10 hours recording time.

28–July 4

> MSC, "Weekly Activity Report for the Office of the Associate Administrator, Manned Space Flight, June 28–July 4, 1964," p. 3.

After acceptance testing, AiResearch Manufacturing Company delivered the first production CM environmental control system to North American.

30

> The Garrett Corporation, AiResearch Manufacturing Division, "Monthly Progress Report, Environmental Control System, NAA/S&ID, Project Apollo, 16 June 1964–15 July 1964," SS-1013-R(26), July 31, 1964, pp. 1, 15.

MSC directed North American to make whatever changes were necessary in the Block I design to make the spacecraft compatible with the Gemini space suit. (See May 7.)

30

> MSC, "ASPO Weekly Management Report, June 25–July 2, 1964."

MSC's Operations Planning Division requested OMSF to revise its spacecraft specifications to (1) delete the requirement for data storage in the LEM (this function would be performed by the CSM data recording equipment via an RF link); and (2) drop the requirement for one portable life support system (PLSS) for each crewman (a third PLSS would only allow the CM pilot to enter the LEM without benefit of a hard dock, and studies had shown that this situation probably would never arise).

During the Month

Early in July, MSC requested OMSF to change two other requirements from tentative to firm: (1) LEM tilt angle at lunar liftoff should not exceed 30 degrees (MSC had accepted this value and Grumman had been asked to design systems to conform [see May 21]); (2) the service propulsion system should include a propellant control so that unused propellants (resulting from mixture ratio shift) would not exceed 0.5 percent of the initial propellant supply. (Studies showed that the North American design already met this requirement.)

> "Apollo Quarterly Status Report No. 8," p. 63; MSC, "ASPO Weekly Management Report, July 2–9, 1964."

ASPO spelled out operational procedures for the space suit emergency oxygen supply (EOS) units. [The primary function of the EOS was as a

July 1

195

1964
July

backup during extravehicular operations, if the portable life support system failed or if suit leakage was excessive. EOS could also be used to back up the spacecraft environmental control system during short-term emergencies such as crew transfer.] The two units, stowed in the CM, would be worn during crew transfer to the LEM, then stored there. After landing on the moon, the crewmen would wear the EOS during the entire lunar stay. Putting on or taking off the units unassisted would not be required. North American and Grumman were directed to provide suitable stowage areas in each spacecraft.

> TWX, C. L. Taylor, MSC, to NAA, Attn: E. E. Sack, July 1, 1964; TWX, W. F. Rector III, MSC, to GAEC, Attn: R. S. Mullaney, July 14, 1964; memorandum, William C. Kincaide, MSC, to Chief, Crew Systems Div., "Apollo Emergency Oxygen Supply Subsystem (EOSS)," July 24, 1964.

2–9

MSC's Operations Planning Division (OPD) examined a 14-day lunar survey mission (a manned Apollo Lunar Orbiter-type of photographic mission). OPD found that the 578-kilowatt-hour capability of the CSM's electrical power system was adequate, provided there were no cryogenic tank failures. If such failures occurred, the maximum mission duration would be 11.8 days (four days in lunar orbit).

> MSC, "ASPO Weekly Management Report, July 2-9, 1964;" interview, telephone, Richard H. Kohrs, Houston, March 11, 1970.

8

Donald K. Slayton, MSC Assistant Director for Flight Crew Operations, announced specific assignments for the astronauts. Alan B. Shepard, Jr., was named Chief of the Astronaut Office, Slayton's former job. This office was now divided into three branches, Apollo, Gemini, and Operations and Training: L. Gordon Cooper, Jr., was head of the Apollo branch, with James A. McDivitt, Charles Conrad, Jr., Frank Borman, and Edward H. White II assisting him; in the Gemini branch, headed by Virgil I. Grissom, were Walter M. Schirra, Jr., John W. Young, and Thomas P. Stafford; the Operations and Training branch was headed by Neil A. Armstrong, assisted by Elliot M. See, Jr., and James A. Lovell, Jr. (M. Scott Carpenter, currently on duty with the U.S. Navy's Project Sealab, was not given a specific MSC assignment.)

The 14 newest astronauts were given individual assignments within the Operations and Training branch: Edwin E. Aldrin, Jr., mission planning (including trajectory analysis and flight plans); William A. Anders, environmental control systems and radiation and thermal protection; Charles A. Bassett II, training and simulators; Alan L. Bean, recovery systems; Eugene A. Cernan, spacecraft propulsion and the Agena; Roger B. Chaffee, communications and the Deep Space Network; Michael Collins, pressure suits and extravehicular experiments; R. Walter Cunningham, electrical and sequential systems and monitoring of unmanned flight experiments in other programs which might relate to MSC programs; Donn F. Eisele, attitude

PART III: DEVELOPING SOFTWARE GROUND RULES

Going over the Apollo Boilerplate 15 command and service module stacking check-off sheet in Hangar AF, Cape Kennedy, Fla., were, left to right, Allen Cave, MSC mechanical systems engineer; Thomas Black, MSC operations engineer; and Orval M. Bradford, Jr., NAA operations engineer. The check-off was made before mating the package to the Saturn SA-7.

and translation control systems; Theodore C. Freeman, boosters; Richard F. Gordon, Jr., cockpit integration; Russell L. Schweickart, future manned programs and inflight experiments in Gemini and Apollo; David R. Scott, guidance and navigation; and Clifton C. Williams, Jr., range operations and crew safety.

<small>MSC News Release 64–125, July 9, 1964; MSC, *Space News Roundup*, July 8, 1964, pp. 1, 3.</small>

Apollo Program Director Samuel C. Phillips called a meeting at NASA Headquarters to discuss disposing of the S-IVB stage and its instrument

1964
July

8

1964
July

unit (IU) during lunar missions. Certain restrictions were considered: (1) the S-IVB/IU must not hit the spacecraft after separation; (2) it was preferable that the S-IVB/IU not impact either the earth or the moon, but in seeking to prevent this no changes would be made to the space vehicle that might result in weight, cost, or schedule penalties; and (3) no special provision would be made for tracking the S-IVB/IU after separation from the spacecraft.

"Minutes of Meeting to Review Disposition of the S-IVB/IU and Related Support Requirements During the Post Injection Phase of Lunar Missions," July 15, 1964.

8-9

MSC representatives attended the second Block I CSM mockup review at North American. (See April 28-30.) Although the crew area was decidedly improved, further changes in the suit umbilicals and the restraint system—and significant ones—still were required.

MSC, "ASPO Weekly Management Report, July 9-16, 1964."

9

ASPO directed Grumman to delete 200 watts, currently appearing on the LEM's power allotment charts, for lighting during television transmission of lunar earthshine scenes. The LEM television camera, which was furnished

The lunar television camera.

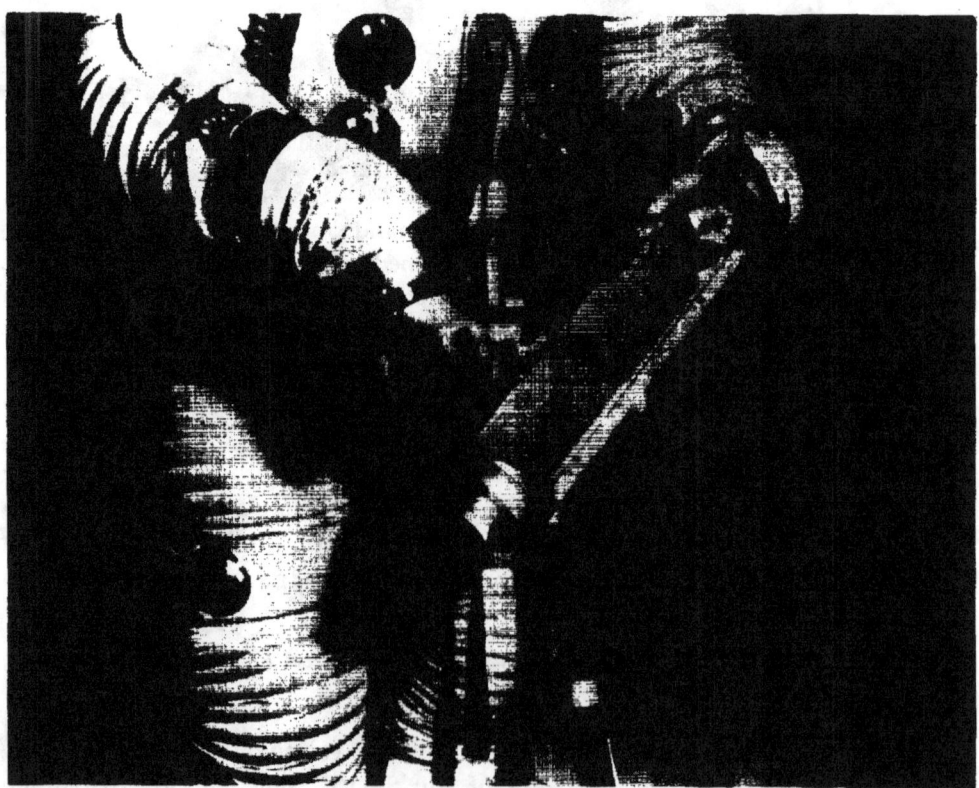

PART III: DEVELOPING SOFTWARE GROUND RULES

The crawler track at Marion, Ohio, in mid-July 1964 before its shipment to Merritt Island. The Marion Power Shovel Company had the contract to build the crawler-transport.

by the government, would be able to televise all lunar scenes during sunshine or earthshine periods.

1964
July

TWX, W. F. Rector III, MSC, to GAEC, Attn: R. S. Mullaney, July 9, 1964.

A NASA–North American Technical Management meeting set the CM control weight (based on an assumed 41 000-kilogram [90 000-pound]-payload capability of the Saturn V) at 5000 kilograms (11 000 pounds). MSC then asked and North American agreed to design, test, and qualify the open ring-sail main parachutes for a CM weighing 5200 kilograms (11 500 pounds).

16

"Minutes of NASA–NAA Technical Management Meeting, July 14, 1964"; MSC, "ASPO Weekly Management Report, July 30–August 6, 1964."

Once the decision was made to use Gemini space suits during Apollo earth-orbital flights, NASA took the next step. The space agency gave to the David Clark Company, manufacturer of the Gemini suit, a program for modifying and testing that suit for use in the Apollo program, and designated it the "Aponi" suit. Formal contract awards were scheduled for late in the year.

16

Memorandum, H. F. Battaglia, MSC, to Chief, MSC Crew Systems Div., "Trip report for visit to David Clark Company, Worcester, Massachusetts concerning Aponi Space Suit Program," July 16, 1964.

Representatives of North American, RCA, and MSC's Instrumentation and Electronic Systems Division held a meeting on the status of the CSM television subsystem. A design review covering all electrical, mechanical, and

16–17

1964

July

optical aspects of the configuration established that the design was complete, subject only to changes growing out of development and qualification tests.

> MSC, "ASPO Weekly Management Report, July 16–23, 1964."

19–25

North American completed a CM-active docking simulation at its Columbus, Ohio, facility to study propellant consumption, engine duty cycles, and stabilization and control system characteristics and performance. Preliminary results showed that sighting aids mounted on the LEM were needed for a satisfactory docking. Furthermore, during transposition docking the S-IVB's roll rate must be no greater than 0.1 degree. North American would prepare a full-scale, three-dimensional study to evaluate differences in lighting and would design sighting aids (to be tested at Langley Research Center).

> MSC, "Weekly Activity Report for the Office of the Associate Administrator, Manned Space Flight, July 19–25, 1964," p. 4; "Apollo Monthly Progress Report," SID 62-300-28, p. 8.

20–21

At Grumman, representatives from MSC's Structures and Mechanics and Systems Engineering Divisions reviewed the design criteria for the LEM's landing gear. The group agreed to study landing stability in various landing conditions. This investigation, and results of MSC Guidance and Control Division's landing simulations, would permit a realistic evaluation of the 406.4-centimeter (160-inch) cantilever gear. (See October 2, 1963.)

> MSC, "ASPO Weekly Management Report, July 23–30, 1964."

21

MSC approved a configuration that Hamilton Standard had recommended for the power supply for the liquid-cooled portable life support system. This configuration embodied an 11-cell secondary battery and separate conversion devices for each electrical load. The total battery capacity required was 108.8 watt-hours.

> TWX, W. F. Rector III, MSC, to GAEC, Attn: R. S. Mullaney, July 21, 1964.

21

Grumman held a portable life support system (PLSS) accessibility test in test mockup 1 for the MSC Crew Systems Division. Subjects were able to put the PLSS on and take it off, unassisted, with the suits pressurized and unpressurized.

> MSC, "ASPO Weekly Management Report, July 23–30, 1964."

21

MSC approved Grumman's subcontract with Allison Division of General Motors Corporation for the LEM descent engine tanks. The amount of the cost-plus-incentive-fee contract was $5.48 million.

> MSC, "Consolidated Activity Report for the Office of the Associate Administrator, Manned Space Flight, July 19–August 22, 1964," p. 41.

21

NASA announced that its Office of Space Science and Applications was inviting scientists to participate in a scientific experiment program for manned

PART III: DEVELOPING SOFTWARE GROUND RULES

1964
July

and unmanned spacecraft. American and foreign scientists from universities, industry, and government were being asked to submit proposals. The earliest Apollo missions that could support this program were anticipated to be the fourth and fifth flights. About 0.06 cubic meter (two cubic feet) of space would be available for instruments and equipment weighing not more than 36 kilograms (80 pounds), but it was expected that additional space and weight would be available in the S–IVB stage during early flights.

NASA News Release 64-177, "NASA Invites World Scientists to Propose Space Experiments," July 21, 1964.

As currently conceived, the LEM's waste management system was designed for direct transfer from the space suit assembly and immediate dumping. If a storage system for the urine were not designed into the LEM, ASPO reported, the spacecraft could be lightened by more than 23 kilograms (50 pounds).

MSC, "ASPO Weekly Management Report, July 23-30, 1964."

23-30

At its Reno, Nev., facility, Rocketdyne conducted the first checkout firing (five seconds) of their LEM descent engine at a simulated altitude of 39 600 meters (130 000 feet). A heavyweight, 20.3-millimeter (.080-inch) thick nozzle extension skirt was used. During the following week, firings of the engine included one of 110 seconds.

MSC, "ASPO Weekly Management Report, July 23-30, 1964"; "ASPO Weekly Management Report, July 30–August 6, 1964."

23-30

Dalmo Victor Company was selected to supply the LEM S-band steerable antenna system to RCA, subcontractor for the LEM communication system.

MSC, "ASPO Weekly Management Report, July 23-30, 1964."

23-30

After comparing capabilities of the space suit assembly with and without the emergency oxygen supply (EOS), the MSC Apollo Portable Life Support Systems Office recommended that the EOS system be retained for crew safety considerations. (See July 1.)

Memorandum, William C. Kinkaide, MSC, to Crew Systems Division, "Apollo Emergency Oxygen Supply Subsystem (EOSS)," July 24, 1964.

24

MSC authorized North American to provide a boost protective cover that would completely enclose the conical portion of the CM during launch. As an integral part of the launch escape system (LES), the cover would be jettisoned after atmospheric exit or during an atmospheric abort. Also the cover would satisfy the requirement for clean windows on the CM after LES separation and would protect the CM's thermal coating and docking mechanism from the launch environment. (See January 15-23 and March 19-26.)

Letter, H. P. Yschek, MSC, to NAA, Space and Information Systems Div., "Contract Change Authorization No. 235," July 24, 1964.

24

1964

July 27

ASPO notified Grumman that spacecraft attitude criteria had been changed to relax thermal design requirements. The former constraints ("worst case orientation") had imposed severe penalties on the design of subsystems and components. The new criteria relieved thermal design problems, but Grumman must ensure that these standards were compatible with other constraints and that they provided adequate operational flexibility.

> Letter, W. F. Rector III, MSC, to GAEC, Attn: R. S. Mullaney, "Contract NAS 9-1100, Apollo spacecraft thermal design mission," July 27, 1964.

28

MSC awarded a $335 791 contract to Lockheed-California Company for transient heat transfer and thermodynamic analyses of the service propulsion system (SPS). Phase I, an analytical study, and Phase II, testing a one-third-scale model of the SPS, were scheduled for completion in January and May. Tests would be run in the Hughes Aircraft Company altitude chamber in a thermal vacuum and under simulated solar radiation.

> MSC, "ASPO Weekly Management Report, August 13-20, 1964."

28

Ranger VII was launched from Cape Kennedy. The 365.6-kilogram (806-pound) spacecraft, carrying six television cameras to take close-up pictures of the moon, was boosted into an earth-parking orbit by an Atlas-Agena launch vehicle. The Agena engines then refired to place the spacecraft on a translunar trajectory. On July 31, *Ranger VII* crashlanded on the moon at 10.7°S, 20.7°W, in the Sea of Clouds. The spacecraft sent back 4316 pictures, beginning at an altitude of about 800 kilometers (500 miles) and ending at impact.

During the next several weeks, MSC's Space Environment Division, ASPO, Grumman, and Bellcomm studied these photographs in great detail. On October 30, ASPO Manager Joseph F. Shea informed Samuel C. Phillips, Deputy Director of the OMSF Apollo Program, that the *Ranger VII* data had eliminated most of the major uncertainties about the lunar surface that could be resolved by photographic techniques.

> *The New York Times*, July 29, 1964; memorandum, John M. Eggleston, MSC, to Shea, "Preliminary analysis of Ranger 7 photographs," August 13, 1964; memorandum, Shea, to NASA Headquarters, Attn: Phillips, "Apollo Mapping and Survey System," October 30, 1964.

30

MSC awarded a cost-plus-fixed-fee contract estimated at $365 000 to the Astronautics Division of LTV for Apollo space suit evaluation and thermal development and qualification testing of Gemini space suits in the company's space environment simulator.

> MSC, "Consolidated Activity Report for the Office of the Associate Administrator, Manned Space Flight, July 19-August 22, 1964," p. 64; memorandum, Robert E. Smylie, MSC, to Chief, Systems Test Branch, "Technical Monitorship of the LTV Space Environment Simulator Contract," August 26, 1964.

PART III: DEVELOPING SOFTWARE GROUND RULES

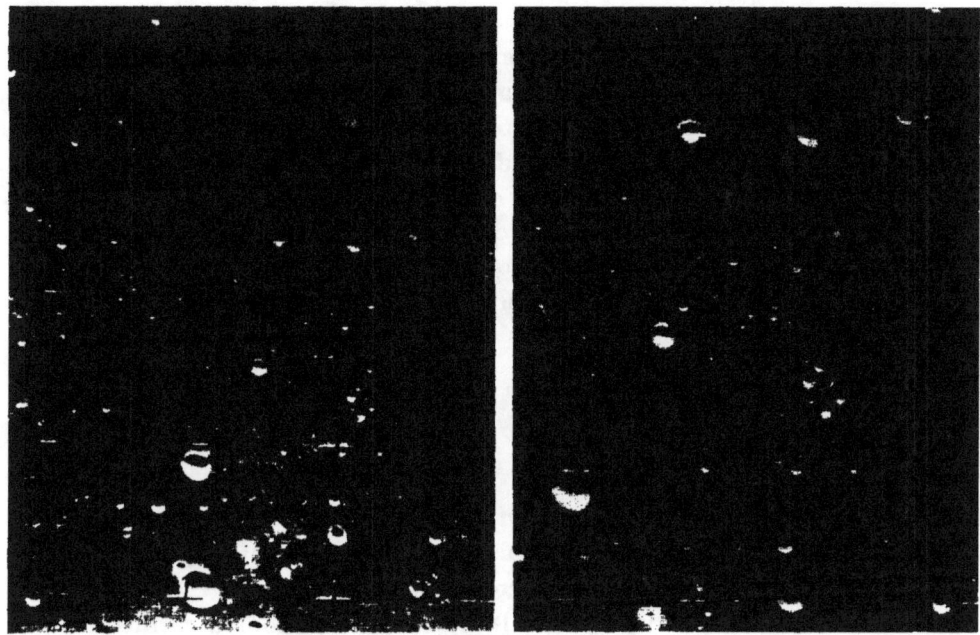

Two of the pictures relayed back to the earth by *Ranger VII* July 31, 1964. The photo at left was taken from an altitude of 124.9 kilometers (77.6 miles) and the photo at right from an altitude of 78.1 kilometers (48.5 miles).

NASA approved Grumman's proposal to use the spacecraft's VHF radios as an "intercom" between the docked LEM and the CM. Early planning had involved the use of a hardline/umbilical arrangement.

TWX, W. F. Rector III, MSC, to GAEC, Attn: R. S. Mullaney, July 30, 1964.

1964
July
30

Technicians in MSC's Operations Planning Division (OPD) studied oxygen storage capacities in the two spacecraft to determine whether those supplies exceeded by 50 percent the levels of consumption anticipated during a normal mission (as required by OMSF specifications). On the basis of current design consumption, they found that mission requirements were exceeded by only 45 and 25 percent for the CSM and LEM, respectively. OPD therefore recommended that OMSF's specifications be revised, because oxygen for the fuel cells as well as for breathing was contained in the same tanks. Rather than the 50 percent reserve, OPD said, Headquarters should instead require the oxygen supplies in both spacecraft to be the maximum amount that would be used for environmental control and for generating power during a lunar mission. And, to allow for safe aborts, some alternate or redundant oxygen storage would be provided in each spacecraft.

MSC, "ASPO Weekly Management Report, July 30–August 6, 1964."

30–August 6

1964

June

During the Month

Members of the National Academy of Sciences' Committee on Lunar Exploration, meeting in Houston, expressed fear about contamination of the lunar surface before Apollo astronauts could secure samples for analysis. Contaminants might come, they noted, from at least two possible sources: (1) air released when the LEM was depressurized, and (2) leakage from the space suits. Elliott S. Harris, head of MSC's Microbiology, Biochemistry, and Hygiene Section, who was present at the meeting, informed Crew Systems Division of the scientists' concern and relayed their recommendations on ways of preventing or controlling such contamination (such as bacteria filters).

> Memorandum, Elliott S. Harris, MSC, to Chief, Crew Systems Division, "Lunar contamination," July 31, 1964.

During the Month

At Hamilton Standard and at MSC, testing continued on early versions of the Hamilton Standard liquid-cooled garment as well as an in-house model developed by the Crew Systems Division. (See February 1 and May 8.) While sweating was not yet completely eliminated, these tests nonetheless confirmed the efficacy of using liquid- rather than gas-cooled garments.

> MSC, *Space News Roundup*, June 24, 1964, p. 7; MSC News Release 64-121, July 8, 1964; MSC, "Consolidated Activity Report for the Office of the Associate Administrator, Manned Space Flight, May 17–June 20, 1964," p. 53; memorandum, Gilbert M. Freedman and Francis J. DeVos, MSC, to Apollo Portable Life Support Systems Office, "Trip Report—Contract NAS 9-723," July 8, 1964; MSC, "ASPO Weekly Management Report, July 2-9, 1964"; "ASPO Weekly Management Report, July 16-23, 1964."

August 3

At its new Magic Mountain, Calif., facility, the Marquardt Corporation began development firings on the LEM reaction control system. By using successively more advanced components, the testing program would gradually build toward a complete prototype. Early in September, MSC's Propulsion and Power Division (PPD) reported that Marquardt had suspended testing temporarily because of problems with monitoring equipment (which, the Division grumbled, could have been checked out before the testing started). Two weeks later, PPD reported that contamination of the thrust chamber had forced Marquardt to halt these developmental firings again. Finally, by mid-October, problems with manufacturing and acceptance checking of the thrust chambers at the company's manufacturing plant portended a twenty-week slippage in delivery of the chambers to the Magic Mountain site.

> MSC, "ASPO Weekly Management Report, July 30–August 6, 1964"; "ASPO Weekly Management Report, August 27–September 3, 1964"; "ASPO Weekly Management Report, September 10-17, 1964"; "ASPO Weekly Management Report, October 8-15, 1964."

4

ASPO tentatively approved Grumman's recommendation to use electroluminescent lighting for controls and display panels inside the LEM's cabin (with backup floodlighting). "Definitive acceptance," of course, was "dependent upon resolution of actual production hardware capabilities." This

PART III: DEVELOPING SOFTWARE GROUND RULES

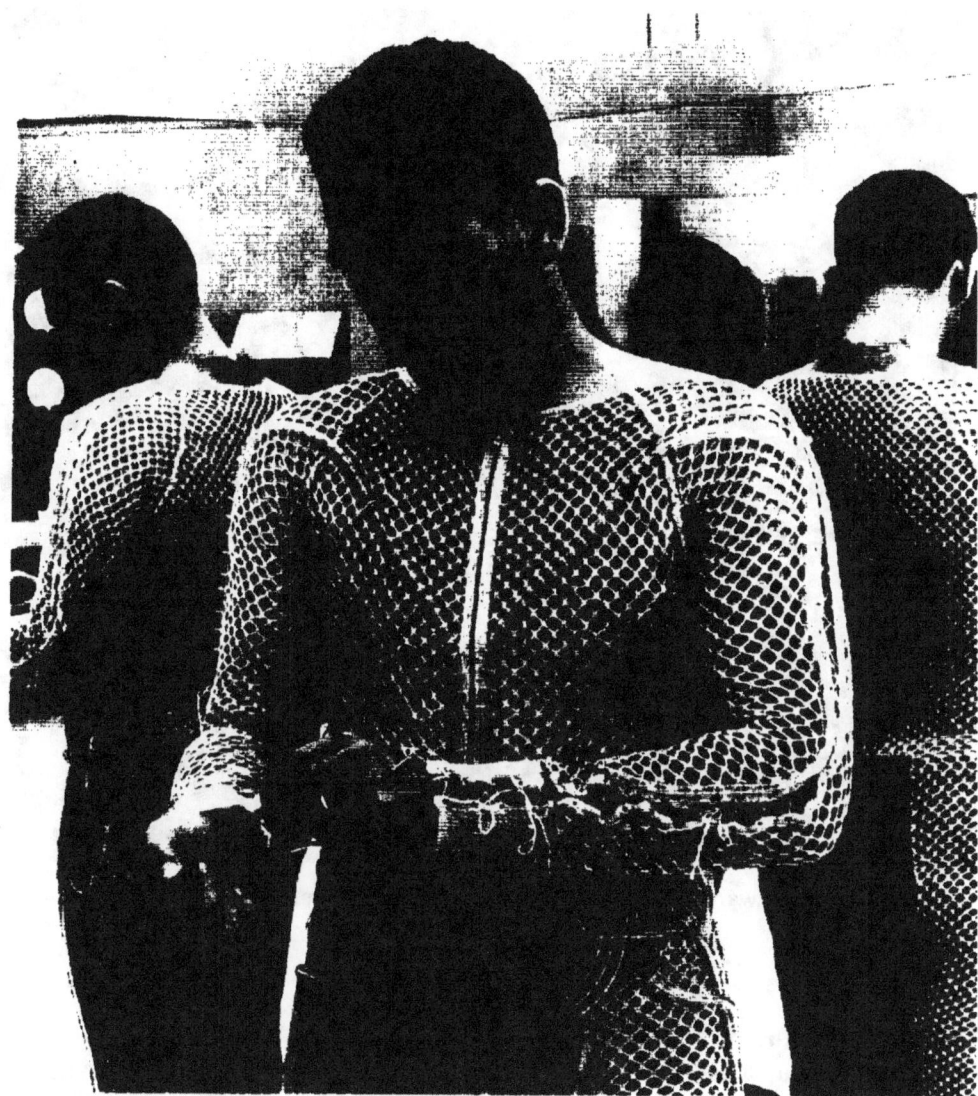

An experimental water-cooled garment, designed to cool space-suit-clad astronauts by water-filled tubes, was delivered to MSC's Crew Systems Division for evaluation. The garments were expected to allow astronauts to work harder and perform more tasks than originally expected.

action followed a July 16 presentation of the electroluminescent concept by Grumman and a review by MSC representatives (among whom were two astronauts, Richard F. Gordon, Jr., and Charles Conrad, Jr.). [Electroluminescence involved the use of a crystalline phosphor to give off light. Advantages of the concept, which was wholly new to manned spacecraft,

1964

August

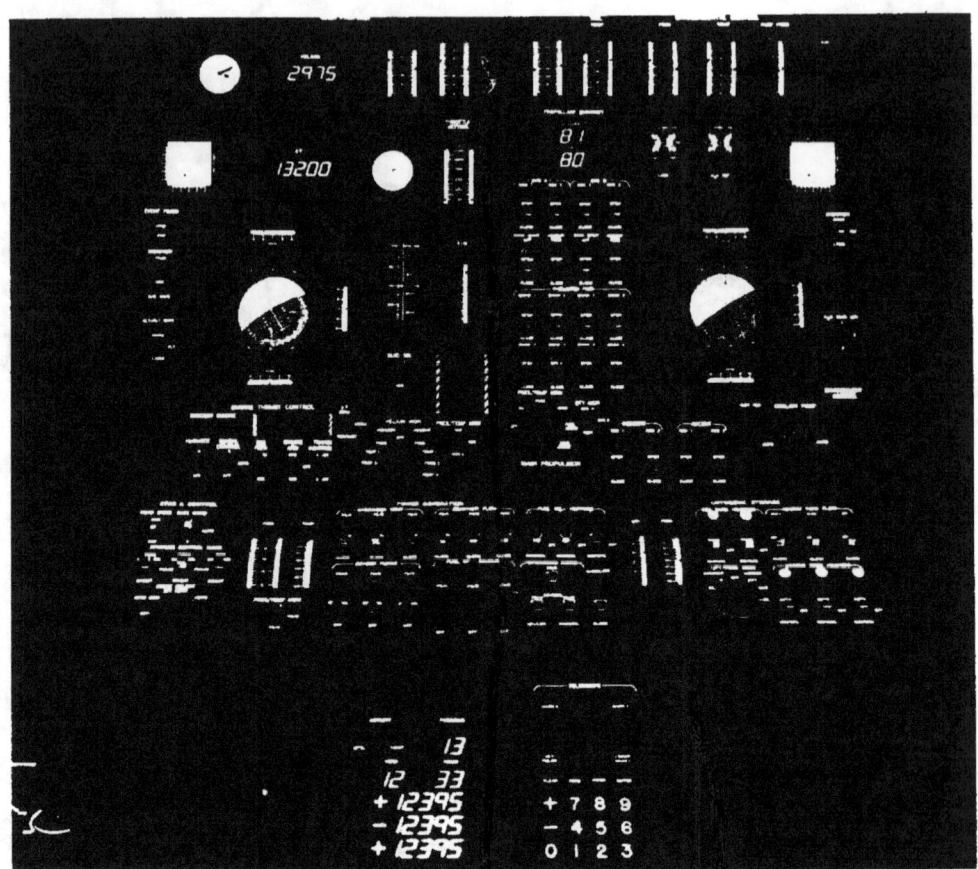

LEM forward display panel showing electroluminescent lighting.

1964

August

were that it used less power and gave off less heat than conventional incandescent bulbs; and, even more significant in the eyes of the astronauts, it was much more even and had an "afterglow" of less than one second.]

> Letter, W. F. Rector III, MSC, to GAEC, Attn: R. S. Mullaney, "Contract NAS 9–1100, Lighting Mockup Review," with enclosure: "Abstract of Proceedings, LEM Crew Integration Meeting, GAEC, Bethpage, L. I., New York, Subject: LEM Interior Lighting Review," July 17, 1964.

4

At a meeting at MSC on July 23, MIT outlined aids and radar display requirements, as well as landing site selection procedures, for lunar landing. This included the recticular patterns on the LEM window that designated where the vehicle was coming down and which enabled the pilot to make touchdown corrections. There was a good deal of concern that, at some time during the final letdown phase, dust might obscure the astronauts' vision and make the radar data unreliable. To overcome this, MSC ordered Grumman to use inertially derived data to monitor automatic touchdown or as a basis for switching to manual control of the descent.

PART III: DEVELOPING SOFTWARE GROUND RULES

Letter, W. F. Rector III, MSC, to GAEC, Attn: R. S. Mullaney, "Contract NAS 9-1100, NASA Coordination Meeting L8A, Implementation of Decisions," August 4, 1964, with enclosure: "Minutes of NASA Coordination Meeting L8A, July 23, 1964."

1964
August

ASPO Deputy Manager Robert O. Piland issued a memorandum concerning the Block II SM, as he put it, "to clear up any confusion which may have existed"—and obviously there was some. (See April 16.) On the basis of revised velocity budget requirements, and as a weight-saving scheme, Piland said, the service propulsion tanks in the Block II SM were being shortened. But he emphasized that the length of the spacecraft per se *"will not be reduced,"* and would thus remain the same as the Block I vehicle.

6

Memorandum, Piland, MSC, to Addressees, "Block II Service Module Length," August 6, 1964.

To investigate problems that might be encountered during the LEM's "blast off" from the moon, Grumman conducted "fire in the hole" tests using a 1/10th-scale model of the spacecraft. (See February and March 11, 1963.) These tests showed that the initial shock of the ascent engine's

6–13

Astronauts Frank Borman, left, and Elliot M. See, Jr., demonstrated prototype thermal overgarments designed to protect men on the moon from the direct rays of the sun unscreened by the thick protection of atmosphere available on the earth. On their backs were mockups of units which would provide life support and communications equipment while astronauts were on the surface of the moon.

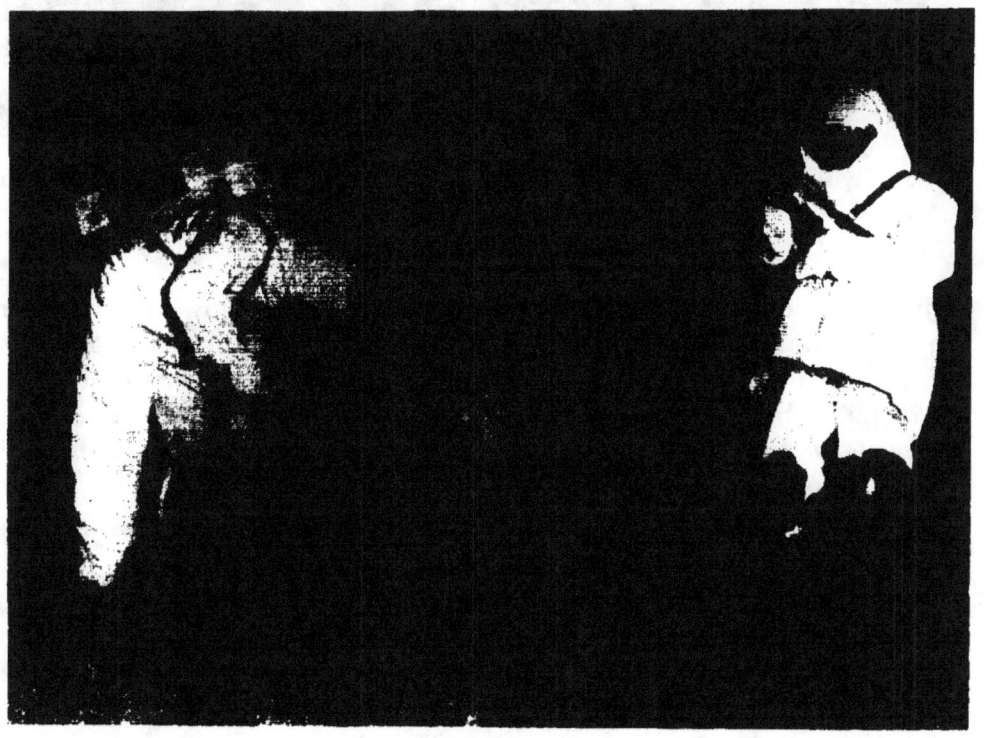

The fully stacked Apollo Boilerplate 15 command module, service module, and adapter section being transported to Pad 37 for mating with the Saturn I.

1964

August

ignition could increase the pressure in the engine nozzle by 2 newtons per square centimeter (3 psi), and that this pressure could vary from one side of the nozzle to the other by as much as 0.53 newtons per square centimeter (0.75 psi). This pressure differential would change the thrust vector and cause an overturning moment on the vehicle. Grumman planned additional testing before actual full-scale firings began at WSMR.

MSC, "ASPO Weekly Management Report, August 6–13, 1964."

7

At North American, engineers from MSC's Crew Systems Division (CSD) reviewed the revised CM couch restraint system. (See May.) CSD still considered the restraint harness unacceptable for use with the pressurized suit. Also the harness attachment gave inadequate restraint when the couch

PART III: DEVELOPING SOFTWARE GROUND RULES

angles were changed and would have to be relocated. North American was asked to install a mirror in the CM to help the astronauts in securing the restraint harness.

1964
August

Ibid.

ASPO's LEM Project Office authorized Grumman to proceed with its subcontractor effort for attitude indicators for the LEM. Until MSC concluded defining the LEM's guidance equipment (anticipated early in November), Grumman should pursue the analog concept (i.e., visual display instruments). (MSC was in the midst of "tradeoff" studies on digital versus analog indicators.) ASPO thus sought to ensure that the manufacturerer did not delay procurement of the devices.

7

Letter, W. F. Rector III, MSC, to GAEC, Attn: R. S. Mullaney, "Contract NAS 9-1100, LEM Attitude Indicator and Gimbal Angle Sequence Transformation Assembly (GASTA)," August 7, 1964.

At its Potrero, Calif., test facility, Lockheed Propulsion Company began qualification testing on the pitch control motors for the launch escape system. Early in September, when the program ended, about two dozen motors had been successfully fired for full duration. Test and reliability results showed that the motors met procurement specifications and had an average specific impulse three percent higher than required.

7

Lockheed Propulsion Company, "Qualification Test Report, Apollo Pitch Control Motor," 588-M-50, December 8, 1964, pp. 1-2, 2-1, 2-2, 2-11.

The modified ring-sail parachutes for the CM's earth landing system demonstrated their potential when Northrop Ventura conducted its first clustered drop using that type of chute.

9–15

MSC, "Weekly Activity Report for the Office of the Associate Administrator, Manned Space Flight, August 9-15, 1964," p. 2.

During late July and early August, MSC and its two spacecraft contractors worked out the dimensions of sample containers and other scientific equipment that would be stowed aboard the spacecraft during lunar missions: 48 by 20 by 29 centimeters (19 by 8 by 11.5 inches). MSC asked Grumman for cost and weight estimates for the containers.

11

Letter, W. F. Rector III, MSC, to GAEC, Attn: R. S. Mullaney, "Contract NAS 9-1100, Results of Meeting on Scientific Equipment Stowage Space," August 11, 1964, with enclosure: "Results of Meeting on Scientific Equipment Stowage Space, July 23, 1964."

In designing batteries for the LEM electrical power system, ASPO ordered Grumman to assume that, if a fuel cell failed, the mission would be aborted.

12

TWX, W. F. Rector III, MSC, to GAEC, Attn: R. S. Mullaney, August 12, 1964.

The U. S. Navy's Air Crew Equipment Laboratory agreed to conduct a series of tests on the water-cooled undergarment. Part I would determine

13–20

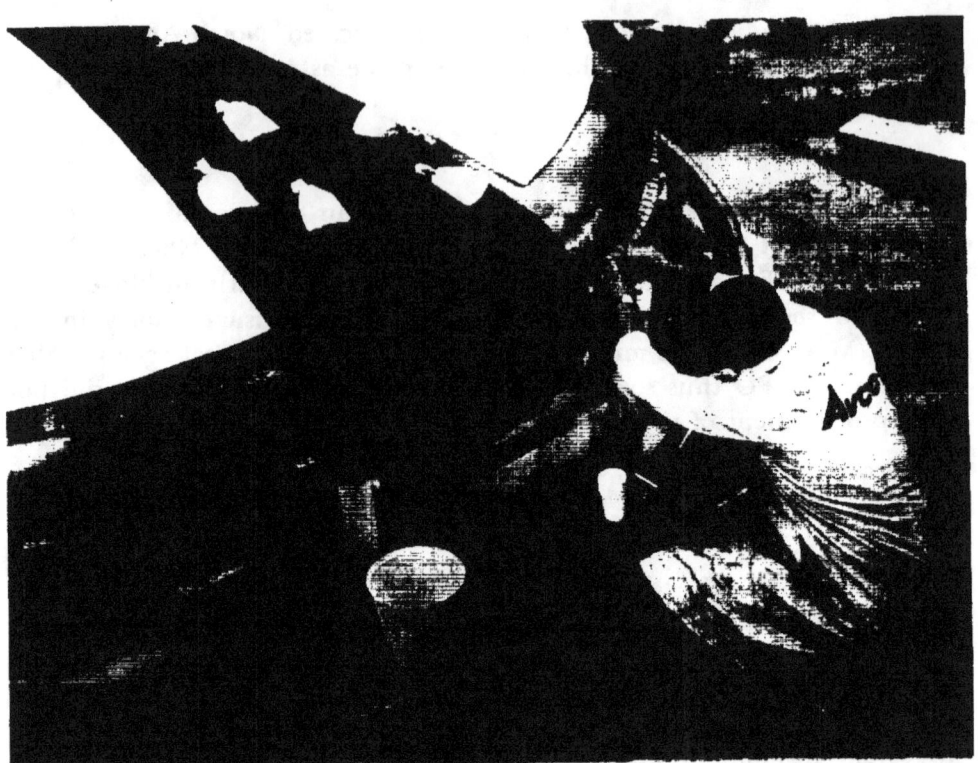

An Avco employee prefitted honeycomb to clamps on edge members of an Apollo command module. The containers held clips that secured edge members to the substructure. The aft compartment faced forward on reentry of the module into the earth's atmosphere and therefore bore the greatest heat.
—Avco Corporation photo.

1964

August

the garment's suitability for the postlanding phase of the mission; Part II would investigate the CM range of temperature that could be tolerated wearing the garment, with and without a space suit.

MSC, "ASPO Weekly Management Report, August 13–20, 1964."

13–20

To save money on the Hamilton Standard contract in Fiscal Year 1965, MSC's Crew Systems Division (CSD) would take over preliminary development of the meteoroid protective garment. Since there was still too little knowledge about the need for meteoroid protection, CSD believed that a concentrated contractor effort was "unwarranted" at that time. (See November 17–December 21, 1963.)

Ibid.

13–September 3

MSC Crew Systems Division engineers evaluated the feasibility of transferring water from the CM to the LEM in lunar orbit. They found that hardware modifications would be needed—either lower water tank pressures

Randolph H. Hester of MSC's Crew Systems Division wore a pressurized space suit and a 13.6-kilogram (30-pound) backpack containing oxygen for coolant, pressurization, and breathing as he traversed a slope at MSC's "moonsite." During the test —part of a dress rehearsal for the crew performance, at Bend, Ore., August 24-28, 1964—Hester used a modified "Jacob's Staff," designed to help him keep his balance.

in the LEM during transfer or a pump added to the water management system in the CM. Six weeks later, Grumman submitted a report confirming that continuous use of CM water from transposition to separation was more desirable than transferring water to the LEM.

> MSC, "ASPO Weekly Management Report, August 13-20, 1964"; "ASPO Weekly Management Report, August 27-September 3, 1964"; "ASPO Weekly Management Report, October 1-8, 1964."

At Baylor University's College of Medicine, investigators presented some results of a joint MSC–Baylor study of human tolerance to low frequency noise (up to 12 cycles per second [cps]). [The study was undertaken because, as launch vehicles for manned spacecraft become larger—i.e., Saturn V and Apollo—they produce higher noise levels, but lower noise frequencies. The possibility of harmful effects upon the crew had to be known.] Audiometry indicated some temporary physiological effect: after three minutes of exposure at levels of about 140 decibels (db), about half of the twenty test subjects suffered some temporary impairment of their hearing. No serious vestibular effects were encountered during noise levels below 12 cps with a maximum of 144 db; heart and respiration rates of the subjects indicated

1964
August

14

THE APOLLO SPACECRAFT: A CHRONOLOGY

1964
August

no severe stresses. Based upon these findings, crew exposure to these noise levels (both frequency and intensity) was considered acceptable.

> MSC, "ASPO Weekly Management Report, August 13-20, 1964;" Burrell O. French et al., *Effects of Low Frequency Pressure Fluctuations on Human Subjects*, NASA TN D-3323, March 1966, pp. 1-2, 7-9.

16—September 15

Studies at North American and at MSC disclosed that, during aborts above 9100 meters (30 000 feet), simultaneous separation of the CM apex cover and the launch escape system (with boost protective cover attached) probably would damage the parachutes or escape hatch. One method of eliminating this hazard was to jettison the apex cover 0.4 second after ignition of the tower jettison motor and firing of the separation bolts. Also being studied were means of sequencing the firing of the jettison motor, the separation bolts, and the heatshield thrusters.

> "Apollo Monthly Progress Report," SID 62-300-29, p. 3; MSC, "Consolidated Activity Report for the Office of the Associate Administrator, Manned Space Flight, August 23–September 19, 1964," p. 63.

16—September 15

North American recommended an uprighting system for the CM composed of three 0.566-cubic-meter (20-cubic-foot) airbags and an inflation system with an electric pump. Using the bags and flooding the aft compartment would maintain a single-point flotation attitude for both Block I and Block II CMs. MSC Structures and Mechanics Division tests of a 1/5-scale model indicated that all three bags were needed to upright the CM. North American contended that any two bags would usually be sufficient, with the third bag providing a redundant capability. The contractor would conduct further

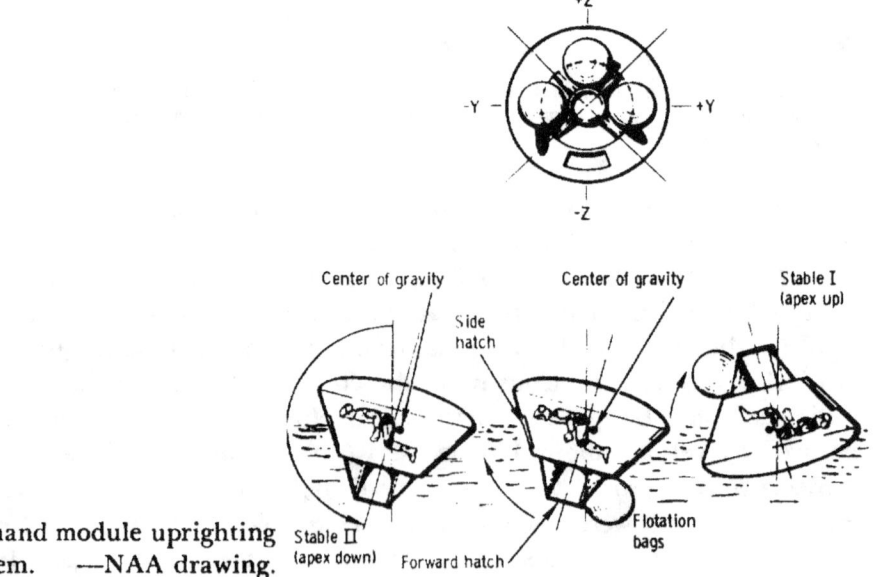

Command module uprighting system. —NAA drawing.

PART III: DEVELOPING SOFTWARE GROUND RULES

tests with inflatable bags (rather than the rigid foam spheres used previously), while MSC would evaluate the use of an extendable boom with two flotation bags.

1964
August

"Apollo Monthly Progress Report," SID 62-300-29, p. 8; MSC, "Consolidated Activity Report for the Office of the Associate Administrator, Manned Space Flight, August 23–September 19, 1964," pp. 45–46.

From Wallops Island, Va., NASA launched another in its series of Scout reentry tests to evaluate the thermal performance of various ablative materials. The material (Avcoat 5026–39, which was being considered for use in the CM's heatshield—see June 10, 1963) was fabricated and bonded in much the same manner as on the actual heatshield. The multi-staged rocket's trajectory propelled the payload into a reentry path that simulated heating loads and shear forces of lunar returns. Though not coming through completely unscathed, the material nonetheless survived.

18

Data on heating, telemetered from the vehicle, established design limits for the ablative material and, thus, were applied to the design of the CM's thermal protection.

James L. Raper (ed.), *Results of a Flight Test of the Apollo Heat-Shield Material at 28,000 Feet Per Second*, NASA TM X-1182, February 1966, pp. 1, 5, 11–12, 23; MSC, "ASPO Weekly Management Report, September 3–10, 1964"; NASA News Release 64-202, "Re-entry Heating Experiment to be Flown by Scout," August 11, 1964.

Thiokol Chemical Corporation began qualification testing on the tower jettison motor. The third motor to be fired in the series, on September 9, experienced a failure of the spot welding on the interstage structure. The motor, now freed, broke apart in the test bay. Analysis of the failure and repairs to the test stand followed, but Thiokol reported that testing could not be resumed until about mid-November—"at the earliest." This foreshadowed a probable delay of about two months in the qualification program.

18

Thiokol Chemical Corporation, Elkton Div., "Apollo Tower Jettison Program, Monthly Progress Report No. 26," A-226, October 14, 1964, pp. ii, 2–12, 32–34; "Apollo Monthly Progress Report," SID 62-300-29, p. 16.

Homer E. Newell, head of NASA's Office of Space Science and Applications, informed MSC Director Robert R. Gilruth that, as NASA had requested (see April 16), the Space Science Board of the National Academy of Sciences had defined the academic requirements for scientist-astronauts for the Apollo program. These requirements demanded graduate studies to the doctorate level, or equivalent.

19

Letter, Newell, NASA, to Gilruth, MSC, August 19, 1964.

MSC's Crew Systems Division (CSD) appraised crew tolerance to SM abort accelerations for Block I spacecraft. Normal mission limits of + •15 g, with total base durations of 50 seconds, were judged tolerable. Under these con-

20–27

Much of the weightless training obtained by the astronauts was gained in Air Force C-135 aircraft following parabolic curves. Here, an Air Force technician braced himself at the side of the cabin as Astronaut Charles A. Bassett II was suspended and Astronaut Theodore C. Freeman balanced himself.

1964

August

ditions, CSD estimated that dizziness or visual disturbance would occur in less than 10 percent of the cases. CSD set emergency limits as + 18 g, with base durations not exceeding 40 seconds.

> MSC, "ASPO Weekly Management Report, August 20–27, 1964."

21

ASPO gave Grumman formal approval to proceed with their concept of a mission programmer for the LEM. The concept, which the contractor had presented in June, involved using the guidance computer as the main sequencing element, with the tape reader as a backup sequencer.

> Letter, W. F. Rector III, MSC, to GAEC, Attn: R. S. Mullaney, "Contract NAS 9–1100, LEM Mission Programer," August 21, 1964; MSC, "ASPO Weekly Management Report, August 20–27, 1964."

23–29

A redesigned thrust chamber (called the "phase C") for the LEM ascent engine was tested in the altitude chamber at Arnold Engineering Development Center. [The "phase C" chamber differed from the "phase B" in that a compression-molded ablative throat section was used.] Firing runs of 60, 380, and five seconds produced only negligible throat erosion. Preliminary

PART III: DEVELOPING SOFTWARE GROUND RULES

data indicated a 2.0-second specific impulse increase over the "phase B" chamber.

1964

August

MSC, "Weekly Activity Report for the Office of the Associate Administrator, Manned Space Flight, August 23–29, 1964." p. 3.

MSC proposed a device affixed to the interior of the spacecraft, called a body-mounted attitude gyro (BMAG), as a backup to the inertial platform in the CM. Should the platform fail during reentry, the pilot could take control of the spacecraft and, using this secondary attitude indicator, fly a safe trajectory. Analog computer analysis indicated the BMAG's feasibility, provided the spacecraft did not maintain a constant roll rate during reentry.

23–September 19

MSC, "Consolidated Activity Report for the Office of the Associate Administrator, Manned Space Flight, August 23–September 19, 1964," p. 49.

MSC completed negotiations with General Electric Corporation (GE) Apollo Support Department for 10 ground stations for spacecraft checkout. (See March 25.) The figure finally agreed upon, $62 244 657 with a $4.1 million fee, was over $20 million less than GE's March quotation.

23–September 19

Ibid., p. 41.

MSC's Technical Services Division (TSD) built a prototype lightweight Apollo couch and test fixture and delivered them to the Crew Systems Division. TSD had designed this couch assembly, as a single unit, to replace previously planned individual couches in the CM, which would save 15.9 kilograms (35 pounds). During subsequent qualification testing, however, the couch did not stand up structurally, and was abandoned. But the concept itself was later useful to North American in the design of their couch arrangement.

23–September 19

Ibid., p. 35; interview, telephone, Ralph Drexel, Houston, March 12, 1970.

At North American, the service propulsion engine was gimbaled during hot firing tests, the first time that the engine had been gimbaled under these conditions. Gimbal operation was satisfactory.

24–28

MSC, "ASPO Weekly Management Report, September 3–10, 1964;" "Apollo Monthly Progress Report," SID 62-300-29, pp. 14–15.

MSC's Crew Systems Division (CSD) conducted mobility tests on lunar-like surfaces near Bend, Oreg. Three types of terrain were used: loose basaltic rubble, low-density pumice with crusty surface and low bearing load, and loose sand. Several CSD engineers and Astronaut Walter Cunningham wore pressurized Apollo prototype space suits and simulated portable life support systems. Climbing steep slopes covered by loose material proved difficult unless aided by ropes. Not surprisingly, how fast they could walk depended upon the terrain. Simple geophysical tasks at the level of the astronaut's feet were easily accomplished, but those requiring good visibility

24–29

Astronaut R. Walter Cunningham climbed a slope near Paulina Lake, about 80 kilometers (50 miles) from Bend, Ore., while wearing a pressurized suit.

1964

August

and dexterity were almost impossible and were better accomplished at a working level of between one and four feet above the ground. The only problems with the space suit were fogging of the visor, inadequate ventilation, and stiffness in the hips and ankles of the suits.

> MSC, "ASPO Weekly Management Report, August 27-September 3, 1964"; "ASPO Weekly Management Report, September 3-10, 1964"; MSC, "Consolidated Activity Report for the Office of the Associate Administrator, Manned Space Flight, August 23-September 19, 1964," p. 65; memorandum, Willis B. Foster, NASA, to Assoc. Adm., Manned Space Flight, "Apollo Field Simulations," September 8, 1964; MSC, *Space News Roundup*, September 2, 1964, p. 1.

25

At a Contractor Coordination Meeting on June 9-10, the point had been made that there existed a single-point failure that would preclude the crew's safe return—a disabled crewman in the CM during LEM operations. MSC demanded unequivocally that, even under these circumstances, the two crewmen in the LEM must be able to complete the mission. Therefore, the CSM must be designed for such a contingency; and to limit hardware impact, this must be done by using onboard equipment as much as possible.

Accordingly William F. Rector III, the LEM Project Officer in ASPO, advised Grumman of two operational requirements:

(1) The radar transponder in the CSM must be turned on before the LEM's ascent from the moon and must be pointed toward the LEM during ascent and rendezvous.

(2) The CSM's attitude had to be stabilized during this phase of the mission.

The two prime contractors, Rector said, should decide on some means of controlling remotely the CSM's transponder and its stabilization and control system. The contractors should, however, use the simplest and most reliable arrangement. To initiate these two functions, the CSM would receive commands from the ground. Finally, Rector informed Grumman of a new ground rule on CSM communications: continuous communications, both telemetry and voice, must be maintained whenever the spacecraft was in view of the earth.

> Letter, Rector, MSC, to GAEC, Attn: R. S. Mullaney, "Contract NAS 9-1100, Operations Groundrule for Disabled CSM Astronaut," August 25, 1964.

Apollo operational radiation protection was divided into two categories: personal dosimeters (attached to the space suit) and a portable, hand-held, radiation survey meter. Grumman was directed to provide a readily accessible stowage location aboard the LEM for the meter, which would weigh about 0.5 kilogram (one pound) and measure approximately 51 x 51 x 191 millimeters (2 x 2 x 7.5 inches).

> Letter, W. F. Rector III, MSC, to GAEC, Attn: R. S. Mullaney, "Contract NAS 9-1100, Space Allocation for LEM Radiation Instrumentation," August 25, 1964.

MSC's Crew Systems Division (CSD) concluded that, in terms of weight and complexity, the "buddy system" concept for supporting two crewmen on a single portable life support system (see July 28–August 3, 1963) was undesirable. An additional emergency oxygen system seemed more practical. The suit assembly already provided at least five minutes of emergency life support; this extra system would afford another five, at a cost of only 1.4 kilograms (three pounds). Consequently CSD redefined the rescue requirement to mean simply "the capability for the crewman remaining in the spacecraft to egress . . . and attend or retrieve the crewman in distress."

> Memorandum, Richard S. Johnston, MSC, to Asst. Chief, Systems Engineering Div., "Portable Life Support System emergency operation," August 26, 1964.

1964

August 30

North American reported that qualification testing had been completed on the launch escape motor. In all, 20 motors had been successfully static fired. (See June 19.)

> MSC, "Project Apollo Quarterly Status Report No. 9 for Period Ending September 30, 1964," p. 17; MSC, "ASPO Weekly Management Report, September 3–10, 1964."

30–September 5

MSC decided to use total mission elapsed time, instead of Greenwich mean time, as the time reference for mission operations. (See February 27, 1963.) North American and Grumman were directed to provide a common format for this display.

> MSC, "Weekly Activity Report for the Office of the Associate Administrator, Manned Space Flight, August 30–September 5, 1964," p. 3.

31

Robert E. Smylie, of MSC's Crew Systems Division (CSD), advised that, as a consequence of MSC's canceling the requirement for inflight maintenance, there were no longer any provisions for tools or for a tool belt inside the spacecraft. Smylie reported that CSD was developing a belt for carrying tools and small equipment needed on the lunar surface, which would be stowed along with the scientific equipment in the LEM's descent stage.

> Memorandum, Smylie, MSC, to Systems Engineering Div., Attn: Lee N. McMillion, "Extravehicular equipment belt," August 31, 1964.

31

Studies of future Gemini and Apollo missions showed that at least four flight directors would be needed. MSC Director Robert R. Gilruth named Christopher C. Kraft, Jr., John D. Hodge, Eugene F. Kranz, and Glynn S. Lunney to these positions. The flight directors would manage all flight operations from launch to recovery. Their responsibilities would include making operational decisions on spacecraft performance, implementing flight plans, and ensuring the safety of the astronauts.

> MSC Announcement 64-120, "Designation of Flight Directors," August 31, 1964; MSC News Release 64-150, September 4, 1964.

During the Month

During zero g tests at Wright-Patterson Air Force Base, subjects wearing pressurized Gemini space suits got into the Apollo crew couch and attached the restraint harness. They entered through a Block II CM tunnel 73.6 centimeters (29 inches) in diameter. One subject made the transfer with a portable life support system (PLSS) strapped on his back and another with the PLSS carried in his hands. One subject also went through the tunnel with an 24.7-meter (81-foot) umbilical hose attached to his suit. These tests demonstrated the feasibility of moving the couch to the earth landing position without readjusting the restraint harness; also they pointed up the need for improving the lap belt.

> MSC, "ASPO Weekly Management Report, September 3–10, 1964."

PART III: DEVELOPING SOFTWARE GROUND RULES

Flight Directors, clockwise from upper left, Glynn S. Lunney, John D. Hodge, Christopher C. Kraft, Jr., and Eugene F. Kranz around the Flight Director's console in Mission Control at MSC.

MSC Crew Systems Division reported that the present water capacity of the LEM (181 kilograms; 400 pounds) was sufficient for either a 35-hour lunar stay with a nine-hour orbital contingency or for a 44-hour lunar stay with no reserve. No excessive weight growths were needed to accomplish this mission flexibility.

1964

September

1

> Memorandum, Richard S. Johnston, MSC, to Asst. Chief, Systems Engineering Div., "LEM ECS Water Provisioning," September 1, 1964; MSC, "Consolidated Activity Report for the Office of the Associate Administrator, Manned Space Flight, August 23–September 19, 1964," p. 19.

NASA and North American signed an amendment to the prime contractor's Apollo contract, extending that agreement to February 15, 1966. The amendment called for production of five additional CSM's (flight articles), three more boilerplate spacecraft, another full-scale mockup, and nine adapters which house the LEM. (See August 14, 1963.) The $496 million amendment increased the estimated value of North American's contract

1

1964

September

(including cost and fee) to over $1.436 billion. Also, the amendment forecast, beyond that February 1966 date, production of 20 more spacecraft.

<blockquote>Oakley, Historical Summary, S&ID Apollo Program, p. 25; MSC, "Consolidated Activity Report for the Office of the Associate Administrator, Manned Space Flight, August 23–September 19, 1964," p. 40; NASA Note to Editors, "Correction on Release No: 64-277 Friday, Sept. 4, 1964," September 11, 1964.</blockquote>

2–9

The alternate mode of escape tower jettison called for firing the launch escape motors. Analyzing the structural integrity of a tower thus jettisoned, MSC Structures and Mechanics Division calculated that it would hold together for 3.5 seconds at least. By that time, it would be 610 meters (2000 feet) away from the flight path of the spacecraft and launch vehicle. This second method for shedding the tower would be tested on the forthcoming AS–102 mission. (See September 18.)

<blockquote>MSC, "ASPO Weekly Management Report, September 3–10, 1964."</blockquote>

A flight kit assembly which would store the equivalent of a 12 000-page library of documents for astronauts was being developed at NAA's Space and Information Systems Division in September 1964. The assembly, a locking case with hinged cover to serve as a lapboard writing surface, had a 10- by 13-centimeter (4- by 5-inch) projection screen in the upper left hand corner of the cover. Its film was coded and indexed so that the astronaut could select any page from 1 to 12 000 and receive it on the display screen in 15 seconds or less. A space-suited engineer held an early model of the flight kit assembly for the photo during a test at NAA.

PART III: DEVELOPING SOFTWARE GROUND RULES

MSC awarded a $2 296 249 contract to Westinghouse Electric Corporation for the LEM television camera. The first test model was scheduled for delivery to Houston in March 1965.

> MSC, "Consolidated Activity Report for the Office of the Associate Administrator, Manned Space Flight, August 23–September 19, 1964," pp. 42, 58.

1964
September
3

MSC issued a definitive contract to Kollsman Instrument Corporation for the LEM optical subsystem. A statement of work had gone into effect on March 10 and had been implemented by technical directives from MIT to Kollsman. The definitive contract covered work until December 31. After that date, Kollsman would become a subcontractor to AC Spark Plug.

> *Ibid.*, p. 40; Kollsman Instrument Corporation, "LEM [Optics] Program Quarterly Technical Progress Report No. 1," September 30, 1964, pp. Kv, K1-1, K2-1.

3

To evaluate lunar surface light, Astronauts Edwin E. Aldrin, Jr., Elliot M. See, Jr., and David R. Scott (accompanied by engineer pilots) began simulated landing approaches over lava flats in southern Idaho. They wore dark glasses that had been modified to permit rapid change to progressively darker (or lighter) filters. Diving in T–33 aircraft from 4600 meters (15 000 feet), they leveled off at 900 meters (3000 feet). See, who had also participated in helicopter exercises earlier in California, believed that the reflected earthshine would be insufficient to allow a LEM pilot to avoid deep surface cracks or large boulders. He also thought that earthshine would limit the crew's visibility to only a short distance. Aldrin, however, felt that this was a pessimistic view. He suggested that the LEM might be equipped with landing lights or flares.

> *The Houston Post*, September 3, 1964; Jim Maloney, *The Houston Post*, September 12, 1964; interview, telephone, Dean F. Grimm, MSC, January 27, 1970.

3

Grumman and the Link Division signed a definitive cost-plus-incentive-fee contract (valued at $7 083 022) for two LEM simulators.

> MSC, "ASPO Weekly Management Report, September 3–10, 1964;" "ASPO Weekly Management Report, September 10–17, 1964."

3–10

North American gave Miineapolis–Honeywell an official go-ahead to begin design work on the Block II CSM stabilization and control system.

> MSC, "ASPO Weekly Management Report, September 3–10, 1964."

3–10

Representatives of Geonautics, Inc., reported on the status of their study of selenodetic experiments for early lunar surface missions. (See June 9.) Results to date indicated that lunar survey measurements could rely heavily on photographic data acquired on the lunar surface.

> MSC, "Consolidated Activity Report for the Office of the Associate Administrator, Manned Space Flight, August 23–September 19, 1964," p. 65.

4

The resident Apollo office at Grumman reported that Pratt and Whitney had achieved reliable 100-hour operation of the LEM fuel cell through the

8–11

1964
September

use of new filling methods. This "apparently" had solved the problem of potassium hydroxide deposits stopping up the cell, the cause of early plugging failures (i.e., after only 10 hours of operation). Some cells, in fact, had run between 200 and 400 hours before failing, the office reported. On the other hand, carbonate plugging was still a problem.

> MSC, "ASPO Weekly Management Report, September 10–17, 1964."

9

Robert E. Smylie, of MSC's Crew Systems Division, asked the Crew Performance Section of the Center's Space Medicine Branch to test the capability of men in space suits to roll over in 1/6 g. In a previous test, using a mockup portable life support system (PLSS), a subject lying on his back had been unable to turn over. Two different PLSS configurations and two kinds of thermal garments would be tested with the Apollo suit. Also an emergency oxygen system mockup would be attached to the helmet.

> Memorandum, Smylie, MSC, to Chief, Space Medicine Branch, "Testing of Apollo SSA roll-over capability in 1/6 g," September 9, 1964.

9

NASA directed North American to add the electronics equipment needed to enable the crew to gimbal the service propulsion engine by using the rotational hand controller.

> Letter, H. P. Yschek, MSC, to NAA, Space and Information Systems Div., "Contract Change Authorization No. 250," September 9, 1964.

11

MSC issued a definitive contract to AC Spark Plug for LEM guidance and navigation equipment. (See October 18, 1963, and June 12.) Estimated cost and fee of the contract was $2.316 million.

> MSC, "Consolidated Activity Report for the Office of the Associate Administrator, Manned Space Flight, August 23–September 19, 1964," p. 40.

14

MSC issued three amendments (worth $6 134 113) to Grumman's LEM contract. These amendments provided funds for data acquisition equipment that MSC formerly was to have furnished; for static test stands at WSMR; and for additional systems engineering studies by Grumman.

> *Ibid.*

14

ASPO issued ground rules for Grumman and MIT to use when defining the LEM guidance and control system. MSC's concerns related to provision for lunar landing aborts and recognition of guidance and control equipment failures. An example of rules during an abort stated that the system should be able to provide information for the astronauts to fire the engines and gain orbital flight on the first effort after initiating an abort. If the first attempt failed, procedures had to specify how the crew could use the system to achieve orbit and then rendezvous and dock with the CM. The second matter concerned investigations to assure that failures in the guidance and

PART III: DEVELOPING SOFTWARE GROUND RULES

A new design of the LEM landing gear foot pad (Aladdin's lamp) was checked by shopmen before the honeycomb core, left, was placed inside for bonding.
—Grumman photo.

control system could be detected and to define what responses the crew must make to those failures.

Letter, W. F. Rector III, MSC, to GAEC, Attn: R. S. Mullaney, "Contract NAS 9-1100, Ground Rules for LEM Guidance and Navigation Operation and Monitoring," September 14, 1964.

North American completed modifications to CM boilerplate (BP) 6, which had been used in Apollo mission PA-1 (see November 7, 1963). The spacecraft, now designated BP-6A, was then delivered to Northrop Ventura for use as a parachute test vehicle.

"Apollo Monthly Progress Report," SID 62-300-29, p. 1.

The first attitude-controlled Little Joe II (see May 1963) was shipped to WSMR. This vehicle would be used for Mission A-002, scheduled for December 1964.

Little Joe II Test Launch Vehicle, NASA Project Apollo, Final Report, p. 1-6.

William A. Lee of ASPO outlined minimum communications requirements for "near-lunar" operations. Those of a general nature included two-way voice communication between spacecraft and ground at any time when a line-of-sight existed with the tracking network. Also there should be

1964
September

14

14

15

1964 **September**	provisions so that the crew could manuever the spacecraft to control antenna position when needing to acquire or reacquire the communication link with the ground.

Requirements for specific phases of the mission—the trip from earth to moon, lunar orbit, and the flight to earth—were also covered:

• Translunar: must be able to transmit, track, and receive telemetry data, television, voice simultaneously at least 50 percent of the time (half-hour on and half-hour off) and, on occasions, as much as two hours at a time.

• Lunar Orbit: (a) continuous voice except when behind the moon and out of sight with the ground network; (b) continuous voice between the LEM and the spacecraft at all times when the LEM was flying—descending or ascending.

• Transearth: the same as translunar.

<small>Memorandum, Lee, MSC, to Addressees, "CSM Lunar Mission Communications Requirements," September 15, 1964.</small> |
| 16 | The Air Force released Launch Complex 16 of its Eastern Test Range to NASA for use as a service propulsion system test facility and static firing stand.

<small>"Apollo Quarterly Status Report No. 9," p. 47.</small> |
| 17 | The first production CM environmental control system was installed in boilerplate 14, and pressurization tests on the water-glycol system were begun. Contamination checks, servicing, and checkout were completed near the end of the month.

<small>MSC, "ASPO Weekly Management Report, September 10–17, 1964"; "ASPO Weekly Management Report, September 24–October 1, 1964"; "Apollo Quarterly Status Report No. 9," p. 47.</small> |
| 17–24 | MSC's Instrumentation and Electronic Systems Division (IESD) advised ASPO that it would probably recommend a second steerable S-band high gain antenna on the CSM. IESD based this assertion upon the operational requirements for communications, the need for reliability, and constraints imposed by the spacecraft's attitude. The division was giving Lockheed Electronics Company the job of analyzing the problems of acquisition and tracking with the high gain antennas on both spacecraft, and thus made the dual-antenna concept for the CSM a part of that study. Also included in Lockheed's study were: an RF (radio frequency) tracking system, comparing it with the current infrared concept; and an inertial reference system for acquisition.

<small>MSC, "ASPO Weekly Management Report, September 17–24, 1964."</small> |
| 18 | Apollo Mission A-102, the second flight of an Apollo spacecraft with a Saturn I (SA-7) launch vehicle, was launched from Complex 37B of the |

PART III: DEVELOPING SOFTWARE GROUND RULES

Eastern Test Range at 11:22:43 a.m., e.s.t. [The first such flight was Mission A-101, with boilerplate (BP) 13, launched on May 28.] A-102 used BP-15, essentially the same configuration as BP-13 except that one of the SM's simulated reaction control system quadrant assemblies was instrumented to measure launch temperatures and vibrations. The mission was intended to demonstrate (1) spacecraft/launch vehicle compatibility, (2) launch and exit parameters to verify design, and (3) the alternate mode of escape-tower jettison (i.e., using the launch escape and pitch control motors).

The launch azimuth was again 105 degrees. The S-I stage shut down at T + 147.4 seconds, only 0.7 second later than planned. The S-I and S-IV stages separated at T + 148.2 seconds, and the S-IV stage ignited 1.7 seconds after that. The launch escape tower was jettisoned at T + 160.2 seconds. S-IV cutoff took place at T + 621.1 seconds, burning 1.3 seconds longer than anticipated. The spacecraft and S-IV were inserted into orbit at 631.1 seconds (2.0 seconds late), at a velocity of 7810.05 meters (25 623.54 feet) per second. The spacecraft weight at insertion was 7815.9 kilograms (17 231 pounds). Orbital parameters were 212.66 and 226.50 kilometers (114.85 and 122.37 nautical miles), and the period 88.64 minutes.

All spacecraft test objectives were met. Satisfactory engineering data verified the launch and exit design criteria. The launch escape and pitch control motors moved the launch escape system safely out of the path of the spacecraft. The Manned Space Flight Network obtained telemetry data into the fifth orbit, at which time the transponders stopped working, but several stations continued to track the vehicle until it reentered over the Indian Ocean on its 59th journey around the earth. As with BP-13, no recovery of the spacecraft was planned.

MSC, "Postlaunch Report for Apollo Mission A-102 (BP-15)," MSC-R-A-64-3 (October 10, 1964), pp. 1-1, 2-1, 3-4, 3-5, 3-6, 5-1, 6-1, 7-15.

1964
September

ASPO asked Grumman to investigate automatic switching mechanisms for LEM VHF and S-band omnidirectional antennas. If such devices were used in manned flights, the crew would need to pay only minimum attention to antenna selection; on unmanned flights, it would improve communication operations and range.

TWX, W. F. Rector III, MSC, to GAEC, Attn: R. S. Mullaney, September 18, 1964.

18

"Fire-in-the-hole" tests of the LEM's ascent engine (see February 1963) were completed at Arnold Engineering Development Center after 18 successful runs. Visual inspection showed no damage to the thrust chamber. Grumman confidently reported to MSC that these tests indicated that "the ascent engine can handle the shock" of ignition with its exhaust nozzle enclosed by the descent stage of the vehicle.

MSC, "Weekly Activity Report for the Office of the Associate Administrator, Manned Space Flight, September 20-26, 1964," p. 3; MSC, "ASPO Weekly Management Report,

20-26

At left Apollo command module boilerplate 15 was checked out at Hangar A-F, Cape Kennedy, before being mated with the SA-7 Saturn I launch vehicle. Below, the total SA-7 vehicle underwent launch preparations on Pad 37.

1964

September

20–30

September 17–24, 1964;" GAEC, "Monthly Progress Report No. 20," LPR-10-36, October 10, 1964, p. 20.

Joseph F. Shea directed that the LEM's television camera built by Westinghouse (see September 3) also be used in the Block II CM. (RCA was the contractor for the Block I's camera.) Engineers from North American and MSC met with Westinghouse representatives to work out the design details (such as mounting, since Westinghouse's camera was larger—and more versatile—than was RCA's).

"Apollo Quarterly Status Report No. 9," p. 2; MSC, "Consolidated Activity Report for the Office of the Associate Administrator, Manned Space Flight, September 20–October 17, 1964," p. 52; MSC, "ASPO Weekly Management Report, October 1–8, 1964"; interview, telephone, Milton G. Kingsley, Houston, March 13, 1970.

PART III: DEVELOPING SOFTWARE GROUND RULES

Rocketdyne conducted its first firing of the prototype LEM descent engine using a new dome manifold injector, called the "Block II" engine (in comparison to the previously tested circumferential manifold type). Rocketdyne reported, in Grumman's words, "no noticeable change in the combustion chamber pattern thrust chamber erosion."

1964
September
20–26

MSC, "Weekly Activity Report for the Office of the Associate Administrator, Manned Space Flight, September 20–26, 1964," p. 3; "Monthly Progress Report No. 20," LPR-10-36, p. 20; interview, telephone, C. Harold Lambert, Jr., Houston, March 19, 1970.

NASA approved Grumman's subcontract with RCA for the LEM attitude and translation control assembly. (See May 1.) The cost-plus-incentive-fee subcontract was valued at $9 038 875.

21

MSC, "Consolidated Activity Report for the Office of the Associate Administrator, Manned Space Flight, September 20–October 17, 1964," p. 39.

North American, MIT, and NASA jointly conducted a series of tests at Wright-Patterson Air Force Base. The tests, in which four astronauts participated, evaluated suit mobility, manipulation of controls, and adjustment of couch and restraints.

21–24

NAA, "Apollo Monthly Progress Report," SID 62-300-30, November 1, 1964, pp. 7–8.

The first SM propulsion engine firing in the F-2 text fixture at WSMR was unsuccessful. Although analysis was incomplete, improper functioning of the engine's main propellant valve might have delayed full combustion until eight seconds after fire signal. In a second test on October 1, the engine was fired for 10 seconds. The engine performed satisfactorily this time, even though oxidizer inlet pressure was below normal.

22

MSC, "ASPO Weekly Management Report, September 17–24, 1964"; "Apollo Monthly Progress Report," SID 62-300-30, pp. 16, 32.

North American and MSC officials negotiated the specifications for the overall Block I CSM system, including special needs for some spacecraft to provide for specific mission objectives. The documents subsequently were incorporated into the North American contract. (See Volume I, July 28 and November 7, 1962; April 28–30, 1964.)

24–27

"Apollo Monthly Progress Report," SID 62-300-30, p. 27.

NASA approved a $14 185 848 contract with North American for spare parts (for Apollo spacecraft and ground support equipment) to expedite repairing of the CSM at WSMR and Cape Kennedy. Spares would include complete electronic packages, hydraulic and mechanical components, reaction control engines, and equipment needed to service the spacecraft.

25

MSC News Release 64-159, September 25, 1964.

MSC Director Robert R. Gilruth approved a Structures and Mechanics Division proposal for three-dimensional dynamic testing of the Apollo

25

A service module propulsion test at WSTF.

1964

September

docking system in a thermal-vacuum environment. Tests were scheduled for late 1965 in the Center's Space Environment Simulation Laboratory.

> MSC, "ASPO Weekly Management Report, September 24–October 1, 1964"; "Apollo Quarterly Status Report No. 9," p. 8.

28

MSC's Crew Systems Division (CSD) advised against increasing the capacity of the portable life support system. CSD contended that the current design was capable of performing a variety of lunar missions (at the maximum design metabolic load of 1600 BTUs per hour) and that the minimum 30 minutes of contingency time was sufficient.

> Memorandum, Richard S. Johnston, MSC, to Systems Engineering Division, "Contingent operation of the Portable Life Support System," September 28, 1964.

PART III: DEVELOPING SOFTWARE GROUND RULES

Richard S. Johnston, Chief of Crew Systems Division, provided Hamilton Standard with some new guidelines and operating procedures formulated by MSC concerning crew transfer from CM to LEM. One major item related to suit umbilicals. A former requirement for end-to-end interchangeability (called the "buddy system") was deleted (see September 19–25, 1963), as was the requirement for quick disconnects at the environmental control system (ECS) outlet. Under MSC's new rules, the crew would transfer with the two cabins unpressurized. Both CM and LEM umbilicals had to be long enough to enable the astronauts to reach the LEM's ECS controls.

1964

September

29

TWX, W. F. Rector III, MSC, to GAEC, Attn: R. S. Mullaney, September 29, 1964; TWX, Richard S. Johnston, MSC, to Hamilton Standard, Attn: R. Breeding, October 8, 1964.

NASA conducted a formal inspection and review of the Block II CSM mockup. [The design resulted from a number of meetings earlier in the year (see April 16 and June 11), a three-month program definition study, and additional investigations requested by NASA.]

30

North American presented mockups of the CM interior, upper deck, lower equipment bay, and the SM with two bays exposed. Actual hardware was

Block II command module's lower equipment bay.

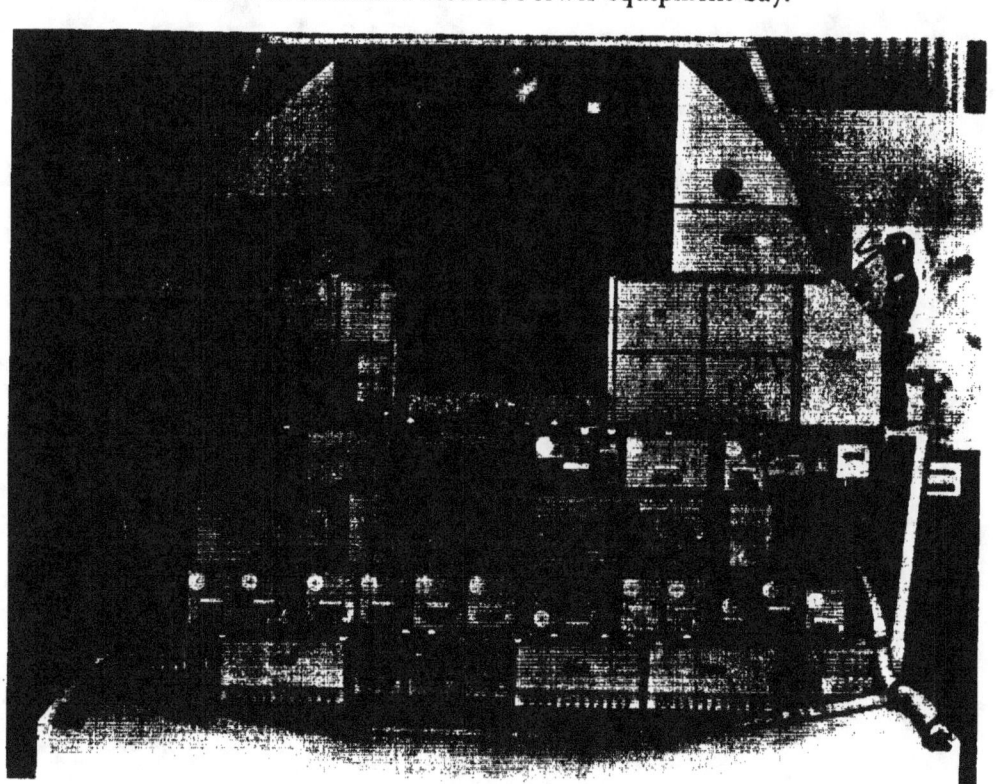

1964
September

simulated. The couches from the Block I review in April were used, with revised harnesses. The Block I inner and outer hatches were displayed, while the CM exterior showed only changes from Block I.

North American explained that this mockup had been designed to depict only volume, space allocations, and arrangements of the CSM. New systems required for Block II were defined only as to maximum size. A detailed mockup, showing actual hardware configuration, of the Block II CSM interior and exterior would be available in February and April, respectively.

> Letter, H. P. Yschek, MSC, to NAA, Space and Information Systems Div., "Contract Change Authorization No. 254," October 1, 1964; MSC, "Command and Service Modules: Project Apollo, Board Report for NASA Inspection and Review of Block II Mockup, September 29–October 1, 1964," pp. 1–4.

APPENDIXES

APPENDIX 1—GLOSSARY OF ABBREVIATIONS

ASPO	Apollo Spacecraft Program Office
BP	Boilerplate
CM	Command module
CSM	Command and service modules
EDD	Engineering and Development Directorate
GAEC	Grumman Aircraft Engineering Corporation
GE	General Electric Company
HF	High frequency
IBM	International Business Machines Corporation
ITT	International Telephone and Telegraph Company
KSC	Kennedy Space Center
LEM	Lunar excursion module
LES	Launch escape system
LEV	Launch escape vehicle
LOC	Launch Operations Center
LTV	Ling-Temco-Vought
MCC	Mission Control Center
MIT	Massachusetts Institute of Technology
MSC	Manned Spacecraft Center
MSF	Manned Space Flight
MSFC	Marshall Space Flight Center
NAA	North American Aviation, Inc.
NASA	National Aeronautics and Space Administration
OMSF	Office of Manned Space Flight
OSSA	Office of Space Sciences and Applications
RASPO	Resident Apollo Spacecraft Program Office
RCA	Radio Corporation of America
RF	Radio frequency
SM	Service module
STL	Space Technology Laboratories, Inc.
VHF	Very high frequency
WSMR	White Sands Missile Range
WSTF	White Sands Test Facility

APPENDIX 2—SPACECRAFT WEIGHTS BY QUARTER

DECEMBER 1962–SEPTEMBER 1964

Item	December 1962			March 1963		
	Control Weight (kgs, lbs)	Target Weight (kgs, lbs)	Current Weight (kgs, lbs)	Control Weight (kgs, lbs)	Target Weight (kgs, lbs)	Current Weight (kgs, lbs)
Command Module	4309 (9500)	3856 (8500)	4246 (9350)	4309 (9500)	3856 (8500)	4067 (8990)
Service Module	5214 (11 500)	4990 (11 000)	4629 (10 205)	4763 (10 500)	4309 (9500)	4336 (9780)
SM Useful Propellant	18 370 (40 500)	15 531 (34 240)	15 744 (34 710)	17 921 (39 730)	16 381 (36 115)	16 860 (37 170)
S–IVB Adapter	1361 (3000)	1361 (3000)	1479 (3260)	1542 (3400)	1361 (3000)	1411 (3110)
Lunar Exc. Module	11 567 (25 500)	11 113 (24 500)	9752 (21 500)	11 961 (26 370)	11 113 (24 500)	11 113 (24 500)
Total Spacecraft Injected	40 823 (90 000)	35 471 (78 200)	35 745 (79 025)	40 823 (90 000)	37 247 (82 115)	38 124 (84 050)

Item	June 1963			September 1963		
	Control Weight (kgs, lbs)	Target Weight (kgs, lbs)	Current Weight (kgs, lbs)	Control Weight (kgs, lbs)	Target Weight (kgs, lbs)	Current Weight (kgs, lbs)
Command Module	4309 (9500)	3856 (8500)	4059 (9170)	4309 (9500)	3856 (8500)	4277 (9650)
Service Module	4763 (10 500)	4309 (9500)	4264 (9620)	4763 (10 500)	4309 (9500)	4291 (9680)
SM Useful Propellant	17 921 (39 730)	11 381 (36 115)	17 060 (37 610)	17 988 (39 900)	16 488 (36 350)	17 958 (39 811)
S–IVB Adapter	1542 (3400)	1361 (3000)	1411 (3110)	1542 (3400)	1361 (3000)	1542 (3400)
Lunar Exc. Module	11 961 (26 370)	11 113 (24 500)	11 521 (25 400)	12 111 (26 700)	11 340 (25 000)	12 916 (28 476)
Total Spacecraft Injected	40 823 (90 000)	37 247 (82 115)	38 471 (85 410)	40 823 (90 000)	37 353 (82 350)	40 285 (91 017)

Item	December 1963			March 1964		
Command Module	4309 (9500)	3856 (8500)	4332 (9770)	4309 (9500)	3856 (8500)	4554 (10 040)
Service Module	4763 (10 500)	4309 (9500)	4408 (9960)	4763 (10 500)	4082 (9000)	4403 (9950)
SM Useful Propellant	22 524 (39 900)	16 488 (36 350)	18 727 (41 285)	16 828 (37 100)	14 662 (32 325)	16 329 (36 000)
S–IVB Adapter	1542 (3400)	1361 (3000)	1542 (3400)	1542 (3400)	1406 (3100)	1542 (3400)
Lunar Exc. Module	12 111 (26 700)	11 340 (25 000)	13 819 (30 465)	13 281 (29 500)	11 567 (25 500)	12 314 (27 149)
Total Spacecraft Injected	40 823 (90 000)	37 353 (82 350)	42 037 (94 880)	40 823 (90 000)	35 573 (78 425)	39 253 (86 539)

APPENDIX 2

Item	June 1964			September 1964		
	Control Weight (kgs, lbs)	Target Weight (kgs, lbs)	Current Weight (kgs, lbs)	Control Weight (kgs, lbs)	Target Weight (kgs, lbs)	Current Weight (kgs, lbs)
Command Module	4309 (9500)	3856 (8500)	4553 (10 030)	4990 (11 000)	*	4576 (10 090)
Service Module	4763 (10 500)	4082 (9000)	4590 (10 120)	4627 (10 200)	*	4559 (10 050)
SM Useful Propellant	16 828 (37 100)	14 662 (32 325)	16 617 (36 635)	17 468 (38 510)	*	16 894 (37 244)
S–IVB Adapter	1542 (3400)	1406 (3100)	1576 (3475)	1724 (3800)	*	1678 (3700)
Lunar Exc. Module	13 281 (29 500)	11 567 (25 500)	12 748 (28 105)	13 281 (29 500)	*	13 250 (29 431)
Total Spacecraft Injected	40 823 (90 000)	35 573 (78 425)	40 082 (88 365)	42 638 (94 000)	*	40 057 (90 515)

* No longer reported.

APPENDIX 3—MAJOR SPACECRAFT COMPONENT MANUFACTURERS

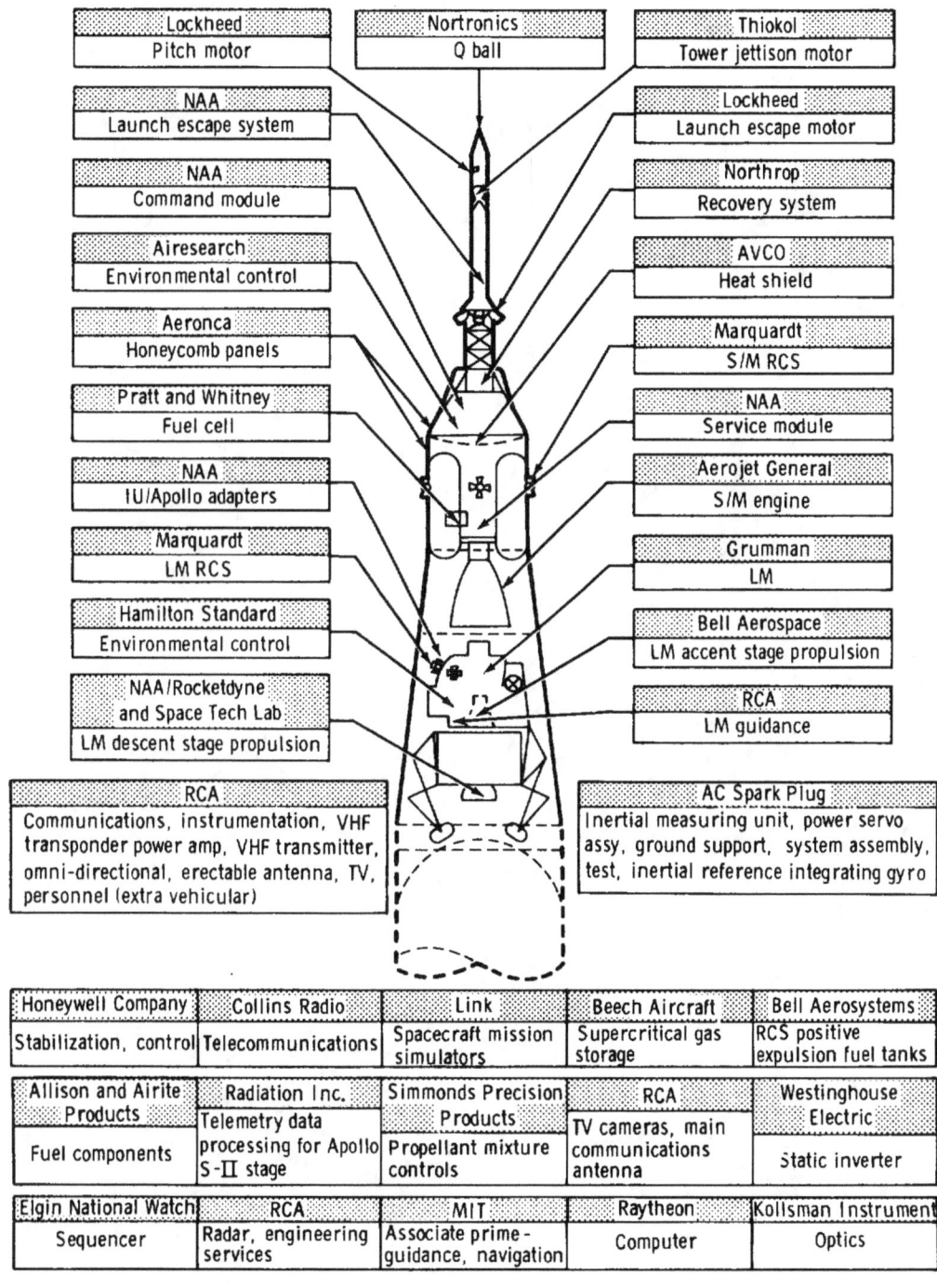

Honeywell Company	Collins Radio	Link	Beech Aircraft	Bell Aerosystems
Stabilization, control	Telecommunications	Spacecraft mission simulators	Supercritical gas storage	RCS positive expulsion fuel tanks
Allison and Airite Products	Radiation Inc.	Simmonds Precision Products	RCA	Westinghouse Electric
Fuel components	Telemetry data processing for Apollo S-II stage	Propellant mixture controls	TV cameras, main communications antenna	Static inverter
Elgin National Watch	RCA	MIT	Raytheon	Kollsman Instrument
Sequencer	Radar, engineering services	Associate prime - guidance, navigation	Computer	Optics

APPENDIX 4—FLIGHT SUMMARY*

[November 8, 1962, through September 30, 1964]

Date	Name	General Mission	Launch Vehicle (site)	Performance Vehicle	Performance Payload	Mission Results
1962						
Nov 16	Saturn (SA-3)	Launch vehicle development test ("Project Highwater")	Saturn C-1 (AMR)	S	S	S
Dec 16	*Explorer XVI* (S-55b)	Scientific micrometeoroid satellite	Scout (WS)	S	S	S
1963						
Mar 28	Saturn (SA-4)	Launch vehicle development test ("engine out" capability test)	Saturn C-1 (AMR)	S	S	S
Apr 2	*Lunik IV*	Lunar probe—reported attempt to soft-land instrument package	Unknown (U.S.S.R.)	S	Unknown	
Apr 2	*Explorer XVII*	Measure atmospheric density, composition, pressure, and temperature at altitudes of 249 to 933 kilometers (155 to 580 miles)	Thor-Delta (AMR)	S	S	S
May 15	*Faith 7* (MA-9)	Project Mercury manned one-day mission—fourth US manned orbital flight	Atlas (AMR)	S	S	S

F—Failure
S—Success

AMR—Atlantic Missile Range
PMR—Pacific Missile Range
WSMR—White Sands Missile Range
WS—Wallops Station

* The launches described in this table include only those related to the exploration of the moon: unmanned lunar probes, unmanned tests of spacecraft designed for later manned missions, and manned spacecraft flights. The table is not intended as a comprehensive summary of all American and Soviet space flights.

THE APOLLO SPACECRAFT: A CHRONOLOGY

Date	Name	General Mission	Launch Vehicle (site)	Performance Vehicle	Performance Payload	Mission Results
June 14	*Vostok V*	Manned orbital space flight—first launch of second tandem flight	Unknown (U.S.S.R.)	S	S	S
June 16	*Vostok VI*	Manned orbital space flight—second launch of second tandem flight	Unknown (U.S.S.R.)	S	S	S
July 20	—	Experimental heatshield reentry (29 934 kilometers per hour) (18 600 mph) test	Scout (WS)	F	F	F
Aug 28	—	Prove capability of Little Joe II as an Apollo spacecraft test vehicle	Little Joe II (WSMR)	S	S	S
Nov 7	Pad Abort–1	Qualification test of Apollo launch escape system to effect a safe pad abort. (Spacecraft BP-6)	— (WSMR)	—	S	S
Nov 26	*Explorer XVIII*	To measure interplanetary magnetic fields, solar wind, and cosmic radiation between earth and moon	Thor–Delta (AMR)	S	S	S
Nov 27	Centaur II	In-space ignition of Centaur's liquid-hydrogen engines (second attempt, first success)	Atlas-Centaur (AMR)	S	S	S
Dec 19	*Explorer XIX*	To measure atmospheric density fluctuations of earth's high latitudes	Scout (PMR)	S	S	S

242

APPENDIX 4

1964							
Jan 29	Saturn (SA-5)	Test structure and performance of 2-stage Saturn; orbit second stage	Saturn I (AMR)	S	S	S	S
Jan 30	*Ranger VI*	Television photographs of moon at close range	Atlas–Agena B (AMR)	S	F	F	F
Apr 8	*Gemini–Titan 1*	Unmanned flight test of structural integrity of Gemini spacecraft and compatibility with launch vehicle	Titan II (AMR)	S	S	S	S
Apr 14	Project Fire	Test of Apollo sample heatshield material at lunar reentry speeds	Atlas D (AMR)	S	S	S	S
May 13	Apollo Mission A-001	Test capability of launch escape system to propel spacecraft from launch vehicle during abort at transonic speed (Spacecraft BP-12)	Little Joe II (WSMR)	S	S	S	S
May 28	Apollo Mission A-101 (SA-6)	First flight of an Apollo-configured spacecraft with a Saturn launch vehicle (Spacecraft BP-13)	Saturn I (AMR)	S	S	S	S
Jul 28	*Ranger VII*	Television photographs of the moon at close range	Atlas–Agena B (AMR)	S	S	S	S
Aug 18	—	Experimental heatshield reentry test	Scout (WS)	S	S	S	S
Aug 25	*Explorer XX*	Map irregularities in topside of earth's atmosphere; obtain electron densities and temperatures near satellite	Scout (PMR)	S	S	S	S
Sep 18	Apollo Mission A-102 (SA-7)	Demonstrate spacecraft-launch vehicle compatibility (Spacecraft BP-15)	Saturn I (AMR)	S	S	S	S

APPENDIX 5—APOLLO PROGRAM FLIGHT OBJECTIVES[*]

Pad Abort 1 (November 7, 1963)

  First Order Objectives:

    (1) Determine aerodynamic stability characteristics of the Apollo escape configuration during a pad abort. (Achieved)

    (2) Demonstrate the capability of the escape system to propel a command module to a safe distance from a launch vehicle during a pad abort. (Achieved)

    (3) Demonstrate launch-escape timing sequence. (Achieved)

    (4) Demonstrate proper operation of the launch-escape tower release device. (Achieved)

    (5) Demonstrate proper operation of the tower-jettison and pitch-control motors. (Achieved)

    (6) Demonstrate earth-landing timing sequence and proper operation of the parachute subsystem of the earth-landing system. (Achieved)

  Second Order Objectives:

    (1) Determine dynamics of command module during jettisoning of escape tower. (Achieved)

    (2) Demonstrate operation of research and development instrumentation and communications equipment to be used on subsequent flights. (Achieved)

    (3) Demonstrate compatibility of prototype handling ground support equipment. (Achieved)

    (4) Determine initial separation trajectory of the launch escape tower. (Achieved)

    (5) Determine escape-tower vibration during pad abort. (Achieved)

Apollo Mission A–001 (May 13, 1964)

  First Order Objectives:

    (1) Demonstrate the structural integrity of the escape tower. (Achieved)

    (2) Demonstrate the capability of the escape subsystem to propel the command module to a predetermined distance from launch vehicle. (Achieved)

    (3) Determine aerodynamic stability characteristics of the escape configuration for this abort condition. (Achieved)

    (4) Demonstrate proper operation of the command module to service module separation subsystem. (Achieved)

    (5) Demonstrate satisfactory recovery timing sequence in the earth-landing subsystem. (Achieved)

---

[*] Apollo spacecraft development flights only

Second Order Objectives:
- (1) Demonstrate Little Joe II-spacecraft compatibility. (Achieved)
- (2) Determine aerodynamic loads caused by fluctuating pressures on the command module and service module during a Little Joe II launch. (Achieved)
- (3) Demonstrate proper operation of the applicable components of the earth-landing subsystem. (Not achieved—a parachute riser chafed against a simulated reaction control subsystem motor. The riser broke after main parachute line stretch, and the command module descended safely on the two remaining main parachutes.)

Apollo Mission A–101 (May 28, 1964)

First Order Objectives:
- (1) Demonstrate physical compatibility of the spacecraft with the launch vehicle under preflight and flight conditions. (Achieved)
- (2) Obtain data to verify design criteria for the launch environment. (Achieved)
- (3) Demonstrate the primary mode of the launch escape tower jettison using the escape tower jettison motor. (Achieved)

Second Order Objectives:
- (1) Demonstrate the structural integrity of the launch escape subsystem under flight-loading conditions. (Achieved)
- (2) Demonstrate the compatibility of the BP–13 communications and instrumentation subsystem with the launch vehicle system. (Achieved)
- (3) Demonstrate the adequacy of ground support handling equipment and procedures. (Achieved)

Apollo Mission A–102 (September 18, 1964)

First Order Objectives:
- None—since Apollo Mission A–101 was successful and the launch and exit environments for the spacecraft were measured satisfactorily.

Second Order Objectives:
- (1) Determine the launch and exit environmental parameters to verify design criteria. (Achieved)
- (2) Demonstrate the alternate mode of spacecraft launch escape system jettison utilizing the launch-escape motor and pitch-control motor. (Achieved)

APPENDIX 6—HARDWARE MANUFACTURE AND ACCEPTANCE

Section A—Boilerplates

No.	Unit	Acceptance Date	Use	Location
BP-1	CM	11-14-62	Land and water impact tests	MSC
BP-2	CM	12-11-62	Land and water impact tests	MSC
BP-3	CM	4-15-63	Parachute recovery	
BP-6	CM	7-01-63	Pad abort	
	LES	7-01-63	Pad abort	
BP-9	CM	3-11-63	Dynamic test	
	SM	3-11-63		
	LES	3-11-63		
	Adapter	3-11-63		
BP-12	CM	2-16-64	Transonic abort	
	SM	2-25-64	Transonic abort	
	LES	2-22-64	Transonic abort	
BP-13	CM	2-17-64	Booster and launch environment compatibility	
	SM	2-15-64		
	LES	2-15-64		
	Adapter	2-14-64		
BP-15	CM	6-14-64	Booster and launch environment compatibility	
	SM	6-05-64		
	LES	6-14-64		
	Adapter	6-05-64		
BP-16	CM	8-17-64	Booster, flight compatibility	
	LES	8-17-64		
	Adapter	8-17-64		
BP-19	CM	2-19-63	Parachute recovery	
BP-23	CM	9-17-64	High-Q abort	
	SM	9-14-64	High-Q abort	
	LES	9-19-64	High-Q abort	
BP-25	CM	10-02-62	Water recovery and handling equipment tests	MSC

No.	Unit	Acceptance Date	Use	Location
BP-26	CM	8-10-64	Micrometeoroid flight	
	SM	8-18-64		
	LES	8-18-64		
	Adapter	8-18-64		
BP-27	CM	9-25-64	Dynamic tests	MSFC
	SM	9-28-64	Dynamic tests	MSFC
	LES	9-25-64	Dynamic tests	MSFC

Section B—Mockups, Trainers, Simulators

No.	Unit	Acceptance Date	Use	Location
M-2	CM	9-29-62	Interior arrangement	KSC
M-3	CM	9-10-62	Interior arrangement	KSC
M-4	SM (partial)	11-14-62	Interface studies	
	Adapter (partial)	11-14-62		
M-5	CM	10-12-62	Exterior arrangement	NAA Storage
M-7	SM	11-04-62	Design studies	MSC
M-9	CM	1-04-63	Handling and transportation studies	Tulsa
	SM	1-04-63		KSC
	LES	1-04-63		
	Adapter	1-04-63		
M-11	CM	1-04-63	Handling and transportation studies	KSC
	SM	1-04-63		Tulsa
	LES	1-04-63		KSC
	Adapter	1-04-63		
M-12	CM (partial)	10-12-62	Crew support studies	
M-22	CM	3-18-64	Interior and exterior arrangement	
M-23	CM (partial)	12-01-64	Umbilical tests	MSFC
	SM (partial)	12-01-64		MSFC
	LES (partial)	12-01-64		MSFC

Excerpted from material compiled by North American Rockwell's Space Division Public Relations Office.

APPENDIX 7—FUNDING

Fiscal Year	Funding Breakdown		
1964			
(Original budget request including Fiscal Year 1963 supplemental)	NASA: $3 926 000 000 Apollo: 2 243 900 000	Command and service modules: Lunar excursion module: Guidance and navigation: Integration, reliability, and checkout: Spacecraft support:	$545 874 000 135 000 000 91 499 000 60 699 000 43 503 000
(Fiscal budget appropriation with Fiscal Year 1963 supplemental)	NASA: $3 974 979 000 Apollo: 2 272 952 000	Saturn I: Saturn IB: Saturn V: Engine development: Apollo mission support:	187 077 000 146 817 000 763 382 000 166 000 000 133 101 000
1965			
(Original budget request including Fiscal Year 1964 supplemental)	NASA: $4 523 000 000 Apollo: 2 818 500 000	Command and service modules: Lunar excursion module: Guidance and navigation: Integration, reliability, and checkout: Spacecraft support: Saturn I:	$577 834 000 242 600 000 81 038 000 24 763 000 83 663 000 40 265 000
(Fiscal budget appropriation with Fiscal Year 1964 supplemental)	NASA: $4 270 695 000 Apollo: 2 614 619 000	Saturn IB: Saturn V: Engine development: Apollo mission support:	262 690 000 964 924 000 166 300 000 170 542 000

Compiled by F. P. Hopson, Program Control and Contracts Directorate

APPENDIX 8—ORGANIZATIONAL CHARTS

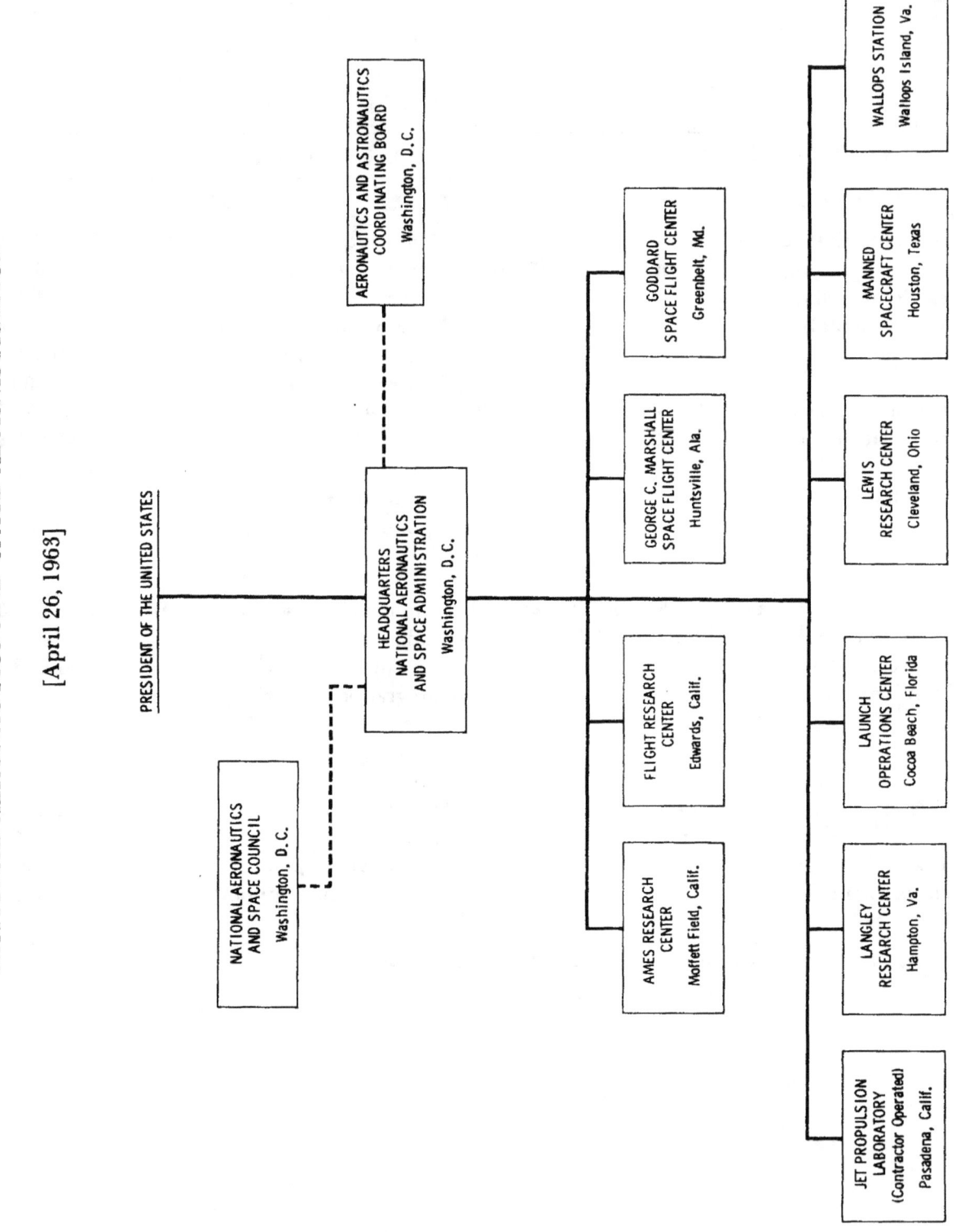

APPENDIX 8

NATIONAL AERONAUTICS AND SPACE ADMINISTRATION

[November 1, 1963]

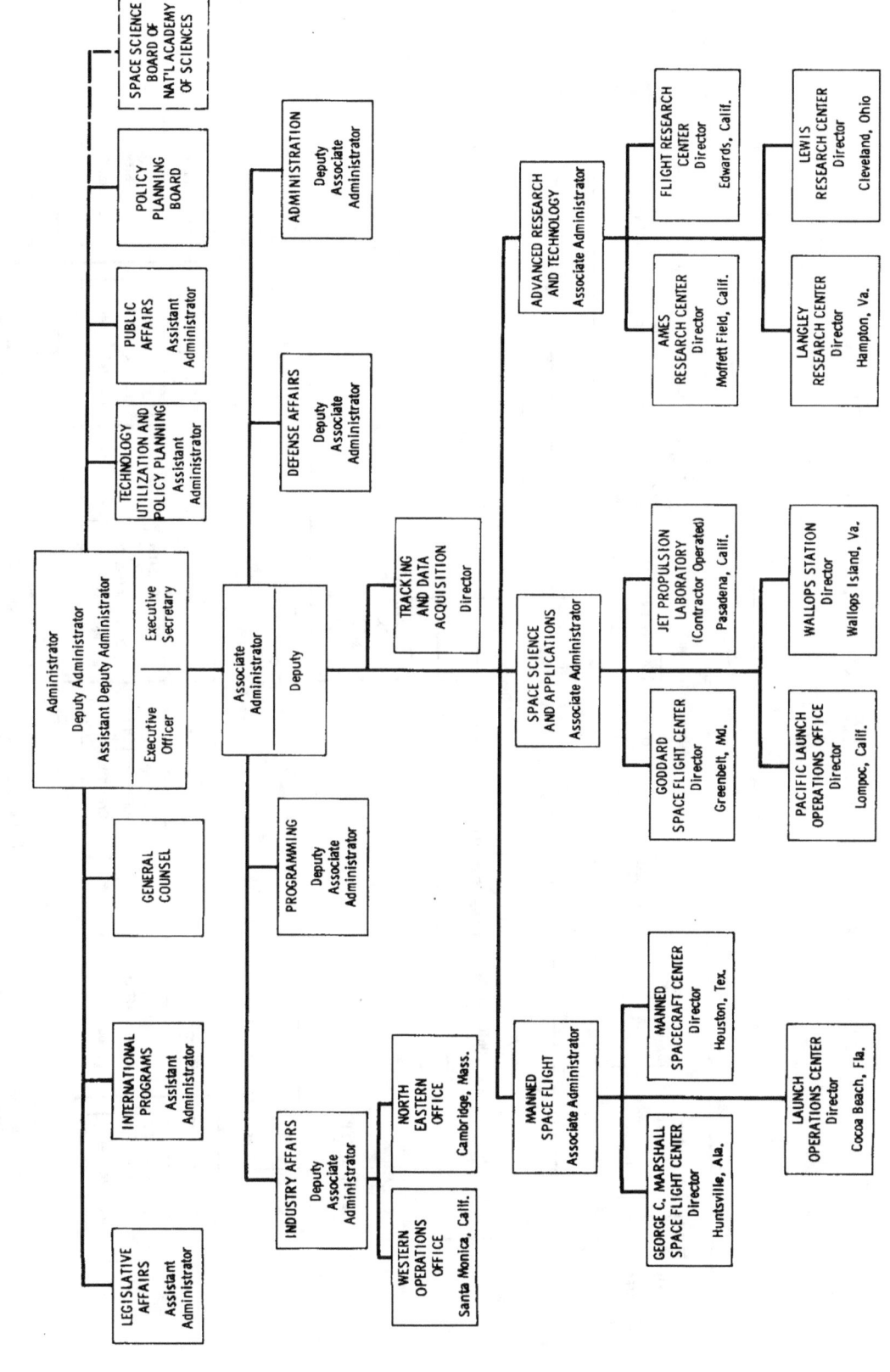

THE APOLLO SPACECRAFT: A CHRONOLOGY

NATIONAL AERONAUTICS AND SPACE ADMINISTRATION
[April 1963]

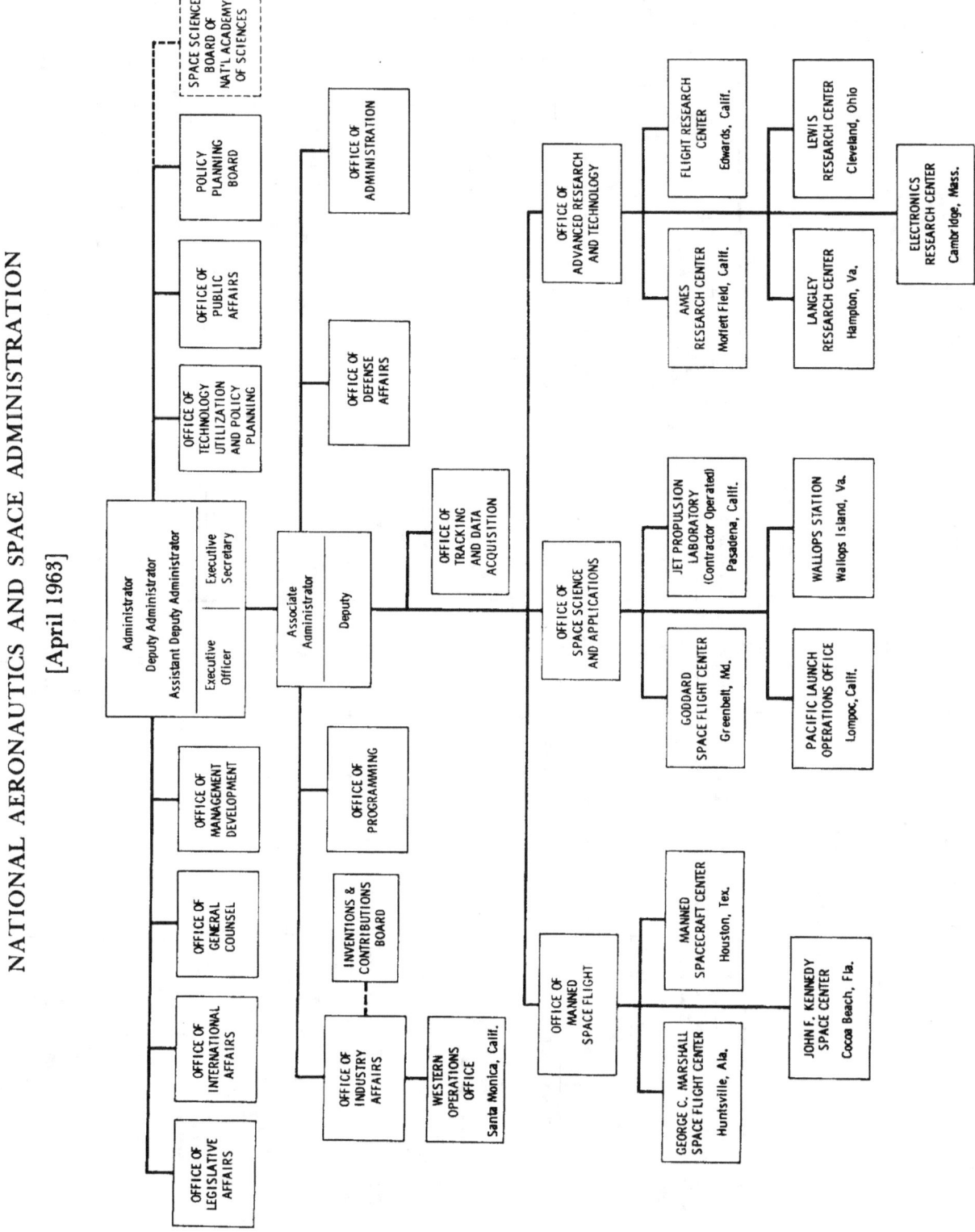

APPENDIX 8

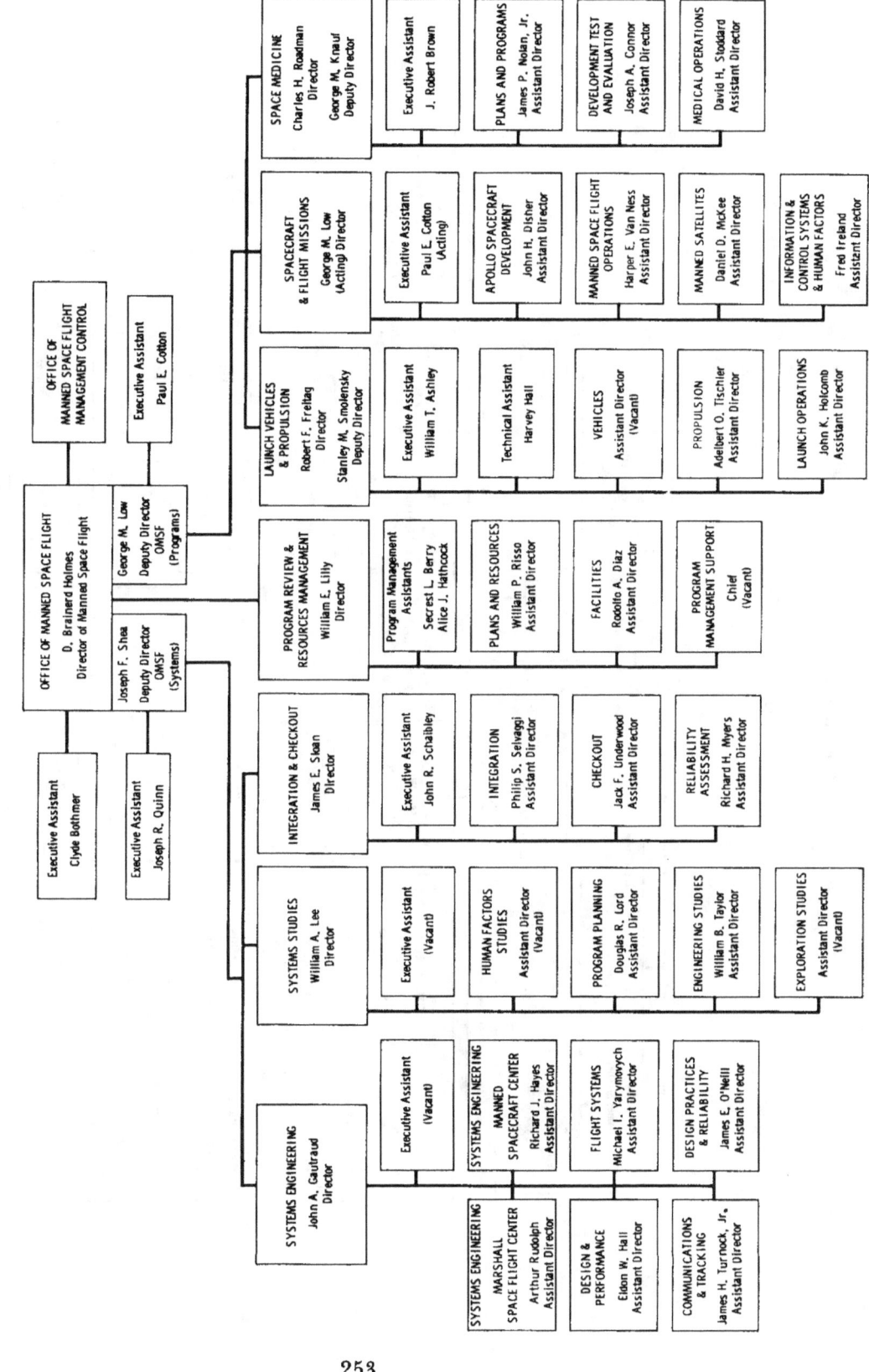

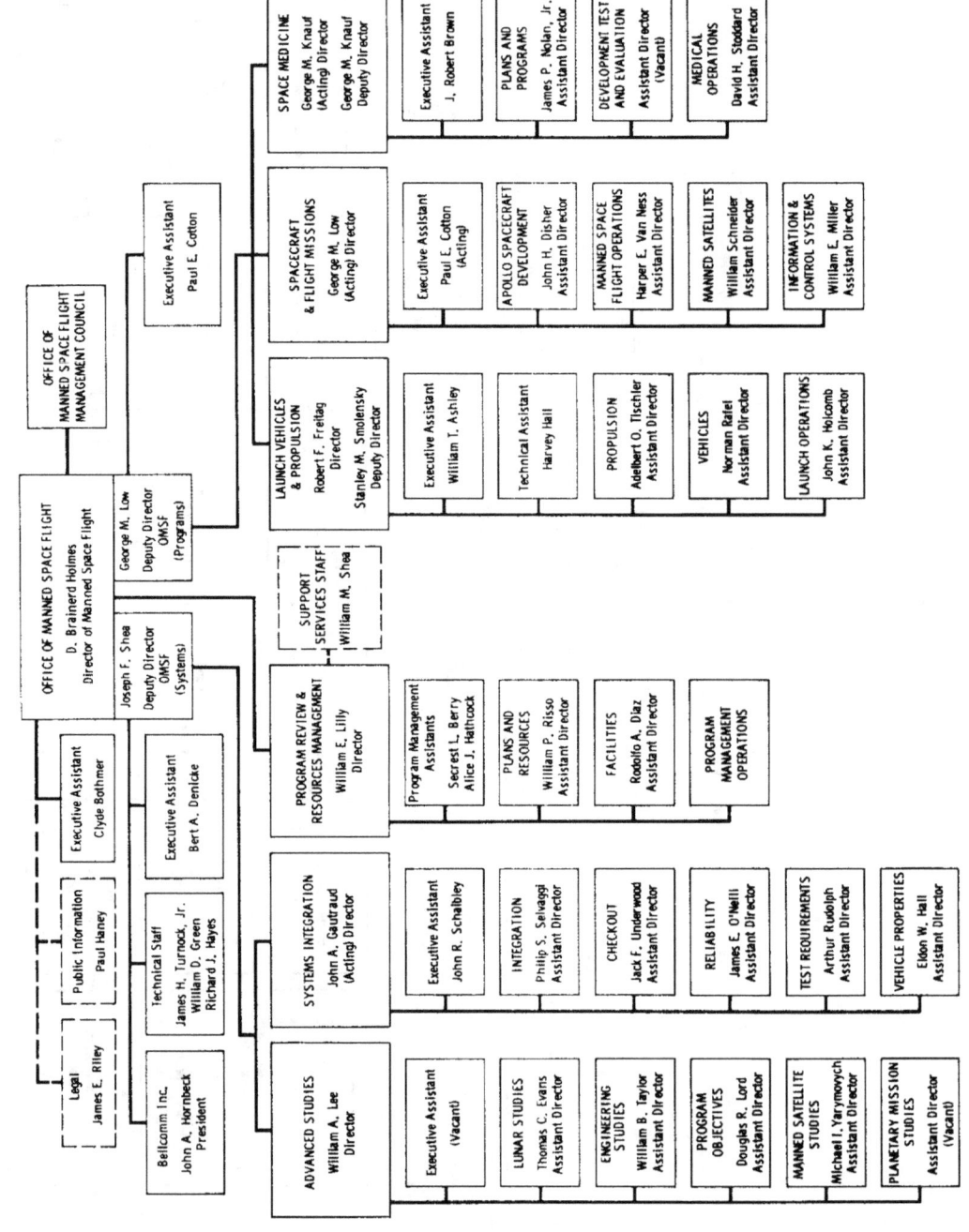

APPENDIX 8

OFFICE OF MANNED SPACE FLIGHT
[January 31, 1964]

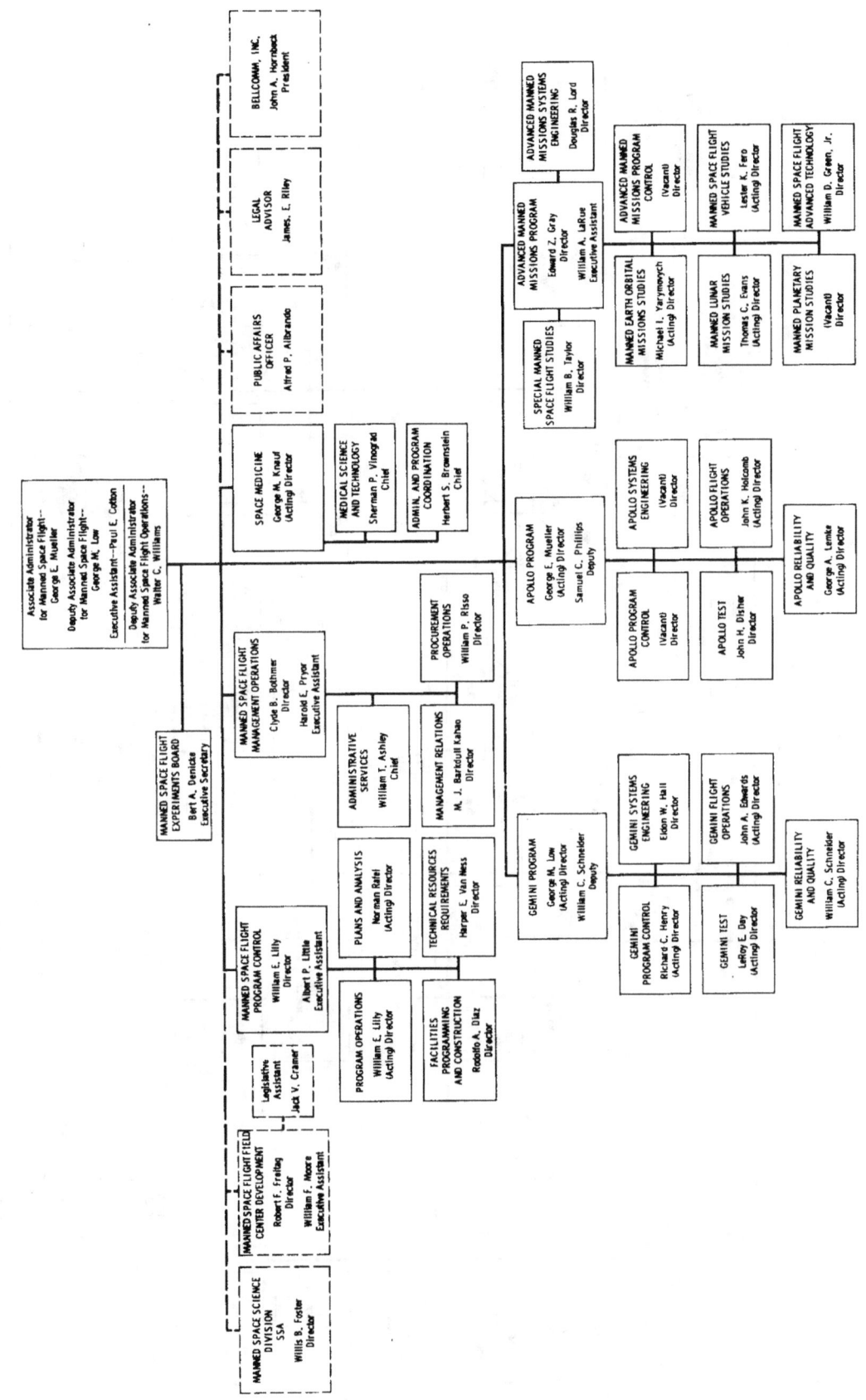

THE APOLLO SPACECRAFT: A CHRONOLOGY

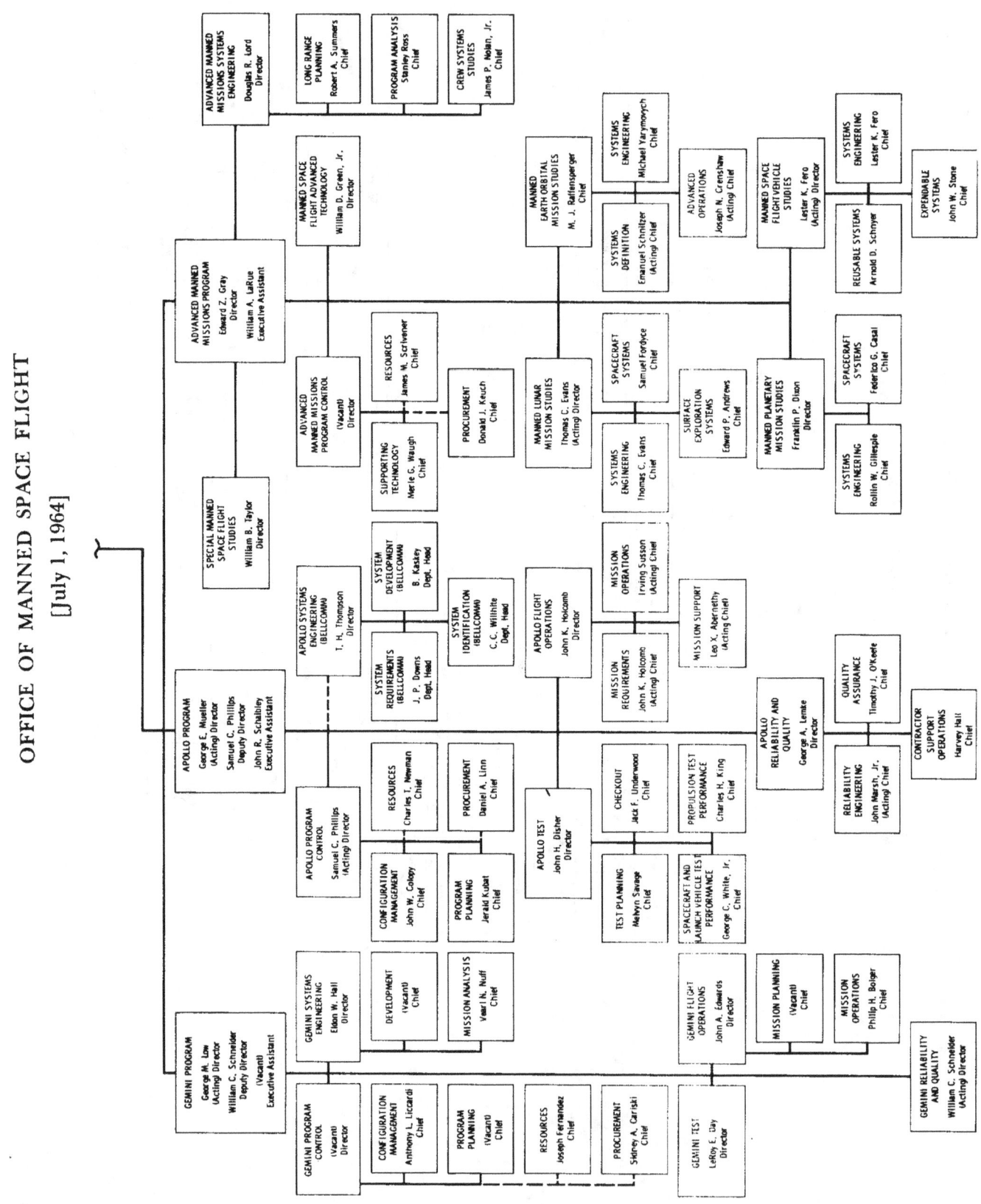

OFFICE OF MANNED SPACE FLIGHT
[July 1, 1964]

APPENDIX 8

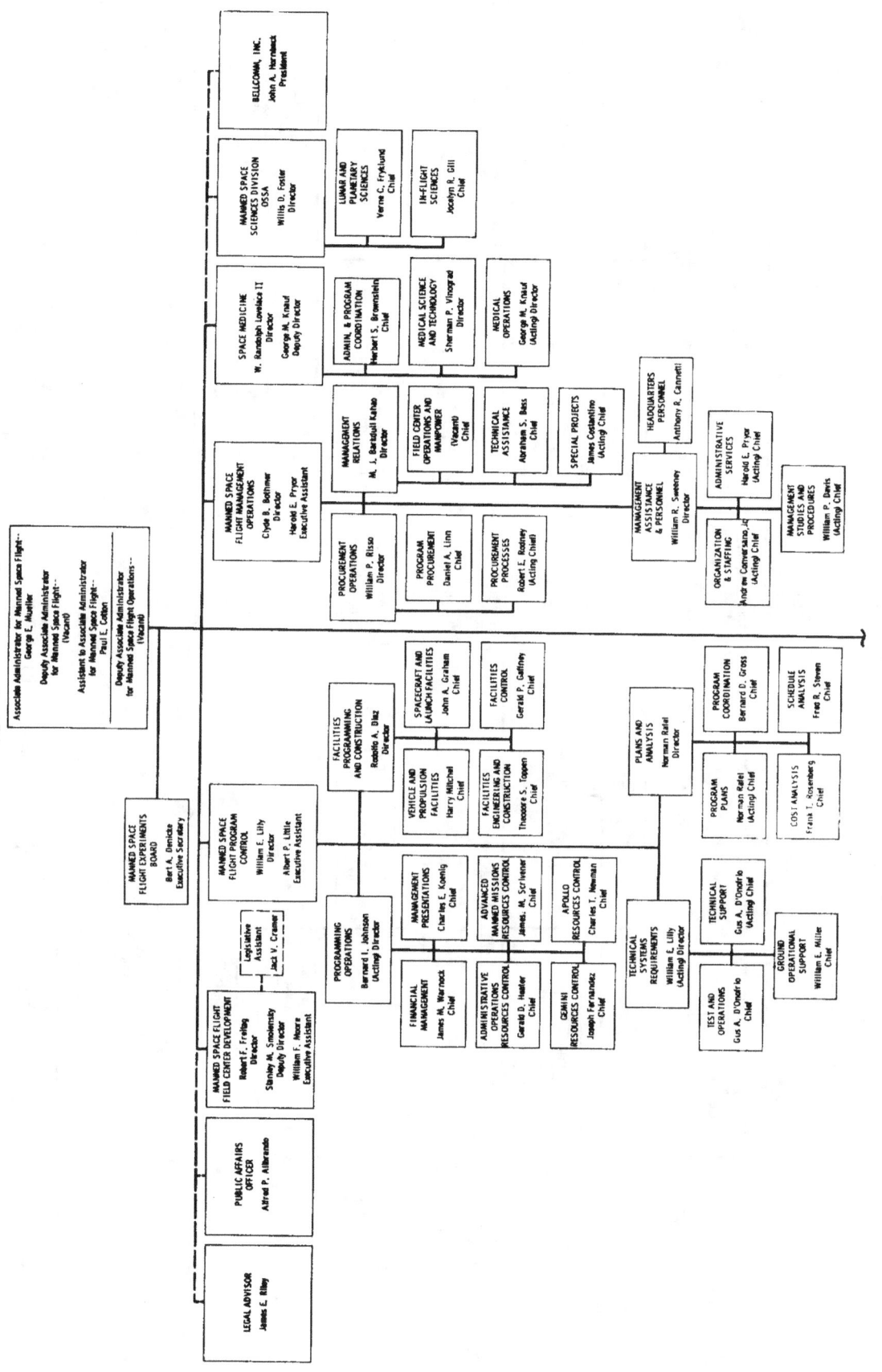

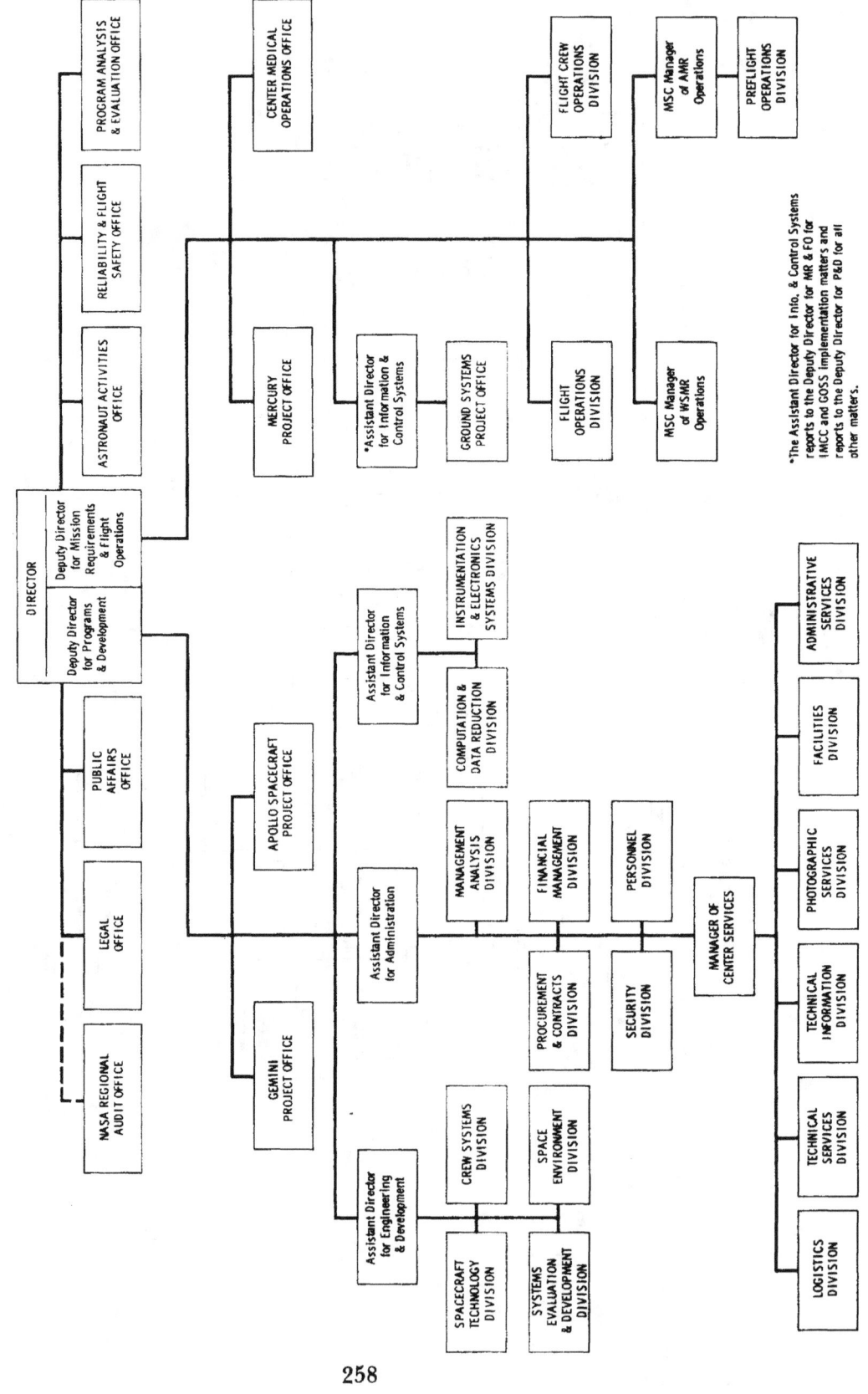

APPENDIX 8

MANNED SPACECRAFT CENTER
[November 1, 1963]

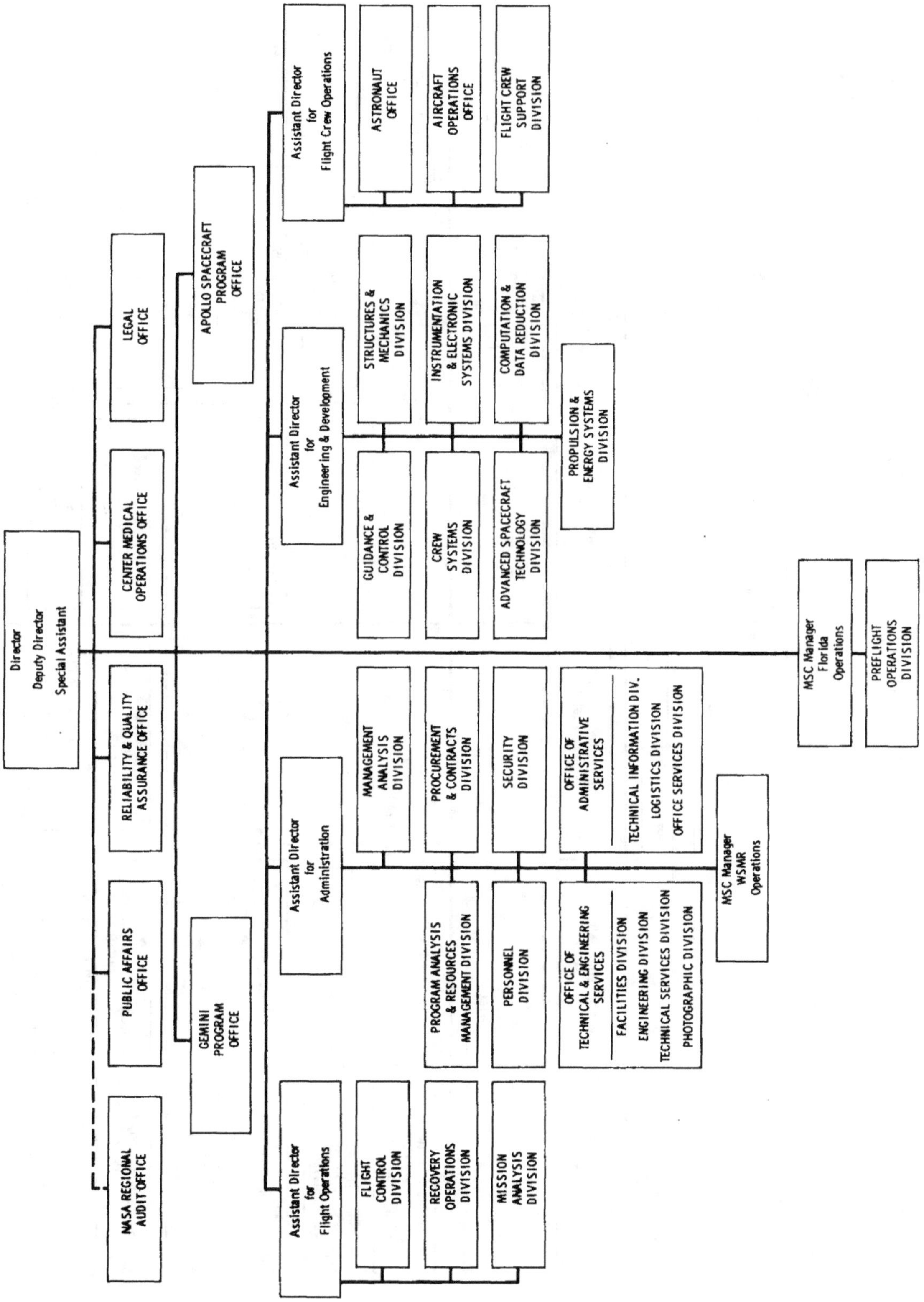

Director / Deputy Director / Special Assistant

Staff offices:
- NASA Regional Audit Office
- Public Affairs Office
- Reliability & Quality Assurance Office
- Legal Office
- Center Medical Operations Office

Program offices:
- Gemini Program Office
- Apollo Spacecraft Program Office

Assistant Director for Flight Operations
- Flight Control Division
- Recovery Operations Division
- Mission Analysis Division

Assistant Director for Administration
- Management Analysis Division
- Procurement & Contracts Division
- Security Division
- Office of Administrative Services
  - Technical Information Div.
  - Logistics Division
  - Office Services Division
- Program Analysis & Resources Management Division
- Personnel Division
- Office of Technical & Engineering Services
  - Facilities Division
  - Engineering Division
  - Technical Services Division
  - Photographic Division

Assistant Director for Engineering & Development
- Structures & Mechanics Division
- Instrumentation & Electronic Systems Division
- Computation & Data Reduction Division
- Guidance & Control Division
- Crew Systems Division
- Advanced Spacecraft Technology Division
- Propulsion & Energy Systems Division

Assistant Director for Flight Crew Operations
- Astronaut Office
- Aircraft Operations Office
- Flight Crew Support Division

MSC Manager WSMR Operations

MSC Manager Florida Operations
- Preflight Operations Division

259

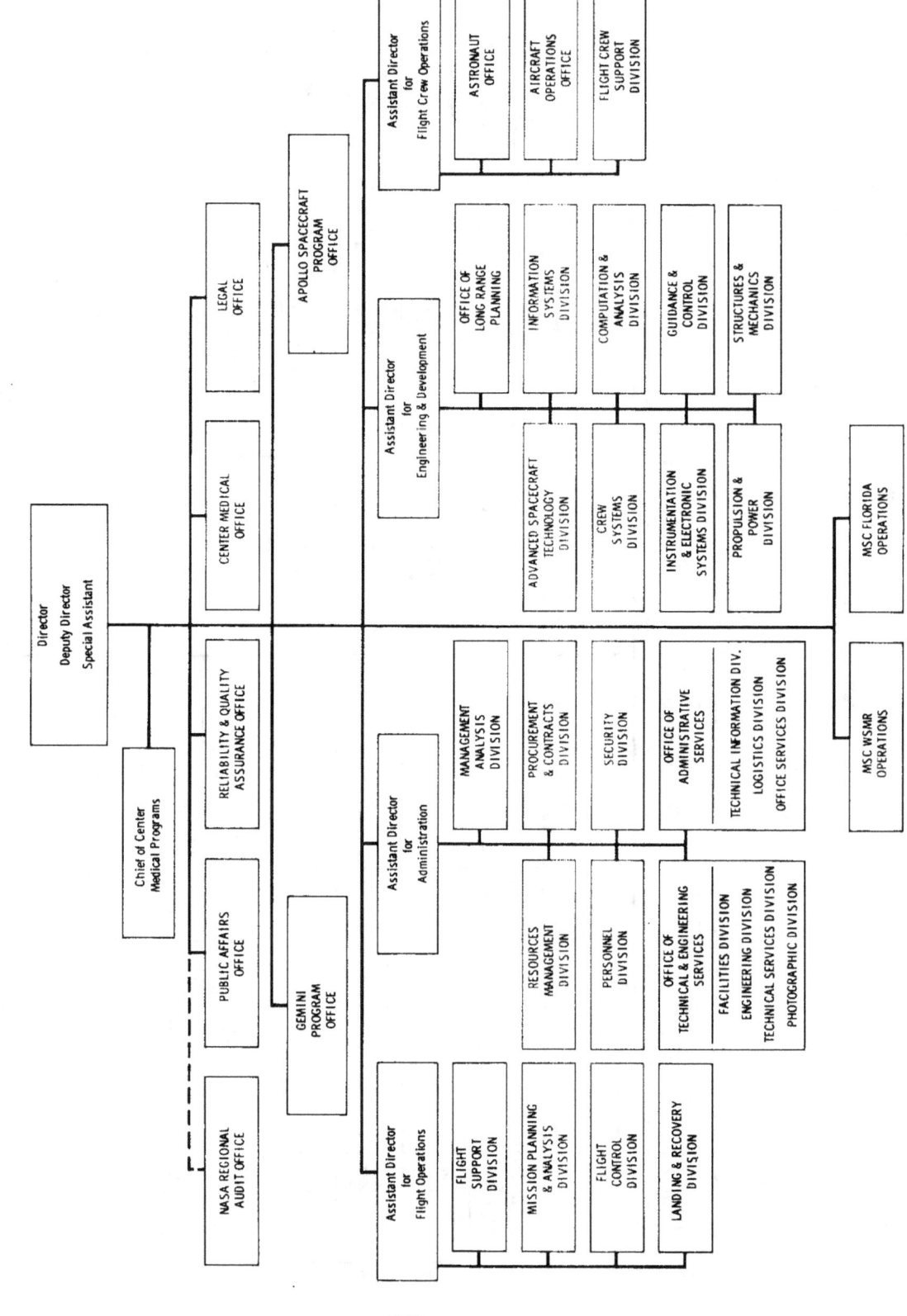

APPENDIX 8

APOLLO SPACECRAFT PROJECT OFFICE
[May 1, 1963]

APOLLO SPACECRAFT PROJECT OFFICE
Acting Manager—R. O. Piland
Special Assistant—T. F. Baker

PROJECT INTEGRATION OFFICE
Manager—J. T. Markley
Asst. Mgr.—A. L. Granfield

Deputy Manager, Spacecraft—R. O. Piland
Asst. Deputy Mgr., C&SM—C. C. Johnson
Asst. Deputy Mgr., Syst. Integ.—A. D. Mardel

- **SYSTEMS INTEGRATION OFFICE**
  Manager—A. D. Mardel
  Asst. Mgr.—J. W. Lanzkron
- **C&SM ADMINISTRATION OFFICE**
  Act. Mgr.—C. L. Taylor
- **C&SM ENGINEERING OFFICE**
  Manager—C. C. Johnson
  Asst. Mgr.—C. L. Taylor
- **RASPO-DOWNEY**
  Res. Mgr.—G. A. Lemke
- **C&SM SYSTEMS TEST OFFICE**
  Manager—O. G. Morris

SPACECRAFT SYSTEMS OFFICE GUIDANCE AND CONTROL
Manager—D. W. Gilbert
Asst. Mgr.—P. Ebersole

- **G&C ADMINISTRATION OFFICE**
  Manager—J. Church
- **RASPO-BOSTON**
  Res. Mgr.—W. J. Rhine
- **G&C ENGINEERING OFFICE**
  Manager—D. W. Gilbert

Deputy Manager, Lunar Excursion Module—J. L. Decker

- **LEM ADMINISTRATION OFFICE**
  Manager—C. E. McCollough
- **RASPO-BETHPAGE**
  Res. Mgr.—J. W. Small
- **LEM SYSTEMS TEST OFFICE**
  Act. Mgr.—J. L. Decker
- **LEM ENGINEERING OFFICE**
  Manager—O. E. Maynard
  Asst. Mgr.—W. F. Rector

APOLLO SPACECRAFT PROJECT OFFICE

[August 26, 1963]

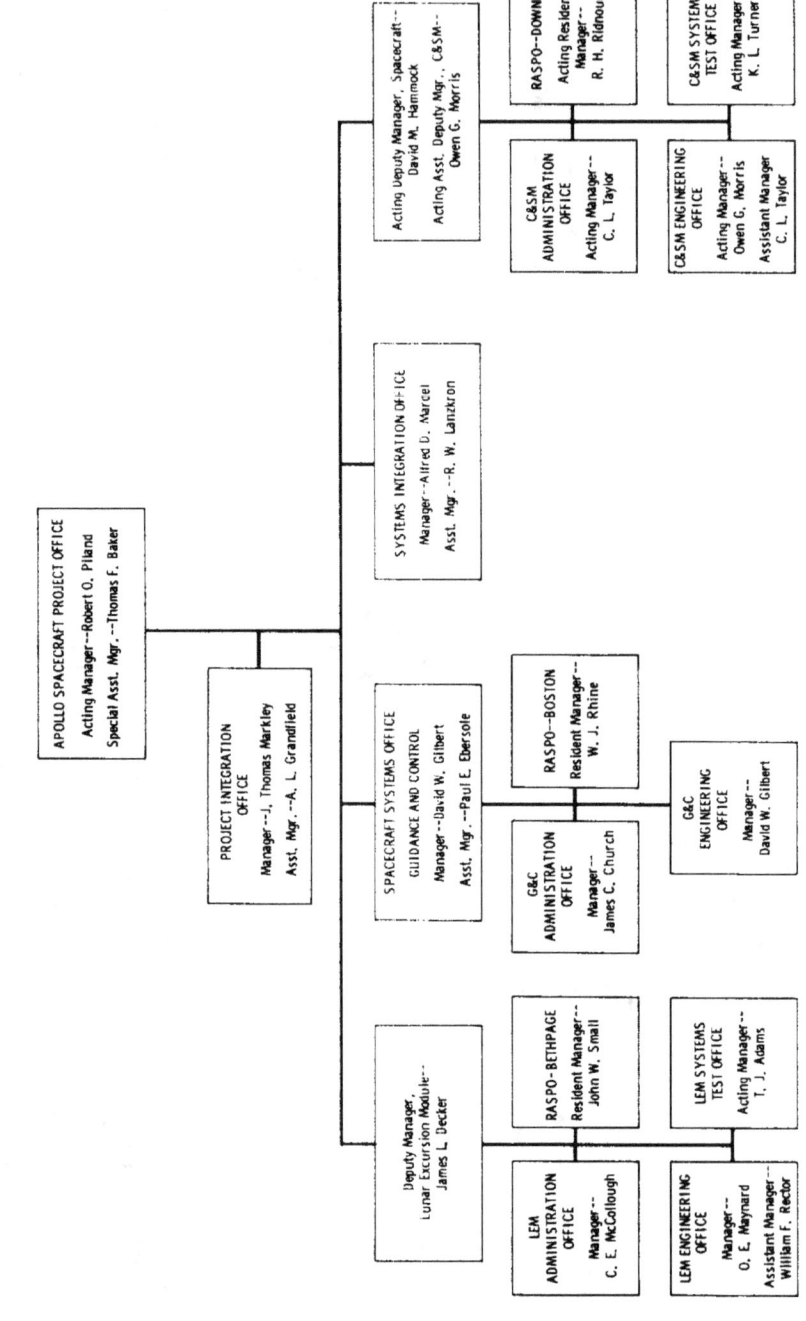

APPENDIX 8

APOLLO SPACECRAFT PROGRAM OFFICE

[August 11, 1964]

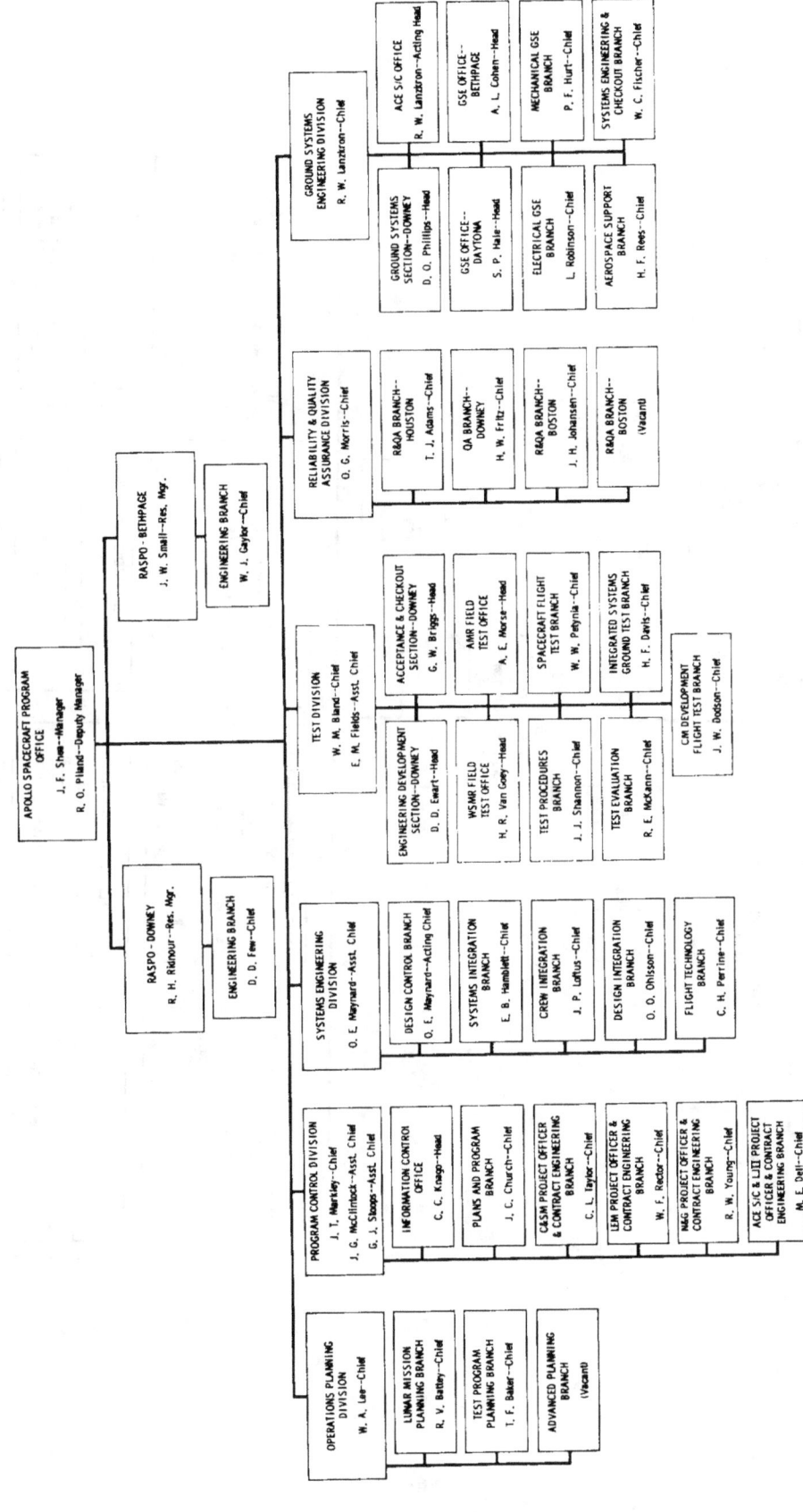

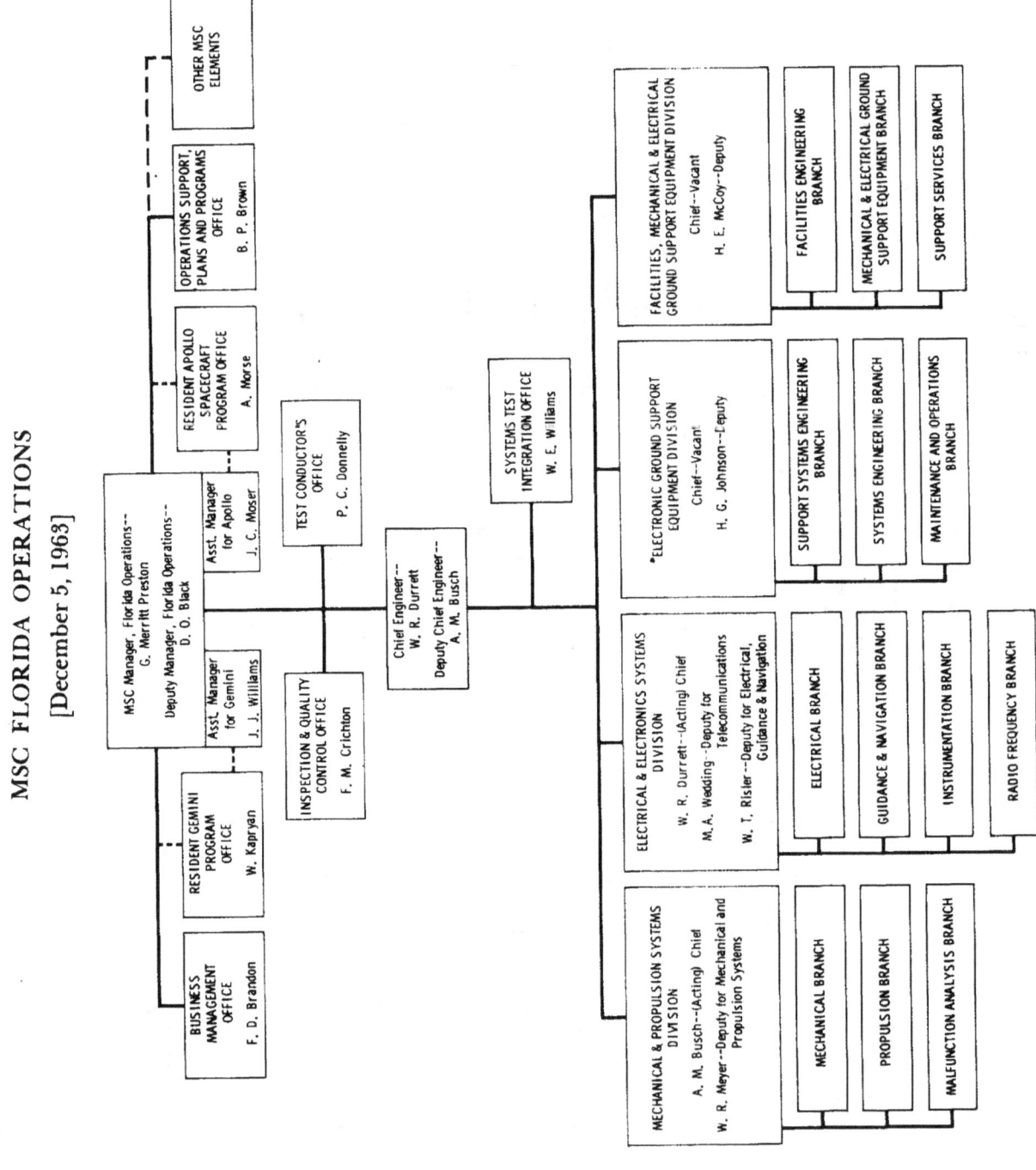

APPENDIX 8

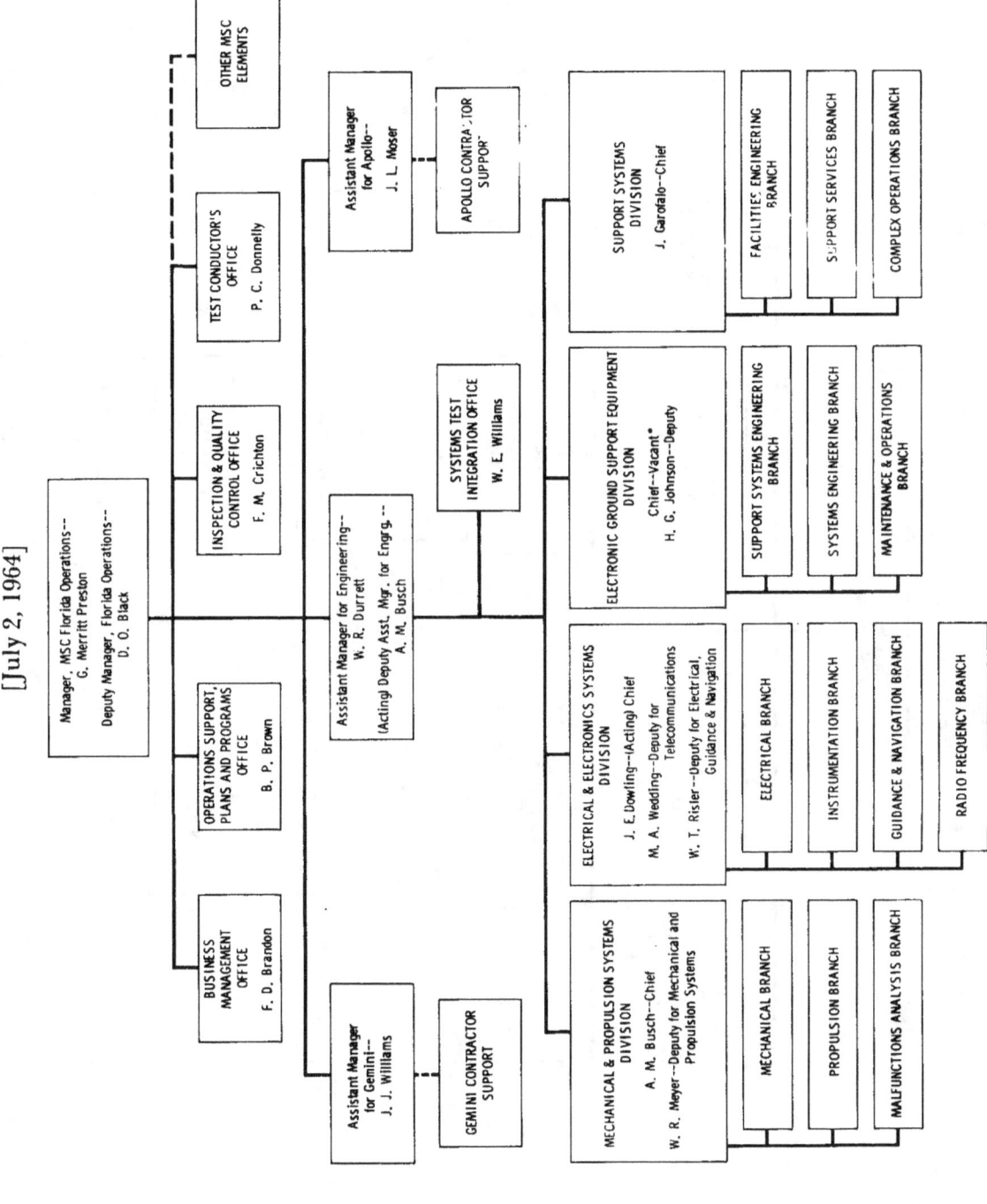

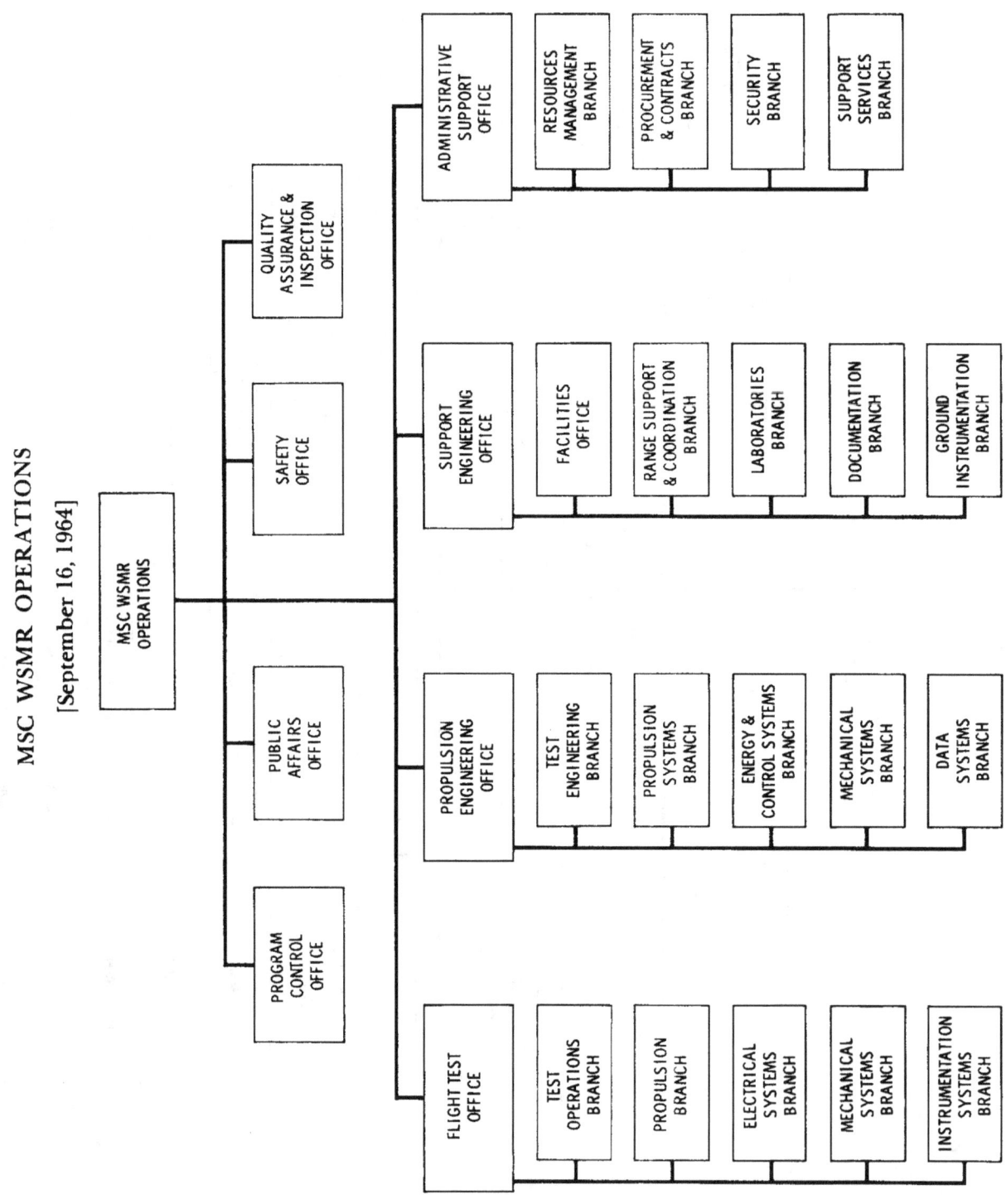

APPENDIX 8

LITTLE JOE II AND APOLLO PAD ABORT TEST OPERATIONS

[June 13, 1963]

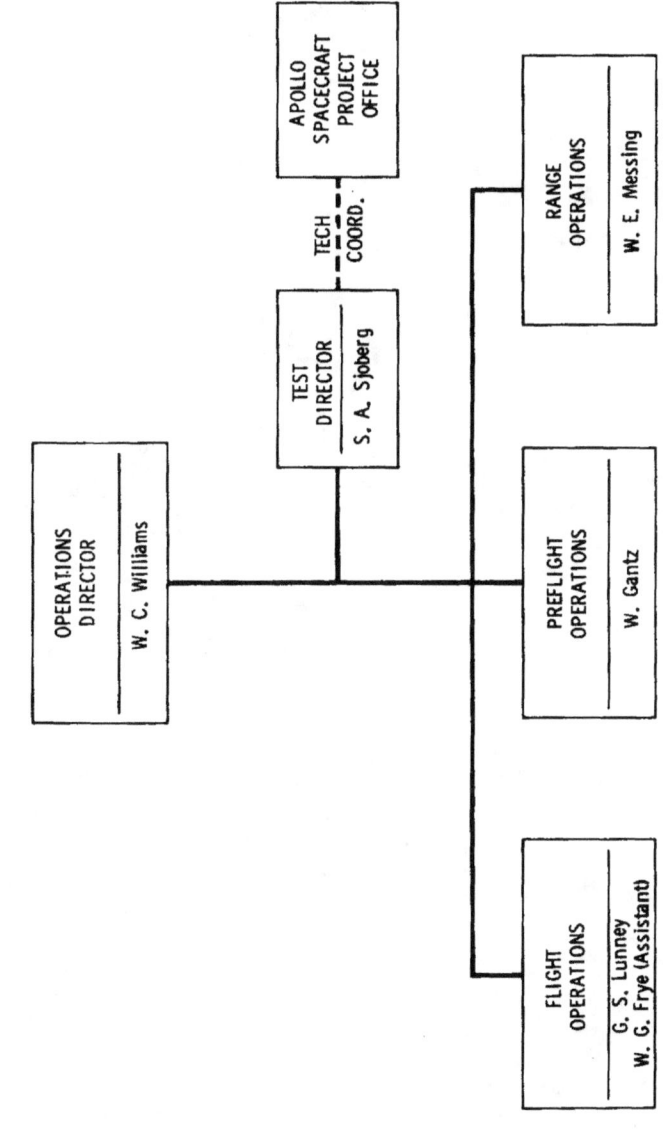

INDEX

A

Ablation. *See* Heatshield.
Abort acceleration, service module, 213–214
Abort, lunar landing, 26, 32, 113, 135–136, 169, 172, 209, 222–223
    simulation, 69
Abort, Saturn launch, 111, 125, 138, 171, 187–188, 212
Accelerometer. *See* Navigation and guidance.
Acceptance checkout equipment. *See* Prelaunch acceptance checkout equipment.
AC Spark Plug Division. *See* General Motors Corporation.
Adapter, spacecraft—Saturn V, 61
Ad Hoc Working Group on Apollo Experiments, 117
Ad Hoc Working Group on Follow-On Surveyor Instrumentation, 103
Aerobee 150 sounding rocket, 105
Aerojet-General Corporation, 3, 27, 39, 62, 145, 157
Aero Spacelines, Inc., 71, 105
AiResearch Manufacturing Company, 41, 88, 111, 115, 127, 129, 130, 146, 154, 195
Air Force Systems Command. *See* United States Air Force.
Airite Products, Inc., 43
Akens, David S., 105
Aldrin, Edwin E., Jr., 101, 196, 221
Alexander, Charles C., 63
"All-up" testing concept, launch vehicle-spacecraft, 104, 105–106, 124
American Car and Foundry Industries, Inc., 19
Ames Research Center, 22, 56, 109, 114–115, 117
    Hypervelocity Ballistic Range, 50–51
Anders, William A., 101, 196
Anderson, Frank W., Jr., xiv
Antennas, 5–6, 22, 103, 124, 131, 225
Apocynthion, 21
Apollo and Gemini mission simulation programs, 189
Apollo Docking Interface Panel, 112
Apollo Executives Group, NASA–Industry, 103
Apollo Mission A–001, 175
Apollo Mission A–002, 100, 223
Apollo Mission A–101, 180–183, 225
Apollo Mission AS–102, 220, 224–225
Apollo Mission AS–201, 104, 106
    heatshield qualification, 132–133
Apollo Mission AS–202, 133
    heatshield qualification, 132–133
Apollo Mission AS–203, 106
Apollo Mission AS–207, 106
Apollo Mission AS–501, 106
Apollo Mission AS–503, 106
Apollo Mission AS–507, 106
Apollo Mission Pad Abort I (PA–1), 108, 223
Apollo mission plan development study, 114
    See also Apollo Mission Planning Task Force.
Apollo Mission Planning Panel, 32–33, 42
Apollo Mission Planning Task Force, 126
    Phase I progress report, 171–172
Apollo Propulsion System Development Facility, 192
Apollo Spacecraft Mission Trajectory Sub-Panel, 50
Apollo Spacecraft Project Office. *See* Manned Spacecraft Center.
Apolune, 21
Armitage, Peter J., 13
Armstrong, Neil A., 25, 53, 196
Arnold Engineering Development Center. *See* United States Air Force.
Ascent engine. *See* Lunar excursion module.
ASPO. *See* Manned Spacecraft Center, Apollo Spacecraft Project Office.
Astronauts
    assignments, 25, 196
    names. *See* under each:
        Aldrin, Edwin E., Jr.
        Anders, William A.
        Armstrong, Neil A.
        Bassett, Charles A., II
        Bean, Alan L.
        Borman, Frank
        Carpenter, M. Scott
        Cernan, Eugene A.
        Chaffee, Roger B.
        Collins, Michael
        Conrad, Charles, Jr.
        Cooper, L. Gordon, Jr.
        Cunningham, R. Walter
        Eisele, Donn F.
        Freeman, Theodore C.
        Glenn, John H., Jr.
        Gordon, Richard F., Jr.
        Grissom, Virgil I.
        Lovell, James A., Jr.
        McDivitt, James A.
        Schirra, Walter M., Jr.
        Schweickart, Russell L.
        Scott, David R.
        See, Elliot M., Jr.
        Shepard, Alan B., Jr.
        Slayton, Donald K.
        Stafford, Thomas P.
        White, Edward H., II
        Williams, Clifton C., Jr.
        Young, John W.
    scientist-astronaut, 159, 213
    selection, 61, 71–72, 101
    third group report to work, 132
Atmosphere, Spacecraft, 14, 118, 151
Australia, 36, 144
Avco Corporation, 61–62, 114
Aviation Medical Acceleration Laboratory. *See* United States Navy.
Avien, Inc., 65, 188

B

Bailey, F. John, Jr., 84
Bassett, Charles A., II, 101, 196
Battaglia, Harold G., 199
batteries, 25, 37, 77, 91, 126, 144, 147, 209
    chargers, 61, 86
    command module, 61, 147
    lunar excursion module, 25, 37, 91, 126, 144, 209
    portable life support system, 77, 86
    See also Electrical system, spacecraft.

Battey, Robert V., 42
Baylor University, 211
Bean, Alan L., 101, 196
Beech Aircraft Corporation, 190
Beilock, Milton, 103
Bell Aerosystems Company, 24, 25, 30, 31, 70, 88, 131, 137, 138, 143, 154, 170
Bellcomm, Inc., 14, 21, 40, 91, 104, 115, 124, 194, 202
Bellows Construction Company, W. S., 120
Bendix Products Aerospace Division, 125
Bertram, Emil P., 77
Bioinstrumentation, 119, 125
Bissett-Berman Corporation, 183
Blaw-Knox Company, 131
Blount Brothers Corporation, 110
Boeing Company, The, 31, 110, 120, 151, 173
Boilerplate spacecraft, 15, 29, 30, 36, 37, 70, 86, 108, 136, 141, 153, 170-171, 175, 180, 184, 186, 223, 224, 225
    BP-2, 186
    BP-3, 15, 29, 86
    BP-6, 37, 70, 86, 108, 223
    BP-6A, 223
    BP-9, 36
    BP-12, 141, 153, 175
    BP-13, 136, 180, 225
    BP-14, 224
    BP-15, 225
    BP-19, 30, 86, 141
    BP-22, 170-171
    BP-25, 184
Borman, Frank, 25, 196
Bothmer, Clyde B., 5, 104
Brandon, George W., 31
British Ministry of Aviation, 132
    Royal Aircraft Establishment, 132
Brockman, Paul R., 103
Brooks, Courtney G., xiv
Bryant, Farris, 114
Bryant, John, 36, 78
Buckley, Edmond C., 6
"Buddy concept." *See* Suit, space.
Budget, NASA, 22
Byers, Bruce K., 40, 84
Bykovsky, Valery F., 64

C

Cameras, 11
Canards, aerodynamic, 22-24, 109
    abort system concept, 125
    versus tower flap configuration, 109-110, 133-135, 138, 140, 148
    See also Launch escape system.
Canberra, Australia, 6, 131, 144
Canning, Frank, 32
Cape Canaveral, 114
Cape Kennedy, 114
Carbee, Robert M., 151
Carbon dioxide measuring system, 36
    sensors, 53, 180
Carpenter, M. Scott, 25, 53, 196
C-band radar, 19, 20, 172
Centrifuge, 32, 89, 121
    Phase I training, 118
Cernan, Eugene A., 101, 196
Chaffee, Roger B., 101, 196
Chamberlin, James A., 84
Chance Vought Corporation, 26
Clark, S. P., 148
CM. *See* Command module, Apollo.
Cohen, Aaron, 195
Collins, Michael, 101, 196
Collins Radio Company, 10, 19, 44, 128, 190

Command module (CM), Apollo
    airbag uprighting system, 212-213
    batteries. *See* Batteries.
    bioinstrumentation, 119
    Block I, 129, 134, 159, 160, 195, 226, 227
        definition, 128
        mockup review, 165, 198
    Block II, 129, 135, 136, 143-144, 154, 159, 160, 161, 174, 186, 189, 190, 226
        definition, 127-128, 154
        mockup review, 229-230
    boilerplate. *See* Boilerplate, spacecraft.
    common use components, 19, 52, 53, 58, 98
    common use ground support equipment, 144
    communications, 124, 169
    computer. *See* Computer, spacecraft.
    contract, 61, 77
    control and display panel, 84
    control weight, 199
    couch. *See* Couch, spacecraft crew.
    crew transfer tunnel mockup, 143-144
    development test program, 113, 121-122
    docking. *See* Docking, spacecraft.
    earth landing system. *See* Landing system, Apollo earth; Landing, spacecraft earth; Landing characteristics, spacecraft; Parachutes; Water landing, spacecraft.
    electrical system. *See* Electrical system, spacecraft.
    environmental control system. *See* Environmental control system.
    first flight, 180-181
    flight schedule, 149
    fuel cell. *See* Fuel cell.
    hatch, 171, 177
    hatch configuration, Block II, 149
    heatshield. *See* Heatshield.
    inflight maintenance concept, deletion of, 187, 218
    landing characteristics. *See* Landing characteristics, command module.
    land versus water landing. *See* Water landing, spacecraft; Landing, spacecraft, earth.
    layout, 10-11, 18
    main display console, 70
    meteoroid hits, 131
    missions. *See* Apollo Mission A-001, A-002, etc.
    navigation and guidance. *See* Navigation and guidance.
    oxygen storage capacity, 203
    parachutes. *See* Parachute system.
    portholes and windows, 129
    postlanding ventilation, 148
    radar, 78, 153, 188
    reaction control engines. *See* Reaction control engines.
    reliability, 69
    scientific equipment weight and volume allocation, 169
    separate operation requirement, 117
    sleeping restraint, 71
    stabilization and control system. *See* Stabilization and control system.
    stowage locations, 126-127
    suit interface. *See* Suit, space.
    temperature and pressure transducer instrumentation, 187
    thermal control system, 114
    upside down floating characteristic, 124, 139, 159, 212-213
    vibration tests, 170-171
    visibility requirements, 129
    warning system, 86
    waste management system, 129-130
    water management system, 48, 137, 159, 210-211
    weight reduction. *See* Weight reduction, spacecraft.

INDEX

Communications, 6, 10, 56, 57–58, 106, 124, 153–154, 169, 223–224, 225
Communications and Tracking Steering Panel, 55
Computer, Mission Control Center, 71
Computer, spacecraft
    command module, 3, 27
    lunar excursion module, 58, 100–101
    up-data link, 22, 62
Conrad, Charles, Jr., 25, 53, 196, 205
Control Data Corporation, 122, 151
Cooper, L. Gordon, Jr., 25, 56, 196
Corps of Engineers. *See* United States Army.
Correale, James V., 7
Couch, spacecraft crew, 14, 44–45, 45–47, 88, 118, 147, 148, 183, 208, 215
    restraint system, 208, 215
Cox, Kenneth J., 191
Crawler transport, 28
Crew safety, 18, 216–217
Crew Systems Division. *See* Manned Spacecraft Center.
Crew transfer between spacecraft, 8, 112, 137, 138, 143, 160, 178–180, 188, 218, 229
    emergency method, 112, 138, 160, 188
    ground rule revisions, 178–180
    zero-g test, 194
Cunningham, R. Walter, 101, 196, 215
Cutler-Hammer, Inc., 22

D

Dalmo Victor Company, 201
Data storage equipment, 69
David Clark Company, 199
Debus, Kurt H., 12
Decker, James L., 28, 53
Deep Space Instrumentation Facility, 6–7, 36, 90, 120, 130, 144–145
Delco-Remy, 37
Denicke, Bert A., 77
Derbyshire, George, 103
Descent engine. *See* Lunar excursion module.
DeVos, Francis J., 204
Disher, John H., 21
Docking, spacecraft, 8, 13, 72, 96, 112, 129, 130, 138, 163–164, 185, 188, 194, 200, 227–228
Douglas Aircraft Company, 76, 117, 170
Drexel, Ralph E., 215
Drop test, spacecraft, 29
Dryden, Hugh L., 3, 6
Duret, Eugene L., 103
Dust, moon, 15, 105

E

Eagle-Picher Industries, Inc., 37, 147
Earthshine, 221
EDD. *See* Manned Spacecraft Center, Engineering and Development Directorate.
Edmiston, Ragan C., 123
Eggleston, John M., 202
Eickmeier, Alfred B., 58
Eisele, Donn F., 101, 196
Electrical Storage Battery Company, 37
Electrical Systems Panel, 56
Electrical system, spacecraft
    command module, 31, 88, 98, 190
    lunar excursion module, 25, 69, 77, 91, 115, 125, 147–148, 172, 198–199, 209
    lunar survey mission, 14-day, 196
    See also Fuel cell; Batteries.
Electro-Optical Systems, Inc., 105
    Micro Systems, 187

Elgin National Watch Company, 33, 141
Elms, James C., 55, 107, 127
Emme, Eugene M., xiv
Environmental control system
    command module, 18, 41, 53, 88, 98, 111, 119, 127, 129, 146, 159, 195, 224
    lunar excursion module, 26, 34, 38, 59–60, 74, 103–104, 120, 163
Ertel, Ivan D., xiv
Experiments, 31, 44, 92, 95–96, 117, 136, 144, 147, 164–165, 200–201, 221
    earth-orbital, 164–165
    incorporation of Apollo-type in Gemini flights, 148
    scientific equipment weight and volume allocation, 169, 171–172, 186, 200–201
Explorer XVI, 139
Extravehicular activity, 149
    equipment belt, 218

F

F-1 engine, 105, 152
Faget, Maxime A., 84, 107, 111, 132, 150
Faith 7 (Mercury-Atlas 9), 56, 98
Farrand Optical Company, 79, 110
Feltz, Charles H., 20, 165
Ferrando, James A., 32
"Fire in the hole" tests, lunar excursion module, 33, 38, 207, 225
Flammability tests, 106
Flight acceleration facility, 32
Flight Research Center (FRC), 24
Flight schedule, Apollo-Saturn, 21, 104, 105, 149
Flight Technology Systems Panel, Lunar Excursion Module, 55
Flotation collar, 184
Food, space, 111, 125, 135, 183, 186
Ford, Bacon, and Davis, Inc., 32
Ford Motor Company, 5
    Philco Corporation, 5, 25, 41, 118, 165
Foss, Ted H., 188
Foster, Willis B., 165, 216
Franklin, George C., 136
Freedman, Gilbert M., 204
Freeman, Theodore C., 101, 197
Free return trajectory, 50, 98, 171–172
Freitag, Robert F., 104
French, Burrell O., 212
Frick, Charles W., 11, 28, 45
Fry, Erika, 113
Fryklund, Verne C., Jr., 95–96, 103, 115, 174
Fuel cell, 3, 25, 37, 73, 91
    command module, 119, 124, 137, 138, 160, 161
    lunar excursion module, 125–126, 138, 147–148, 154, 172, 189, 209, 221–222

G

GAEC. *See* Grumman Aircraft Engineering Corporation.
Garrett Corporation, The. *See* AiResearch Manufacturing Company.
Gates, Sally D., xiv
Gault, Donald E., 51
Gautrand, John A., 30
Gemini Flights Experiments Review Panel, 148
Gemini-Titan I, 155
General Dynamics Corporation, 118
General Dynamics/Convair, 18, 29, 41, 47, 51, 60, 72, 100, 137, 177
General Electric, 12, 34, 37, 122, 175–177, 215
    Policy Review Board, 12

General Motors Corporation, 8
    AC Spark Plug Division, 8, 27, 63, 85, 137, 141, 189, 191, 222
    Allison Division, 43, 65, 115, 200
General Precision, Inc.
    Kearfott Products Division, 145
    Link Division, 47, 64, 110
Geonautics, Inc., 186, 221
Gerstle, John E., Jr., 76
Gilbert, David W., 28, 54
Gilruth, Robert R., 7, 12, 14, 55, 58, 66, 84, 89, 107, 213, 218, 227
Glenn, John H., Jr., 25
Goddard Space Flight Center (GSFC), 5, 44, 144
Goldstone, California, 6, 131
Goodman, Clark, 103
Gordon, Richard F., Jr., 101, 197, 205
Graves, Claude A., 195
Graves, G. Barry, 84
Gresser, Angela C., 114
Grimm, Dean F., 221
Grimwood, James M., xiv, 56, 63, 155
Grissom, Virgil I., 25, 196
Grumman Aircraft Engineering Corporation (GAEC), *passim*
    contract, 5, 38, 222
    flight support engineering, 192
    lunar excursion module. *See* Lunar excursion module.
Guidance and Control Panel, 76
Guidance and navigation. *See* navigation and guidance.
Guidance and Performance Sub-Panel, 98
Gulton Industries, Inc., 37
Gurley, John R., 63

H

H-1 engine, 91, 108-109
Hacker, Barton C., 155
Hamilton Standard Division. *See* United Aircraft Corporation.
Hammock, David M., 107, 110, 111, 116
Handcontroller, spacecraft, 58
Harris, Elliott S., 204
Heatshield, 10, 26, 61-62, 84, 124, 163, 186
    design, 10, 84, 146
    development program, 26
    material, 10, 61-62, 73-74, 75, 146, 213
    meteoroid impact study, 114-115
    termination of backup, 61-62
    test plan, 124, 132-133
Hess, Harry H., 103, 159
Hjornevik, Wesley L., 107
Hodge, John D., 218
Holmes, D. Brainerd, 30, 63, 71, 74
Hornbeck, John A., 40
Howard, B. T., 115
Hughes Aircraft Company, 103, 188-189, 202
Hunt, Gerald L., 55
Hurt, J. B., 47
Huss, Carl R., 133, 195
Huzar, Stephen, 21

I

IBM. *See* International Business Machines.
ILC. *See* International Latex Corporation.
Impact test facility, 29
Inertial measuring unit, 75
Inertial reference integrating gyro (IRIG), 8, 27, 141
Instrumentation and Electronic Systems Division. *See* Manned Spacecraft Center.

International Business Machines (IBM), 71, 162
International Latex Corporation (ILC), 7, 12, 24-25, 42, 118, 138, 161
International Telephone and Telegraph Corporation (ITT), 7
    Industrial Products Division, 61
    Kellogg Division, 7, 41, 77
IRIG. *See* inertial reference integrating gyro.
ITT. *See* International Telephone and Telegraph Corporation.

J

J-2, engine, 113, 170, 192
James, Dennis, 103
Jarrell, A. Ruth, 105
Jet Propulsion Laboratory (JPL), 6, 36, 44, 120, 129, 171
Johnson, Caldwell C., 28, 53, 165
Johnson, Lyndon B., 114
Johnston, Richard S., 5, 217, 219, 228, 229
Jones, Enoch M., 136
Jones, Leo L., 105

K

Karegeannes, Carrie E., xiv
Kelley, Albert J., 14
Kelley Board, 14
Kelly, Tom J., 151
Kennedy, John F., 22, 89, 114
Kennedy Space Center (KSC), 114
Kincaide, William C., 196, 201
Kingsley, Milton G., 226
Kleinknecht, Kenneth S., 84
Kohrs, Richard H., 123, 177
Kollsman Instruments Corporation, 85, 101, 137, 191, 194, 221
Koons, Wayne E., 50
Kraft, Christopher C., Jr., 19, 22, 62, 85, 86, 106, 107, 113-114, 139, 165, 218
Kranz, Eugene F., 218
KSC. *See* Kennedy Space Center.
Kuettner, Joachim P., 77

L

Landing characteristics, command module, 10, 15, 88, 109, 151-152, 186
Lambert, C. Harold, Jr., 227
Landing gear. *See* Lunar excursion module.
Landing, spacecraft, earth, 18-19, 27, 141
    water versus land, 18-19, 27, 36, 139
Landing system, Apollo, earth, 53, 141, 209
Langley Research Center (LRC), 22, 39, 96, 116, 120, 200
Largent, H. R., 147
Launch escape system, 13, 17, 18, 22, 30, 47-48, 50, 73, 86-87, 109-110, 125, 138, 140-141, 171, 175, 187-188, 191, 201, 209, 212, 213, 218, 220
    Apollo Mission Pad Abort I (PA-1), 108
    Block I design, 140-141
    Block II design, 140-141
    canards versus tower flap, 109-110, 133-135, 138, 140-141, 148
    design criteria clarification, 187-188
Launch Operations Center (LOC), 47, 71, 110
    redesignated, 114
    See also Kennedy Space Center.
Launch Complex 39, Saturn V, 110
Leach Corporation, 10
Lee, William A., 30, 332
LEM. *See* Lunar excursion module.

INDEX

LEM spacecraft and test articles
    LEM-5, 84
    LTA-8, 84
    LTA-9, 84, 114
    TM-1, 149-150, 159, 163-164, 200
    TM-2, 90-91
    test vehicle interrelationship, 92
Lewis Research Center (LeRC), 104
Lighting conditions, spacecraft, 41
Lilly, William E., 30
Linde Company, 130
Lineberry, Edgar C., Jr., 32
Ling-Temco-Vought, Inc. (LTV), 6, 69, 88, 96, 107, 164, 194, 202
    Astronautics Division, 202
    Electronics Division, 107
Little Joe II, 18, 41, 47, 51, 60, 62, 63, 100, 137, 145, 153, 177, 223
    Algol motors, 62, 45
    attitude control system, 60, 223
    contract, 29, 51, 60, 100, 177
    Design Engineering Inspection, 100
    instability with lunar excursion module, 116, 135
    launch, 175
    launch complex inspection, 63
    launch site decision, 29
    qualification test vehicle, 72, 80
LOC. *See* Launch Operations Center.
Lockheed-California Company, 202
Lockheed Propulsion Company, 17, 50, 73, 191, 209
Lockheed Electronics Company, 224
Loftus, Joseph P., Jr., 164
Logsdon, John M., 13
Lovell, James A., Jr., 25, 196
Low, George M., 30, 47, 95, 127
LTV. *See* Ling-Temco-Vought, Inc.
Lunar excursion module (LEM), Apollo
    abort capability design, 32, 123-124, 125, 135-136
    abort guidance system, 135-136
    abort, lunar landing. *See* Abort, lunar landing.
    abort simulation, 69
    adapter, 116
    adapter installation, 61, 94
    all-welded to partially riveted construction, 180
    ascent engine, 25-26, 31, 33-34, 43-44, 70, 88, 110, 119, 121, 131-132, 137, 143, 172, 207-208, 214-215, 225
    attitude translation and control assembly, 171, 227
    batteries. *See* Batteries.
    bioinstrumentation, 125
    cockpit layout, 84, 87, 204-205
    common use ground support equipment, 144
    communications system, 44, 69, 108, 140, 153, 169, 201
    computer. *See* Computer.
    contract, 5, 38, 222
    controllers, translation and descent engine thrust, 123
    data storage, 195
    decision for crew to stand, 65, 80, 87, 136, 162, 183
    deletion of front docking, 163-164
    descent engine, 25-26, 28, 32, 39, 53, 55, 59, 70, 112, 115, 118, 121, 126, 137, 153, 163, 174, 175, 200, 201, 227
    development test program, 113, 121-122
    docking. *See* docking, spacecraft.
    dual-control requirement, 123-124
    electrical system. *See* Electrical system.
    electroluminescent lighting, 204-205
    emergency detection system, 185
    environmental control system. *See* Environmental control system.
    first progress report, 38

Lunar excursion module, *Cont.*
    flight attitude director indicator ("eight-ball"), 133, 159, 209
    flight program schedule, 133
    "front porch," 183-184
    fuel cell. *See* Fuel cell.
    harness, crew, 136
    helicopter landing simulation, 65
    helium pressurization system, 130
    ingress and egress, 183-184
    instability with Little Joe II, 116-117, 135
    "intercom" with command module, 203
    landing, 15, 21, 30-31, 85, 103, 123-124, 175-177, 206
    landing gear, 33-34, 48, 56, 66, 67, 84, 85, 94-95, 103, 116, 125, 175-177, 200
    landing simulation test, 45
    lunar landing aids, 206
    lunar landing mission. *See* Lunar landing mission, 7-day; 10-day; and 14-day.
    lunar landing mission design. *See* Lunar landing mission design.
    Lunar Landing Research Facility. *See* Lunar Landing Research Facility.
    lunar landing research vehicle. *See* Lunar landing research vehicle.
    lunar surface water requirements, 219
    M-1 mock-up, 60, 65, 87
    meteoroid hits, 131
    missed rendezvous with CSM after launch, 127
    mission programmer, 214
    mock-up review, 123-124
    navigation and guidance. *See* Navigation and guidance.
    onboard checkout concept, 192-193
    operation requirements, 42, 202, 216-217
    overhead window, 163-164, 173, 177
    oxygen storage capacity, 203
    postlanding tilt angle, 175-176, 195
    preliminary configuration freeze, 48-49
    Project Christmas Present Report, 121-122, 133
    quality control program, 55
    radar, 44, 69, 78, 108, 121, 139, 141-142, 145-146, 177-178, 188, 201, 206-207
    reaction control engines. *See* Reaction control engines.
    reliability, 59
    scientific equipment weight and volume allocation, 169, 171-172, 186, 200-201
    service module failure backup, 76, 145
    spacecraft and test articles. *See* LEM spacecraft and test articles.
    stabilization and control. *See* Stabilization and control.
    suit and cabin pressure control, 69
    suit interface. *See* Suit, space.
    television cameras, 121
    terminal rendezvous maneuvers, 120
    thermal test, 90-91
    upper docking hatch, 163-164, 177
    use of command module components, 19, 52, 53, 58, 98, 172-173
    visibility, 60, 65, 73, 80, 87, 163-164, 206
    weight reduction study, 110-111
    wiring insulation, 155-156
Lunar landing mission
    7-day, 123
    10-day, 159-160
    14-day, 58-59, 88, 137, 159-160
Lunar landing mission design, 18, 21, 30-31, 32-33, 42, 76, 78, 98, 123-124, 126, 171-172
    abort criteria, 32, 123-124
    Apollo Mission Planning Task Force, 126
    fuel budget, 113

Lunar landing mission design, *Cont.*
    ground rules, 102-103, 123-124, 171-172
    land two astronauts on moon, 171-172
    operation requirements, 42, 76, 123-124, 171-172
    planning constraints, 78
Lunar Landing Research Facility, 96
Lunar landing research vehicle (LLRV), 24, 55-56, 154
Lunar orbit rendezvous, 13, 76
Lunar Orbiter program, 83, 117, 120, 173, 174
Lunar Orbiter spacecraft, 39-40, 83, 115, 120, 151, 173
Lunar sample containers, 209
Lunar specimen handling, 189
Lunar surface
    contamination, 204
    equipment, 103, 136, 144
    exploration, 44, 89, 92, 95-96
        proposals, 136, 144, 147, 148
    microrelief, 103
    model training area, 188
Lunar Surface Experiments Panel, 31, 147
Lunar surface model training area, 188
Lunar survey mission, 14-day, 196
Lunik IV, 45
Lunney, Glynn S., 219

M

McAdams, R. E., 133
McDivitt, James A., 25, 53, 196
McMillion, Lee N., 42, 86, 107, 218
Madrid, Spain, 130, 131
Maloney, Jim, 221
Maloney, Philip R., 77
Manager, Apollo subsystem, 118
Manned Spacecraft Center (MSC), *passim*
    Apollo Spacecraft Project Office (ASPO), *passim*
        reorganization, 28, 53
    Center Medical Office, 125
    Crew Systems Division, 77, 91, 103, 106, 111, 132, 135, 141, 148, 159, 162, 183, 200, 204, 210, 213, 215, 217, 219, 228, 229
    Engineering and Development Directorate (EDD), 118, 123, 133, 163
    Flight Crew Operations Directorate, 125, 129, 141, 162
    Flight Operations Division, 18-19, 21, 22, 36, 59, 61, 63, 78, 85, 102
    Landing and Recovery Division, 184
    Preflight Operations Division, 13-14
    reorganization, 55, 107
Manned Spacecraft Criteria Board, 84
Manned Space Flight Management Council, 7, 12, 14, 58-59, 66-67, 70-71, 79, 104
Maps, moon, 174, 192
Mardel, Alfred D., 14, 53
Marion Power Shovel Company, 28
Markley, J. Thomas, 20, 28, 54
Marquardt Corporation, The, 26, 38, 71, 73, 105, 204
Marshall Space Flight Center (MSFC), 4, 30, 61, 71, 98, 104, 105, 132
Martin Company, The, 38, 68, 123
    lunar landing study, 123
Massachusetts Institute of Technology (MIT), 12, 27, 36, 44, 101, 121, 137, 163, 188, 190, 192, 206, 221, 227
Mathews, Charles W., 107
Mayer, John P., 113
Maynard, Owen E., 28, 91, 114, 116, 118, 120, 150, 154, 165, 188, 195
Mercury. *See* Project Mercury.
Messing, Wesley E., 47
Michel, Edward L., 5
Michoud Operations Plant, 32

Micrometeoroid impact simulation, 50
    danger on the moon, 50
Minneapolis-Honeywell Regulator Company, 15, 133, 190, 221
Mission and Training Facility, 120
Mission Control Center, 5, 25, 41, 58, 62, 71
    computer complex, 71, 118
    contract to build, 5, 25, 41
    space flight control support, 165
Mission planning, 32-33, 42, 50, 78, 91, 102, 149, 171, 183
Mission Trajectory and Control Program, 189
Mississippi Test Operations, 32
MIT. *See* Massachusetts Institute of Technology.
Moore, Henry J., 51
Morris, Corinne L., xiv
Morrison-Knudson Company, 12
Moser, Jacob C., 84
Motorola, Inc., 10, 120, 128, 137
MSC. *See* Manned Spacecraft Center.
MSFC. *See* Marshall Space Flight Center.
Mueller, George E., 74, 89, 102, 103, 104, 105, 121, 138, 188
Mullaney, R. S., 91, 114, 116, 118, 119, 120, 123, 124, 130, 132, 136, 145, 156, 159, 169, 174, 180, 185, 186, 194, 196, 199, 200, 202, 203, 206, 207, 209, 214, 217, 223, 225, 229
Murrah, E. D., 133
Myers, Dale D., 158

N

NAA. *See* North American Aviation, Inc.
NASA Headquarters
    Manned Space Science Division, 164
    Office of Advanced Research and Technology (OART), 14
    Office of Manned Space Flight (OMSF), 21, 30, 91, 115, 149
        Apollo Program Office, 121
        reorganization, 30
    Office of Space Sciences (OSS), 39, 95, 117, 148, 159, 174, 200-201, 213
    Office of Space Sciences and Applications (OSSA). *See* Office of Space Sciences.
    Office of Tracking and Data Acquisition (OTDA), 6
    Space Sciences Steering Committee, 148
    reorganization, 79, 96
National Academy of Sciences, 139, 159, 204, 213
    Committee on Lunar Exploration, 204
National Rocket Club, 89
Navigation and guidance
    accelerometer, 36, 184
    Bissett-Berman contract, 183
    body-mounted attitude gyro, 215
    command module, 3, 8, 27, 85, 184, 190, 191, 215
    computer, *See* Computer, spacecraft.
    contractor relationship, 137, 191-192
    equipment testing, 63
    flight director attitude indicator ("eight-ball"), 133, 159, 209
    inertial measuring unit, 75, 100-101
    lunar excursion module, 65, 75, 100-101, 135, 145, 171, 173, 186, 189, 194, 222
    optical devices, 12, 75, 85, 221
    radar. *See* Command module; Lunar excursion module.
    reentry trajectories, 163
    unmanned earth-orbital mission, 119
Nelson, Richard D., 132
Newell, Homer E., 14, 39, 117, 144, 148, 159, 213
noise, low-frequency, 211-212
noise tests, spacecraft parts, 107
North American Aviation, Inc. (NAA), *passim*
    Apollo spacecraft contract, 61, 77, 219-220
    modification and construction of facilities, 129

North American Aviation, *Cont.*
    Rocketdyne Division, 17, 25, 28, 39, 53, 59, 105, 108, 112, 113, 137, 152, 163, 170, 192, 201, 227
    spare parts contract, 227
Northrop Corporation
    Ventura Division, 13, 15, 30, 35, 53, 209

O

Oakley, Ralph B., 17
Oller, W. E., 13
Optical devices. *See* navigation and guidance.
Osbon, H. G., 76, 165
Oxygen environmental tests, 5, 14, 106, 118, 151, 155
    fire, 5, 14, 151

P

PACE. *See* Prelaunch acceptance checkout equipment.
Panel Review Board, 70–71, 76–77
Parachute system, 12, 15, 35, 53, 68, 85, 86, 88, 108, 141, 175, 199, 209
Paul Hardeman, Inc., 12
Paup, John W., 158, 161
Payne, Joe D., 76
Pearce, J. L., 47
Pericynthion, 21
Perilune, 21
Perini Corporation, 12
Perkin-Elmer Corporation, 36, 180
Perrine, Calvin H., 111, 135, 163
Peter Kiewit and Sons, Inc., 120
Philco Corporation. *See* Ford Motor Company.
Phillips, Samuel C., 121, 188, 197, 202
Piland, Robert O., 28, 45, 47, 53, 76, 95, 124, 165, 174, 207
Pioneer Parachute Company, 35, 68–69
PLSS. *See* Portable life support system.
Portable life support system, 7, 45, 59–60, 69, 73, 77, 86, 91, 101–102, 163, 172–173, 174, 177, 180, 195, 200, 215–216, 228
Pratt and Whitney Aircraft, 37, 73, 119, 124, 138, 189, 221
Preflight Operations Division. *See* Manned Spacecraft Center.
"Pregnant Guppy," 71, 105
Prelaunch acceptance checkout equipment, 34, 122, 151, 215
Preston, G. Merritt, 107
Project Apollo, *passim*
Project Christmas Present Report, 121–122, 133
Project Fire, 6, 124, 157–158
Project High Water, 4
Project Luster, 105
Project Mercury, 22, 56, 63
Propellant budget, 113
Propellant measuring system, 47
Pulse code modulation. *See* Telemetry.
Purser, Paul E., 21
Putnam, William D., xiv
Pyrotechnics, 31, 189
    reclaimed explosives, 189

Q

Qualification test vehicle. *See* Little Joe II.

R

Radar. *See* Command module; Lunar excursion module.
Radiation altimeter system, 30
Radiation, cosmic, 56, 120, 138–139, 217
Radiation instrumentation, 107, 217
    warning system, 107
Radiation, Inc., 16, 151

Radio Corporation of America (RCA), 19–20, 43, 69, 108, 121, 140, 142, 145, 171, 188, 199, 226, 227
Ranger program, 14, 86, 117
    cancellation of flights, 117
Ranger V, 14
Ranger VI, 14, 131
Ranger VII, 202
Raper, James L., 213
Raytheon, 3, 27, 58, 101, 137, 191
RCA. *See* Radio Corporation of America.
Reaction control engines
    command module, 17, 30, 114, 119, 127
    lunar excursion module, 25, 52, 73, 138, 161, 170, 204
    service module, 71, 105, 136, 160, 161, 192
Recovery aids, spacecraft, 13, 22, 36, 50, 59, 61, 85
    flotation collar, 184
Recovery rules and procedures, 50
Rector, William F., III, 119, 123, 124, 130, 132, 145, 150, 156, 159, 162, 169, 174, 180, 185, 186, 194, 196, 199, 200, 202, 203, 206, 207, 209, 214, 217, 223, 225, 229
Reentry, 22, 44–45, 63, 124, 163, 194–195
    automatic modes, 119
    loads, 44–45, 194–195
    range, 194–195
    "skip" phenomenon, 63
    test plan, 124
    trajectories, 163
    ultraviolet emissions, 186
Reentry heating, 22, 73–74, 84, 157–158, 186, 194–195, 213
    Scout tests, 73–74, 213
    See also Project Fire.
Reliability, spacecraft, 18, 59, 69
Rendezvous study, 88, 96, 120
Rendezvous Working Group, Ad Hoc, 78
Repair concept, spacecraft, 102–103
Retriever, 35
Richard, Louie G., 136
Richter, Henry L., Jr., 131
Roadman, Charles H., 30
Rocketdyne. *See* North American Aviation, Inc.
Rosen, Milton W., 30
Rosholt, Robert L., 89
Rowell, Billie D., xiv
Russia. *See* Soviet Union.
Russian-American space cooperation, 89
Ryan Aeronautical Company, 121, 141, 145, 177
Ryker, Norman J., Jr., 155

S

S–64A Skycrane helicopter, 114
Sack, E. E., 76, 107, 111, 116, 188
Saegesser, Lee D., xiv
Saturn
    "engine out" capability, 30, 42
    hard-over engine failure, 111
    manufacture, 31–32
    simplification of names, 27
    tower jettison, 86–87
Saturn I, 104, 224–225
    hard-over engine failure, 111
Saturn IB, 91, 104, 105–106, 108–109, 117, 124, 132–133, 162, 192
    flight schedule, 149
    hard-over engine failure, 111
Saturn V, 12, 98, 105–106, 110, 117, 124, 132–133, 152–153, 162, 171, 192
    flight schedule, 149
    hard-over engine failure, 111
Saturn SA–3, 4
    Project High Water, 4
Saturn SA–4, 42
Saturn SA–5, 38–39, 77, 131

Saturn SA-6, 180-181
Saturn SA-7, 224-225
Saturn SA-10, 104, 105-106
    end of Saturn I mission, 104
Saturn SA-201. *See* Apollo Mission AS-201.
Saturn SA-203. *See* Apollo Mission AS-203.
Saturn SA-207. *See* Apollo Mission AS-207.
Saturn SA-501. *See* Apollo Mission AS-501.
Saturn SA-503. *See* Apollo Mission AS-503.
Saturn SA-507. *See* Apollo Mission AS-507.
Saturn S-IC stage, 110
Saturn S-IV stage, 76, 86-87, 91, 117, 192, 197-198
    disposal during lunar mission, 197-198
Saunders, J. F., Jr., 42
Sawyer, Ralph S., 7
S-band radar, 5, 10, 44, 56, 57, 65, 78, 120, 128-129, 137-138, 144-145, 188, 190, 201, 224
Scherer, Lee R., 151, 174
Schirra, Walter M., Jr., 25, 53, 196
Schweickart, Russell L., 101, 197, 221
Scott, David R., 101, 197
Scout rocket, 73-74, 124
Seamans, Robert C., Jr., 55, 83, 89
Sea of Tranquillity, 131
See, Elliot M., Jr., 25, 53, 163, 196, 221
Service module, Apollo
    addition of micrometeoroid protection, 188
    contract, 61, 77
    engine, 67-68, 76, 98, 157, 215, 222, 223-224, 227
    failure procedure, 76, 145
    heat transfer and thermodynamic analysis, 202
    propulsion tanks, 207
    reaction control engines. *See* Reaction control engines.
Sextant. *See* Navigation and guidance, optical devices.
Shea, Joseph F., 21, 30, 70, 95, 107, 113, 124, 133, 135, 136, 139, 140, 154, 159, 171, 173, 202, 226
Shepard, Alan B., Jr., 25, 196
Shoemaker, Eugene M., 51
Simmonds Precision Products, Inc., 47
Simulators, training, 13, 47, 59, 64, 110, 221
    CSM part-task trainer, 115
    dynamic crew procedures, 194
    extravehicular, 68
    lunar excursion module, 221
    mockup review, 64
    visual displays, 78-79, 110
Sites, Apollo lunar landing, 39-40, 103, 112, 113, 115, 171, 174, 186-187
"Skip" phenomenon. *See* Reentry.
Skopinski, Ted H., 133
Slayton, Donald K., 25, 107, 150, 165, 196
Sloan, James E., 30
Smith, George B., Jr., 5
Smith, Norman F., 84
Smylie, Robert E., 202, 218, 222
Sonett, Charles P., 117
Sonett Committee, 117
Soviet Union, 45, 64, 89
Space Business Daily, 189
Space Technology Laboratories (STL), 39, 55, 70, 126, 137, 153, 175, 189
Space Vehicle Review Board, 70-71
Spain, 130
Sperry Gyroscope Company, 36, 101, 184
    Sperry Phoenix, 86
Stabilization and control system
    command and service modules, 5, 15-16, 65, 190-191
        Block II, 221
    lunar excursion module, 69, 130, 170, 171, 173
    lunar orbit requirement, 130
Stafford, Thomas P., 25, 196
Stanford Research Center, 183

STL. *See* Space Technology Laboratories.
Strakes, aerodynamic, 22-24
Suit, space, 7, 24, 45, 121, 138, 153, 159, 173-174, 180, 190, 195, 202
    Aponi suit, 199
    "Buddy concept," 75, 88, 122-123
    command module interface, 24-25, 42, 58, 77, 88, 101-102, 118, 122-123, 126-127, 161-162, 165, 177, 190, 195, 198, 218
    communications and telemetry system, 7, 24, 41, 77, 101, 123, 153-154
    emergency oxygen supply unit, 195-196, 201, 217
    flight-wear ground rules, 162-163
    gas-cooled versus liquid-cooled undergarment, 173-174, 204
    improved version, 141
    lunar excursion module interface, 58, 88, 91, 122-123, 132
    lunar surface, 7, 88
    micrometeoroid protective garment, 111, 210
    mobility, 77, 101-102, 126-127, 159, 172-173
    mobility test at Bend, Oregon, 215-216
    PLSS interface, 91, 180
    Project Gemini suit, 141, 159-161, 165, 177, 195, 199, 202, 218
    second prototype, 101-102
    waste management, 162
    water-cooled undergarment, 132, 173-174, 204, 209-210
Sundt Construction Company, 110
Surveyor, 31, 40, 86, 103, 115, 117, 171
Survival equipment, postlanding, 61, 85-86, 135
Swenson, Loyd S., Jr., xiv, 63
Swetnick, Martin J., 103
Sword, C. D., 69

T

Tanks, gas storage, 43, 59, 65, 92, 115, 121, 130, 145, 190, 200
Tape recorders, 175, 195
Taylor, Clinton L., 165, 196
Taylor, William B., 174
Telemetry, 5-6, 10
    pulse code modulation, 16-17
    up-data link, 22, 62
Telescope. *See* Navigation and guidance, optical devices.
Television, 5-6, 10, 52, 53, 90, 121, 131, 173, 198-199, 200, 221, 226
Tereshkova, Valentina V., 64
Texas Instruments, Inc., 92, 147
Thiokol Chemical Corporation, 13, 213
    Reaction Motors Division, 39
Thompson, Floyd L., 84
Timing equipment, 33, 86, 141, 223
Tinsley, James A., xiv
TLI. *See* Translunar injection.
"Tower flap." *See* Canards.
Tracking network, 5-6, 6-7, 19, 20, 25, 36, 55, 62, 69, 90, 106, 120, 128-129, 130, 131, 144-145, 153-154, 169, 172, 190, 223-224
Translunar injection, 50, 98
Transposition and docking. *See* Docking, spacecraft.
Truszynski, Gerald M., 7
Tunnel, crew transfer. *See* Crew transfer between spacecraft.

U

Up-data link. *See* Computer, spacecraft; Telemetry.
United Aircraft Corporation, 7
    Hamilton Standard Division, 7, 12, 24, 26, 34, 41, 60,

INDEX

United Aircraft Corporation, *Cont.*
    74, 101, 111, 120, 121, 132, 141, 153, 162, 174, 180, 190, 200, 204, 210, 229
    United Technology Center, 39
United Nations General Assembly, 89
United States Air Force
    Air Force Systems Command, 10
    Aerospace Medical Division, 118
    Arnold Engineering Development Center, 10, 67, 157, 214
    Holloman Air Force Base, 183
    School of Aerospace Medicine, 14
    Wright-Patterson Air Force Base, 101, 218, 227
United States Army, 11, 35
    Corps of Engineers, 11, 120
    Map Service, 192
United States House of Representatives, Committee on Science and Astronautics, 40
United States Navy, 5, 12
    Aircrew Equipment Laboratory, 5, 10, 209
    Aviation Medical Acceleration Laboratory, 89, 101, 118
    Instrumentation Ships Projects Office, 172
    Naval Air Facility, El Centro, 12, 15
U.S.S.R. *See* Soviet Union.

V

VAB. *See* Vertical Assembly Building.
Vacuum chamber, test facility, 13–14, 19–20
Vertical Assembly Building (VAB), 11–12
Vibration test, spacecraft part, 107
von Braun, Wernher, 12
Vorzimmer, Peter J., 155
Vostok V, 64
Vostok VI, 64

W

Wallops Island, 73, 213
Water landing, spacecraft, 13, 27, 36, 85
Webb, James E., 13, 22, 63, 77, 96, 104
Weight reduction, spacecraft, 98, 116, 119, 145
Westinghouse Electric Corporation, 3, 56, 186, 226
White, Edward H., II, 25, 53, 163, 196
White Sands Missile Range (WSMR), 29, 41, 47, 63, 70, 72, 80, 137, 184, 192, 227
Williams, Clifton C., Jr., 101, 197
Williams, Walter C., 12, 27, 47, 55, 102, 104
Wiseman, Donald G., 126, 127
Wright-Patterson Air Force Base. *See* United States Air Force.
Wyle Laboratories, 107
WSMR. *See* White Sands Missile Range.

X

X-band radar, 78

Y

Yardney Electric Corporation, 37, 144
Young, John W., 25, 53, 101, 196
Yschek, H. P., 11, 22, 26, 31, 35, 73, 86, 88, 90, 136, 145, 147, 169, 183, 187, 188, 190, 201, 222, 230

THE AUTHORS

Mary Louise Morse has been a Research Associate with the Department of History of the University of Houston since the fall of 1966, currently serving as an assistant to the institution's Director of Research. Born in Beverly, Massachusetts, she received her B.S. in Education from Salem State College, Salem, Massachusetts (1947), and her M.A. in History from Columbia University (1950). She was a senior editor with the MIT Press, Cambridge, Massachusetts, before moving to Houston.

Jean Kernahan Bays was a Research Associate with the Department of History of the University of Houston from August 1967 to June 1969. Born in Pittsburgh, Pennsylvania, she received a B.A. in History from Denison University (1965) and an M.A. in History from the University of Houston (1967), where she served as a teaching fellow. She has also worked as a Research Librarian for the Houston, Texas, Public Library. She is married and lives in Houston.

NASA HISTORICAL PUBLICATIONS

HISTORIES

- Robert L. Rosholt, *An Administrative History of NASA, 1958–1963*, NASA SP–4101, 1966, GPO, $4.00.*
- Loyd S. Swenson, James M. Grimwood, and Charles C. Alexander, *This New Ocean: A History of Project Mercury*, NASA SP–4201, 1966, GPO, $5.50.
- Constance McL. Green and Milton Lomask, *Vanguard—A History*, NASA SP–4202, 1970; also Washington: Smithsonian Institution Press, 1971, $12.50.
- Alfred Rosenthal, *Venture into Space: Early Years of Goddard Space Flight Center*, NASA SP–4301, 1968, GPO, $2.50.
- Edwin P. Hartman, *Adventures in Research: A History of the Ames Research Center, 1940–1965*, NASA SP–4302, 1970, GPO, $4.75.

HISTORICAL STUDIES

- Eugene M. Emme (ed.), *History of Rocket Technology*, Detroit: Wayne State University, 1964, $8.50.
- Mae Mills Link, *Space Medicine in Project Mercury*, NASA SP–4003, 1965, NTIS, $6.00.**
- *Historical Sketch of NASA*, NASA EP–29, 1965 and 1966, NTIS, $6.00.
- Katherine M. Dickson (Library of Congress), *History of Aeronautics and Astronautics: A Preliminary Bibliography*, NASA HHR–29, NTIS, $6.00.
- Eugene M. Emme (ed.), *Statements by the Presidents of the United States on International Cooperation in Space*, Senate Committee on Aeronautical and Space Sciences, Sen. Doc. 92–40, 1971, GPO, $0.55.
- William R. Corliss, *NASA Sounding Rockets, 1958–1968: A Historical Summary*, NASA SP–4401, 1971, GPO, $1.75.
- Helen T. Wells with Susan Whiteley, *Origins of NASA Names*, NASA SP–4402 (1973).
- Jane Van Nimmen and Leonard C. Bruno with Robert L. Rosholt, *NASA Historical Data Book, 1958–1968*, NASA SP–4012 (1973).

CHRONOLOGIES

- *Aeronautics and Astronautics: An American Chronology of Science and Technology in the Exploration of Space, 1915–1960,* compiled by Eugene M. Emme, Washington: NASA, 1961, NTIS, $6.00.

- *Aeronautical and Astronautical Events of 1961,* published by the House Committee on Science and Astronautics, 1962, NTIS, $6.00.

- *Astronautical and Aeronautical Events of 1962,* published by the House Committee on Science and Astronautics, 1963, NTIS, $6.00.

- *Astronautics and Aeronautics: A Chronology on Science and Technology, 1963,* NASA SP–4004, 1964, NTIS, $6.00.

- ———, *1964,* NASA SP–4005, 1965, NTIS, $6.00.

- ———, *1965,* NASA SP–4006, 1966, NTIS, $6.00.

- ———, *1966,* NASA SP–4007, 1967, NTIS, $6.00.

- ———, *1967,* NASA SP–4008, 1968, GPO, $2.25.

- ———, *1968,* NASA SP–4010, 1969, GPO, $2.00.

- ———, *1969,* NASA SP–4014, 1970, GPO, $2.25.

- ———, *1970,* NASA SP–4015, 1972, GPO: domestic mail, $3.10; bookstore, $2.75.

- ———, *1971,* NASA SP–4016 (1973).

- James M. Grimwood, *Project Mercury: A Chronology,* NASA SP–4001, 1963, NTIS, $6.00.

- James M. Grimwood and Barton C. Hacker, with Peter J. Vorzimmer, *Project Gemini Technology and Operations: A Chronology,* NASA SP–4002, 1969, GPO, $2.75.

- Ivan D. Ertel and Mary Louise Morse, *The Apollo Spacecraft: A Chronology,* Vol. I, *Through November 7, 1962,* NASA SP–4009, 1969, GPO, $2.50.

- R. Cargill Hall, *Project Ranger: A Chronology,* JPL/HR–2, 1971, NTIS, $6.00.

---

* GPO: Titles may be ordered from the Superintendent of Documents, Government Printing Office, Washington, D.C. 20402.

** NTIS: Titles may be ordered from National Technical Information Service, Springfield, Va. 22151.

www.ingramcontent.com/pod-product-compliance
Lightning Source LLC
Chambersburg PA
CBHW081720170526
45167CB00009B/3638